Materials Horizons: From Nature to Nanomaterials

Series Editor

Vijay Kumar Thakur, School of Aerospace, Transport and Manufacturing, Cranfield University, Cranfield, UK

Materials are an indispensable part of human civilization since the inception of life on earth. With the passage of time, innumerable new materials have been explored as well as developed and the search for new innovative materials continues briskly. Keeping in mind the immense perspectives of various classes of materials, this series aims at providing a comprehensive collection of works across the breadth of materials research at cutting-edge interface of materials science with physics, chemistry, biology and engineering.

This series covers a galaxy of materials ranging from natural materials to nanomaterials. Some of the topics include but not limited to: biological materials, biomimetic materials, ceramics, composites, coatings, functional materials, glasses, inorganic materials, inorganic-organic hybrids, metals, membranes, magnetic materials, manufacturing of materials, nanomaterials, organic materials and pigments to name a few. The series provides most timely and comprehensive information on advanced synthesis, processing, characterization, manufacturing and applications in a broad range of interdisciplinary fields in science, engineering and technology.

This series accepts both authored and edited works, including textbooks, monographs, reference works, and professional books. The books in this series will provide a deep insight into the state-of-art of Materials Horizons and serve students, academic, government and industrial scientists involved in all aspects of materials research.

Review Process

The proposal for each volume is reviewed by the following:

1. Responsible (in-house) editor
2. One external subject expert
3. One of the editorial board members.

The chapters in each volume are individually reviewed single blind by expert reviewers and the volume editor.

Amirhossein Mosaffa

Phase Changing Materials

An Approach to Storing Thermal Energy

Springer

Amirhossein Mosaffa
Department of Mechanical Engineering
Azarbaijan Shahid Madani University
Tabriz, Iran

ISSN 2524-5384 ISSN 2524-5392 (electronic)
Materials Horizons: From Nature to Nanomaterials
ISBN 978-981-95-3741-9 ISBN 978-981-95-3742-6 (eBook)
https://doi.org/10.1007/978-981-95-3742-6

This Springer imprint is published by the registered company Springer Nature Singapore Pte Ltd.
The registered company address is: 152 Beach Road, #21-01/04 Gateway East, Singapore 189721, Singapore

"For LEILI, who turns life's ordinary moments into magic;

To PANAH, my reason to dream in color;

And for my PARENTS, who gifted me the courage to begin.

This book is yours, as am I."

Preface

The exploration of phase change materials (PCMs) commenced in the early 20th century, notably through the groundbreaking investigations of Alan Tower Waterman at Yale University. During his studies on the thermionic emission of select hot salts, Waterman identified anomalous conductivity behaviors in molybdenite (MoS_2), elucidating that the conductivity of this chalcogenide could be systematically modulated. This finding established a fundamental basis for the evolution of PCMs.

PCMs, classified as latent heat storage materials, are recognized for their exceptional efficacy in thermal energy storage. These materials facilitate energy absorption during phase transitions, specifically from solid to liquid, and subsequently release energy upon solidification, maintaining a constant temperature throughout these processes. The energy transferred during these transitions significantly enhances the vibrational modes of the constituent atoms or molecules. Upon melting, atomic bonds weaken, leading to a phase transition from solid to liquid. Conversely, energy is liberated during crystallization, resulting in a gain in order as the material solidifies.

PCMs can be categorized into three primary classes: organic (paraffin and non-paraffin), inorganic (salt hydrates and metallic alloys), and eutectic (combinations of organic and inorganic components). Recent research has honed in on PCMs that leverage the latent heat of fusion, which has been instrumental in minimizing the volume required for heat storage units. The urgent need for efficient energy utilization, spurred by dwindling fossil fuel reserves, escalating greenhouse gas emissions, and rapid energy demand, has made the advancement of alternative energy solutions critical. In this context, PCMs present a cost-effective and efficient approach to thermal energy storage, enhancing the performance of energy systems while mitigating peak demand and overall energy consumption.

Thermal energy storage (TES) systems include both latent heat storage and thermochemical heat storage. Unlike sensible heat storage, which necessitates considerable volumes due to its lower energy density, latent heat storage systems offer superior cost-effectiveness and manageable logistics. PCM-based cooling systems stand out for their capability in thermal regulation, underpinned by their high latent heat capacity, negligible volume transition during phase changes, substantial specific heat capacity, and chemical inertness, along with endurance throughout extensive cycling.

Moreover, PCMs are characterized by high thermal conductivity, non-corrosiveness, non-flammability, non-toxicity, and relative economic feasibility.

To enhance heat transfer efficiency in thermal control units utilizing PCMs, various methods have been developed. Innovations include the incorporation of conductive materials or pathways, such as metal matrices or foams, nano- and microscale metal or metal oxide fillers, discrete elements like pins and fins, graphite, carbon nanotubes or fibers, and advanced materials such as exfoliated graphite or graphene.

The application of PCMs has proliferated across diverse sectors, including solar cooling and power generation, photovoltaic systems, electronics thermal management, waste heat recovery, agricultural solar drying, domestic hot water systems, pharmaceuticals, food preservation, and even in space exploration endeavors. In the realm of building design, PCMs play a pivotal role in optimizing energy performance and thermal comfort, particularly in the context of net zero energy buildings. For optimal thermal storage application, PCMs necessitate containment in specialized vessels tailored to their properties, such as embedding them in construction materials like wood, concrete, or plasterboard.

This book is designed to equip researchers and practitioners with an in-depth understanding of the properties of PCMs, emphasizing their technological applications. It reviews the phase change dynamics examined by various scholars and explores the integration of PCMs in information storage systems. Additionally, it surveys the current landscape and future directions of PCM research, providing insights into prospective advancements and applications. Join us in this comprehensive examination of PCMs, where we will uncover their transformative potential in thermal energy storage and their pivotal contribution to a sustainable energy future.

Tabriz, Iran
June 2025

Amirhossein Mosaffa

Competing Interests The author has no competing interests to declare that are relevant to the content of this manuscript.

Contents

1 Energy Storage Systems ... 1
1.1 Introduction ... 1
1.2 Thermal Energy Storage Systems ... 3
1.3 Methods of Storing Thermal Energy ... 4
1.3.1 Storage of Sensible Heat ... 4
1.3.2 Storage of Latent Heat ... 8
1.3.3 Passive Phase Change Material Heat Transfer Concepts ... 11
1.3.4 Active Phase Change Material Heat Transfer Concepts ... 13
1.3.5 Storage Integration and Applications ... 13
1.3.6 Thermochemical Storage ... 14
References ... 16

2 Phase Change Materials ... 19
2.1 Introduction ... 19
2.2 The Marvel of Phase Change Materials ... 19
2.3 Characteristics of Phase Change Materials ... 21
2.4 Drawbacks of Phase Change Materials ... 23
2.5 Thermal Stability of Phase Change Materials ... 23
2.5.1 Thermal Stability Test ... 24
2.6 Classification of Phase Change Materials ... 25
2.6.1 Advantages and Disadvantages of Organic and Inorganic PCMs ... 27
2.7 Summary ... 28
References ... 28

3 Shortcomings of PCMs in LHTS ... 31
3.1 Introduction ... 31
3.2 Incongruent Melting/Phase Separation ... 31
3.3 Subcooling ... 32

3.4 Low Thermal Conductivity and Heat Transfer Rate . . . 34
3.4.1 Larger Heat Exchanger Surface . . . 34
3.5 Dynamic PCM System . . . 39
3.5.1 Insufficient Long-Term Stability . . . 40
3.6 Leakage Prevention and Encapsulation Techniques . . . 41
3.6.1 Microencapsulation Techniques . . . 41
3.6.2 Macroencapsulation Techniques . . . 42
3.6.3 Composite Matrices for Leakage Prevention . . . 42
3.6.4 Impact on Thermal Performance . . . 43
3.7 Economic Viability of PCM Systems . . . 43
3.7.1 Material Costs: Organic Versus Inorganic PCMs . . . 43
3.7.2 Manufacturing Expenses . . . 44
3.7.3 Lifecycle Cost Comparisons with Conventional Systems . . . 44
3.8 Summary . . . 46
References . . . 46

4 Analyses of the Phase Change Problems . . . 51
4.1 Introduction . . . 51
4.2 Mathematical Formulation of the Phase Change Problem (Stefan Problem) . . . 52
4.2.1 General Assumptions . . . 52
4.2.2 Heat Transfer Equations . . . 53
4.2.3 Boundary Conditions at the Moving Interface . . . 54
4.2.4 Stefan Condition . . . 56
4.3 Nonlinear Nature of the Energy Equation at the Interface . . . 57
4.4 Exact Solution to the One-Phase Stefan Problem . . . 57
4.4.1 Similarity Solution . . . 58
4.5 Approximate Solutions to Phase Change Problems . . . 59
4.5.1 Quasi-Steady Method . . . 59
4.5.2 Megerlin Method . . . 60
4.5.3 Comparison of Approximate Methods . . . 61
References . . . 63

5 Semi-analytical Solutions for the Stefan Problem in Finned Rectangular Storage Tanks . . . 65
5.1 Introduction . . . 65
5.2 Problem Definition: Freezing in Finned Rectangular Storage Tanks . . . 65
5.2.1 Problem Definition in Region#1 . . . 67
5.2.2 Problem Definition in Region#2 . . . 68
5.3 Semi-analytical Solution of the Freezing Problem in Region#1 . . . 70
5.3.1 Solution of the Freezing Problem in Region#1 for the First-Type Boundary Condition (T_w = constant) . . . 70

5.3.2 Solving the Freezing Problem in Region#1 for a Second-Type Boundary Condition ($q''_w = \text{cte}$) 71
5.3.3 Solving the Freezing Problem in Region#1 for a Third-Type Boundary Condition ($h, T_\infty = \text{cte}$) 72
5.4 Semi-analytical Solution for the Freezing Problem in Region#2 72
5.4.1 Solving the Freezing Problem in Region#2 for a First-Type Boundary Condition 72
5.4.2 Solving the Freezing Problem in Region#2 for a Second-Type Boundary Condition 73
5.4.3 Solving the Freezing Problem in Region#2 for Third-Type Boundary Conditions 75
5.5 Numerical Solution of the Freezing Problem in a Rectangular Container Based on the Enthalpy Method 76
5.6 Illustrative Examples 79
5.6.1 Example 1: Solidification of Climsel C28 in a Rectangular LHTS System with Horizontal Fins 79
5.6.2 Example 2: Solidification with Imposed Heat Flux 82
5.7 Semi-analytical Solution for 1st Type of Boundary Condition 85
5.7.1 Material Selection 86
5.8 Semi-analytical Solution for 2nd Type of Boundary Condition 89
5.8.1 Summary and Conclusions 94
5.9 Semi-analytical Solution for 3rd Type of Boundary Condition 95
5.9.1 Summary of Findings 99
References 99

6 Semi-analytical Solutions for the Stefan Problem in Finned Cylindrical Storage Tanks 101
6.1 Introduction 101
6.2 Defining the Freezing Problem in Finned Cylindrical Storage Tanks 101
6.2.1 Defining the Problem in Region#1 103
6.2.2 Defining the Problem in Region#2 104
6.3 Semi-analytical Solution of the Freezing Problem in Region#1 106
6.4 Solving the Steady-State Part of the Freezing Problem in Region#1 107
6.4.1 Analytical Solution for Temperature Distribution for 1st and 2nd Type of Boundary Conditions 107

6.4.2 Analytical Solution for Temperature Distribution for 3rd Type of Boundary Condition 109
6.5 Semi-analytical Solution of the Freezing Problem in Region#2 111
6.5.1 Mathematical Model Description 111
6.5.2 Analytical Solution via Green's Function 113
6.6 Numerical Solution of the Freezing Problem in a Cylindrical Storage Tank Based on the Enthalpy Method 116
6.7 Results and Discussion 118
6.7.1 Numerical Analysis of PCM Solidification in Rectangular and Cylindrical Shell Thermal Storages with Internal Fins 118
6.7.2 Analytical and Numerical Comparison of Fin Temperature Distributions 121
6.7.3 Solid–Liquid Interface Predictions for Cases 1 and 2 125
6.7.4 Effect of Geometry on Solidification for Case 3 126
6.7.5 Summary 129
References 133

7 A Survey of Analytical Solutions for Phase-Change Problems: Theory and Applications 135
7.1 Introduction 135
7.2 Analytical Modeling of Phase Change in a Composite Wall with Two PCMs in Series 136
7.2.1 Problem Description 136
7.2.2 Governing Equations 137
7.3 Analytical Modeling and Thermal Performance Analysis of a Flat Plate Latent Heat Storage Unit (LHSU) 138
7.3.1 Governing Equations and Assumptions 139
7.4 Analytical and Numerical Modeling of PCM and Hybrid PCM-Heat Sinks for High-Power Wireless EV Charging 141
7.4.1 Analytical Model Development 142
7.4.2 Numerical Modeling 143
7.5 Enhancements in Thermal Energy Storage Systems Using Fins 145
7.5.1 Theoretical Framework 145
7.5.2 Governing Equations 145
7.5.3 Results and Design Insights 147
7.6 Analytical Exploration of PCM Solidification in Annular Systems 148
7.6.1 Theoretical Model 148
7.6.2 Results 150

7.7 AI-Driven Optimization of PCM Systems AI-Driven Optimization of PCM Systems ... 151
7.7.1 AI-Driven Phase-Change Prediction ... 151
7.7.2 Material Selection via Machine Learning ... 152
7.7.3 System Design Optimization ... 153
7.8 Summary ... 154
References ... 155

8 Latent Heat Thermal Storage and Its Applications ... 157
8.1 Introduction ... 157
8.2 Harnessing PCM Technology for Thermal Efficiency in Buildings ... 158
8.2.1 Building Model and Methodology ... 158
8.2.2 Results and Seasonal Analysis ... 161
8.2.3 Criticism and Comparative Analysis ... 163
8.2.4 Conclusion ... 165
8.3 Innovations in Sustainable Building Heating Systems ... 165
8.3.1 The Proposed System Design ... 166
8.3.2 Key Results and Findings ... 167
8.3.3 Economic Evaluation ... 168
8.3.4 Challenges and Mitigation ... 169
8.3.5 Criticism and Comparative Analysis ... 170
8.3.6 Conclusion ... 173
8.4 Advanced Thermal Insulation: PCM-Integrated Foam Concrete Panels ... 173
8.4.1 Materials and Methodology ... 173
8.4.2 Results and Findings ... 176
8.4.3 Criticism and Comparative Analysis ... 177
8.4.4 Conclusion ... 179
8.5 Packed Bed Cold Storage Using Composite PCM Capsules for High-Temperature Solar Cooling ... 180
8.5.1 Experimental Setup and Procedures ... 180
8.5.2 Mathematical Model ... 181
8.5.3 Results and Discussion ... 182
8.5.4 Criticism in Comparison to Similar Works ... 184
8.5.5 Recommendations for Improvement ... 186
8.5.6 Conclusion ... 187
8.6 Experimental Evaluation of Low-Temperature PCMs for Refrigeration and Air Conditioning ... 187
8.6.1 Material Selection ... 188
8.6.2 Compatibility Tests ... 188
8.6.3 Experimental Setup ... 188
8.6.4 Governing Equation: Carnot Efficiency ... 190

8.6.5 Results and Discussion ... 190
8.6.6 Criticism Compared to Similar Works ... 191
8.6.7 Conclusion ... 191
8.7 Characterization of Nano-Mixed PCM for Solar Cooker Applications ... 192
8.7.1 Materials and Characterization of Aluminum Oxide Nanoparticles ... 192
8.7.2 Experimental Results and Discussion ... 193
8.7.3 Governing Equations ... 193
8.7.4 Criticism Compared to Similar Works ... 194
8.7.5 Conclusion ... 195
8.8 Application of Active PCM Storage Systems for Building Energy Efficiency ... 195
8.8.1 Experimental Setup ... 195
8.8.2 Governing Equations ... 196
8.8.3 Results and Analysis ... 197
8.8.4 Discussion ... 197
8.8.5 Criticism Compared to Similar Works ... 198
8.8.6 Conclusion ... 199
8.9 Numerical Modelling of High-Temperature Latent Heat Thermal Storage for Solar Applications ... 199
8.9.1 Governing Equations ... 199
8.9.2 System Design and Operation ... 201
8.9.3 Performance Evaluation ... 204
8.9.4 Criticism and Comparison to Similar Works ... 205
8.9.5 Conclusion and Future Directions ... 205
8.10 Advanced Thermal Energy Storage: Cascaded PCM Configurations ... 206
8.10.1 Cascaded PCM Overview ... 206
8.10.2 Governing Concepts and Equations ... 207
8.10.3 Cascaded Configurations and Thermal Performance ... 211
8.10.4 Governing Equations for Cascaded Optimization ... 212
8.10.5 Criticism and Comparison ... 213
8.10.6 Conclusion and Future Directions ... 214
8.11 Scalability of Composite PCM Capsules ... 214
8.11.1 Fabrication Challenges in Mass Production ... 214
8.11.2 Material and Cost Considerations ... 215
8.11.3 Integration with Solar Cooling Systems ... 215
8.11.4 Long-Term Stability and Environmental Impact ... 216
8.12 Summary ... 217
References ... 217

9 Phase-Change Materials for Free Cooling Application 221
9.1 Introduction 221
9.2 Thermal Performance Optimization of Multiple PCM Thermal Storage Units for Free Cooling 222
9.2.1 System Description 222
9.2.2 Mathematical Modeling of PCM Heat Transfer and System Performance 223
9.2.3 Results and Discussion 227
9.2.4 Conclusion 231
9.3 Optimizing Free Cooling Systems Using Multiple PCM-Based Thermal Storage Units 231
9.3.1 System Description 232
9.3.2 Mathematical Formulation 233
9.3.3 Model Assumptions 235
9.3.4 Results and Discussion 235
9.3.5 Conclusions 240
9.4 Energy and Exergy Evaluation of Multiple-PCM Thermal Storage for Free Cooling Applications 242
9.4.1 System Description 242
9.4.2 Mathematical Formulation 243
9.4.3 Results and Discussion 245
9.4.4 Conclusions 249
9.5 Advanced Exergy Analysis of Air-Conditioning Systems with Latent Heat Thermal Storage 250
9.5.1 System Description 251
9.5.2 Mathematical Formulation 253
9.5.3 Advanced Exergy Analysis 254
9.5.4 Results and Discussion 258
9.5.5 Conclusions 260
9.6 Exergoeconomic and Environmental Evaluation of Air Conditioning Systems with Latent Heat Thermal Storage 263
9.6.1 System Description 263
9.6.2 Mathematical Formulation 264
9.6.3 System Specifications 267
9.6.4 Results and Discussion 268
9.6.5 Conclusions 278
References 278

10 Economic and Environmental Dimensions of PCMs 281
10.1 Introduction 281
10.2 Life-Cycle Assessment of PCM Systems 281
10.2.1 Operational and Manufacturing Impacts 283
10.2.2 Comparative Analysis of PCM Applications 284
10.2.3 Impact of System Design and Application Scenarios 284

10.2.4 Challenges and Future Directions 285
10.3 Cost Drivers: Material Sourcing, Manufacturing, and Recycling 285
10.3.1 Material Sourcing 286
10.3.2 Manufacturing Processes 286
10.3.3 Recycling and Life-Cycle Considerations 287
10.3.4 Economic Feasibility and Future Directions 287
10.4 Policy and Regulatory Frameworks for PCM Adoption 288
10.4.1 Financial Incentives and Market Mechanisms 289
10.4.2 Technical Standards and Grid Integration 289
10.4.3 Environmental and Safety Regulations 290
10.4.4 Regional Case Studies and Policy Models 290
10.4.5 Challenges and Recommendations 290
10.5 Environmental Impact of Nano-PCMs 291
10.5.1 Toxicity of Nanoparticles in Nano-PCMs 291
10.5.2 Recycling Challenges of Nano-PCMs 292
10.5.3 Economic Viability 293
10.6 Cost–Benefit Analysis for Building Integration 294
10.6.1 Long-Term Energy Savings and Life-Cycle Cost Analysis 294
10.6.2 Challenges in Cost–Benefit Realization 295
References 296

Chapter 1
Energy Storage Systems

1.1 Introduction

Thermal energy storage (TES) technology captures heat or cooling potential for later utilization, addressing discrepancies between when energy is available and when it's needed across changing temperatures, locations, or power requirements. These systems function through three distinct phases: charging, storage, and discharging, constituting a complete cycle (Fig. 1.1).

TES implementations fall into two primary categories: active and passive concepts (Fig. 1.2). Active systems rely on forced convection, circulating the storage material itself through components like heat exchangers or solar receivers to transfer thermal energy. This approach can be direct (using the heat transfer fluid, HTF, as the storage medium) or indirect (employing a separate material for storage). Passive systems, conversely, typically utilize a dual-medium design where the HTF flows through the storage material only during the energy addition (charging) and extraction (discharging) processes.

Key design considerations for TES systems revolve around several critical factors. First, the storage medium should exhibit high energy density to maximize storage capacity. Efficient thermal conductivity between the HTF and the storage medium is essential for optimal performance. Additionally, the storage material must demonstrate robust mechanical and chemical stability to withstand operational conditions. Compatibility between the storage medium and its containment system is crucial, ensuring longevity and reliability. The system must also allow for complete reversibility over multiple charge–discharge cycles to maintain efficiency. Minimizing thermal losses during the storage process is another key requirement, alongside a user-friendly control mechanism. Further design criteria include the operational strategy, maximum load demands, desired nominal temperature ranges, and managing enthalpy gradients during storage. Integrating these aspects into the overall application system is fundamental for achieving effective TES operation.

Implementing TES in an energy system offers several benefits:

A. Mosaffa, *Phase Changing Materials*, Materials Horizons: From Nature to Nanomaterials, https://doi.org/10.1007/978-981-95-3742-6_1

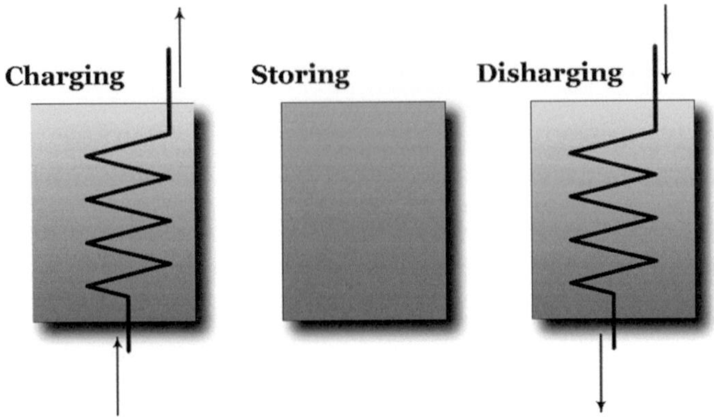

Fig. 1.1 TES complete storage cycle [5]

Fig. 1.2 Classification scheme of various storage systems based on storage concepts [5]

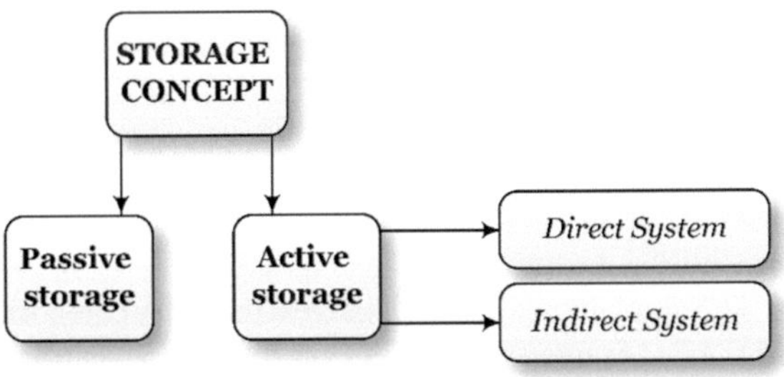

- Improved economics through reduced capital and operational costs
- Enhanced efficiency by making better use of energy
- Reduced environmental pollution and lower CO_2 emissions
- Better system performance and reliability.

Research by Arce et al. [2] assessed the energy-saving potential of TES across Europe, examining applications in both building and industrial contexts. The building sector analysis covered seasonal solar thermal installations, district/central heating networks, short-term solar thermal systems, and passive cooling technologies. For industry, the evaluation included Combined Heat and Power (CHP) systems, industrial heating/cooling processes, power generation facilities, and concentrated solar power (CSP) plants.

The study determined TES potential by analyzing key energy and CO_2 emission parameters. Two primary metrics have been established to evaluate the effectiveness of energy storage solutions. The first metric, thermal load reduction, refers to the decrease in capacity requirements that would be necessary in the absence of storage. This indicates how much less energy is needed due to the presence of storage systems. The second metric focuses on thermal and electrical energy savings, which measure the reusable heat or cold obtained from storage. By effectively utilizing this stored energy, we can significantly reduce the demand for newly generated energy, thereby enhancing overall efficiency.

Collectively, reusing stored thermal energy lowers CO_2 emissions by reducing dependence on fossil fuels and other greenhouse gas-emitting sources.

In general, thermal energy storage tanks are an excellent solution for correcting the mismatch between energy production and demand. They can play a significant role in optimizing energy consumption and meeting the environmental needs of society efficiently. An ideal thermal energy storage tank should have minimal energy losses and be capable of storing or releasing the maximum possible energy when needed. Therefore, extensive studies have been conducted to optimize these systems for the best performance during operation.

1.2 Thermal Energy Storage Systems

Energy requirements across commercial, industrial, and institutional facilities fluctuate daily, weekly, and seasonally. TES systems offer a viable solution for effectively managing these variable demands. Recently, significant interest has developed in applying TES technologies for water heating and cooling, air conditioning, and related applications. As illustrated in Fig. 1.3, integrating energy storage tanks into air conditioning systems can substantially cut energy consumption. This approach avoids the need to run AC units during peak daytime hours when electricity costs are highest. Instead, the energy storage cycle is completed overnight during off-peak periods when energy is more affordable.

In recent decades, various techniques for thermal energy storage have emerged. Such systems have a high potential for use in thermal energy-consuming equipment and, from an economic perspective, are suitable alternatives to support large thermal energy-consuming units. Energy can be stored in various ways. However, since in many countries, energy is produced and transmitted as heat, the role of thermal energy storage systems becomes increasingly evident. Some of the advantages of using thermal energy storage tanks include:

- Reducing energy costs and consumption
- Lowering initial and maintenance costs
- Decreasing the size of energy equipment
- Utilizing more effective equipment with better and higher efficiency
- Reducing fossil fuel consumption by using clean energy
- Decreasing environmental pollution.

Energy storage tanks may store energy short-term, over a few hours, or long-term, over several months. However, in most applications, short-term energy storage systems are used. For example, solar energy is only available during the day, and therefore, thermal energy storage systems must be highly efficient to store the maximum possible amount of heat during sunlight hours to be used at night. A

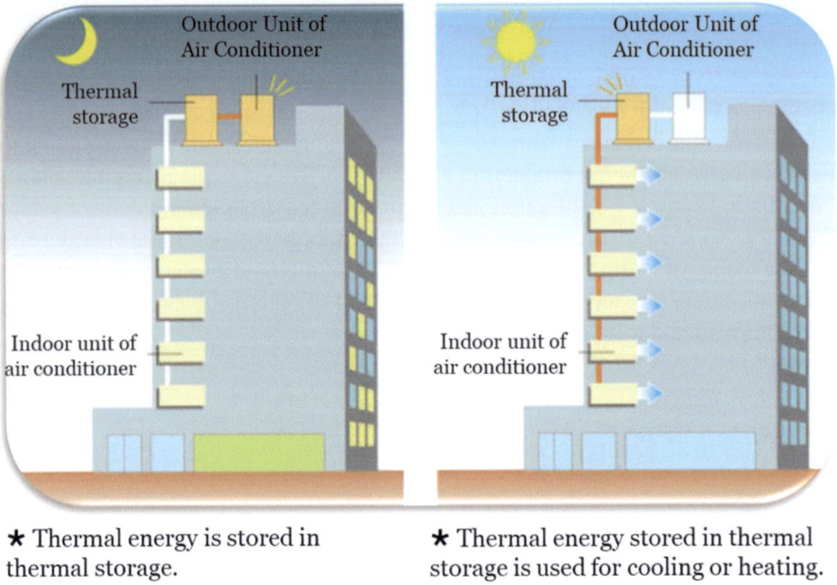

Fig. 1.3 Using thermal energy storage in air conditioning to reduce the cost of energy consumption [14]

similar phenomenon can be observed in heat recovery systems, where the wasted energy varies at different production times.

1.3 Methods of Storing Thermal Energy

Thermal energy storage technologies are fundamentally classified into three categories: sensible heat storage, latent heat storage, and thermochemical storage. Table 1.1 provides a comprehensive comparison of these TES types alongside electrical storage, detailing their key characteristics. Figure 1.4 further defines the three primary TES concepts, specifying their associated physical phase changes or chemical reactions, and identifies the materials typically best-suited for high-temperature applications (above 300 °C).

1.3.1 Storage of Sensible Heat

Sensible heat storage involves changes in the temperature of a material. All materials can absorb and store heat due to their mass and specific heat capacity at constant pressure (c_p). As temperature increases, so does the heat capacity, a concept explained

Table 1.1 Comparison of the storage density of different TES methods [5]

Type of storage technology	Material	Stored energy (MJ/m^3)	Energy density (kJ/kg)	Comments
Sensible heat	Water	84	84	$\Delta T = 298$ K
	Granite	50	17	$\Delta T = 298$ K
Latent heat	Paraffin	180	200	$T_{melting} =$ 278–403 K
	Water	306	330	$T_{melting} =$ 273 K
	Salt hydrates	300	200	$T_{melting} =$ 278–403 K
	Salt	600–1500	300–700	$T_{melting} =$ 573–1073 K
Chemical reactions	H_2 gas (oxidation)	11	120,000	1 bar, 300 K
	H_2 gas (oxidation)	2160	120,000	200 bar, 300 K
	H_2 liquid (oxidation)	8400	120,000	1 bar, 20 K
	Fossil gas	32	–	1 bar, 300 K
	Gasoline	33,000	43,200	–
Electrical storage	Pb battery	–	70–180	–
	Zn/Mn oxide battery	–	180	–

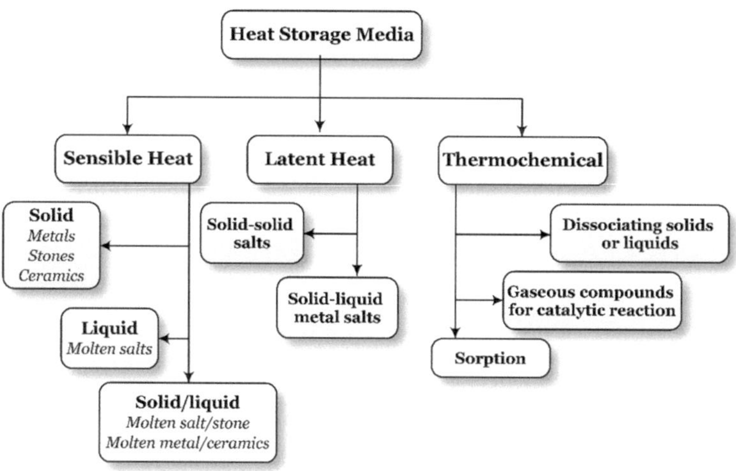

Fig. 1.4 Classification of the three thermal energy storage concepts based on physical phase and thermochemical reaction type for temperatures exceeding 300 °C [7]

by the Debye model. For a temperature difference ($\Delta T = T_2 - T_1$), the heat (or enthalpy) stored is given by:

$$\Delta H_m = mc_p(T_2 - T_1) = mc_p \Delta T \tag{1.1}$$

In this context, T_2 represents the material's final temperature after heat absorption (charging), while T_1 denotes its temperature after heat release (discharging). Systems engineered with a greater temperature differential ΔT achieve significantly lower specific storage costs because storage capacity scales approximately with the temperature difference between hot and cold states.

For pure solid materials at high temperatures, especially heavy elements, the molar specific heat approximates $3R$ (Dulong–Petit law), where R is the universal gas constant (8.31441 J/mol K). Nonporous solids and liquids typically exhibit volumetric heat capacities ($\rho \cdot c_p$) between 1.5 and 6 MJ/m^3 K [6].

1.3.1.1 Storage Materials

Liquids as Storage Materials

Liquids provide a dual-functional advantage in TES systems, serving as both the storage medium and HTF. Beyond previously discussed properties, the following thermochemical and physical characteristics are essential for liquid TES media:

- Minimum Operating Temperature (T_{min}): Defined as the liquidus or melting point temperature (T_m).
- Maximum Operating Temperature (T_{max}): Constrained by thermal stability limits, vapor pressure, or corrosion risks.
- Dynamic Viscosity (μ): Expressed in Pa s; must be sufficiently low to minimize pumping power requirements.

At lower temperatures, water is commonly used for storing and transferring heat. This includes uses like hot water tanks in homes and industries, storing solar energy during the sunny months, and using aquifers that can handle temperatures up to 150 °C without high pressure. However, when water is under high pressure and heated above 374 °C, it can create challenges due to safety limits. The best technology for TES in high-temperature processes is steam accumulators, which are expensive because of the strong pressure tanks needed. For this reason, this discussion will not focus on water-based systems anymore.

Another common fluid used for heat transfer is thermal oil. Mineral oil works at normal pressure and temperatures up to around 300 °C. Synthetic oils can handle even higher temperatures, up to 400 °C, but are more costly and need strong pressure tanks, making them less practical over 300 °C. For temperatures above 150 °C, molten salts are a good option for storing heat in liquid form. They are affordable, stable at high temperatures, and do not need pressure tanks, which simplifies storage design. Molten salts are used in industries for heat treatment and other processes, but they come with a challenge: they can freeze if they get too cold, which means extra heating systems are needed. Their upper temperature limit depends on how stable they are and how they react with metals.

Using mixtures of salts instead of single salts can lower melting points and maintain thermal stability. These mixtures, like eutectics, allow for a broader range of temperatures. In solar thermal power plants, salts made from alkali nitrates, and sometimes alkali nitrites, are preferred for thermal energy storage. A common mixture is Solar Salt, which consists of 60% sodium nitrate and 40% potassium nitrate. This mixture can stay liquid up to about 250 °C and is stable up to around 560 °C. Other important salt mixtures for higher temperatures include combinations like K_2CO_3-Li_2CO_3-Na_2CO_3, which works from 400 to 650 °C, KCl-$MgCl_2$, which operates between 430 and over 700 °C, and KF-LiF-NaF, effective from 460 to over 700 °C [4].

In high-temperature storage systems, molten metals like sodium and lead are also good options. These metals, often used in nuclear reactors, have great heat stability and transfer qualities. Sodium, in particular, has very good thermal characteristics. However, because these metals react strongly with air and water when hot, they need careful handling and special designs. Other possible materials for storing heat in liquid form include molten glass, which can function between 400 and 1200 °C, and molten sulfur, which works from 200 to 600 °C [3]. For temperatures above 300 °C, systems that combine both liquids and solids are also possible. Table 1.2 lists some commonly used material systems.

Solids as Storage Materials

Solid materials can be used at many different temperatures and can be heated to very high levels. For example, ceramics found in Cowper regenerators can handle temperatures above 1000 °C. These materials are usually chemically stable and have low vapor pressure, making it easier to contain them compared to liquids. Solid storage materials fall into two main groups: metals and nonmetals, and they can also be natural or manufactured. Natural materials, like rocks and pebbles, are common and inexpensive. Stones like granite, basalt, and quartzite can work well in high-temperature situations.

Manufactured solid materials, especially different types of ceramics, are often used for heat storage. Refractory bricks made from substances like silica, alumina,

Table 1.2 Material systems for combined heat transfer fluid with filler

Filler material as a solid storage medium	Heat transfer fluid and liquid storage
Cast iron and natural stone (granite, silica and quartzite)	Thermal oil [8]
Graphite	Halogen salt [9]
Natural stone (quartzite), graphite, and alumina	Sodium lead-lead-bismuth eutectic [10, 15]
Natural stone (basalt and quartzite)	Nitrate/nitrite salts [8, 13, 16]
Fuller's earth	Heat transfer fluid—liquid metal (NaK) [15]

magnesia, and iron oxide are used in things like Cowper regenerators, night storage heaters, and tiled stoves. Slag is another low-cost option for storage. Concrete is another option for storing heat, working at temperatures up to 500 °C. It is cheap, easy to find, and can be easily formed into shapes [12]. Graphite can also be heated to very high temperatures but needs an inert atmosphere above about 400 °C to stop it from burning away.

Metals usually have higher thermal conductivity than nonmetals but are not as cost-effective. However, their high conductivity helps heat storage systems charge and discharge quickly. High-conductivity metals like aluminum and copper are used in electronics for heat sinks, while cast iron has good thermal conductivity, high heat capacity, and is affordable. Common solid salts, like sodium chloride and sodium sulfate, can also be used for high-temperature storage. Important things to consider for these materials include creep (the tendency to deform over time), strength, how they are made, and how well they work with heat transfer fluids.

1.3.2 Storage of Latent Heat

When we talk about storing latent heat, we refer to materials called Phase Change Materials. These materials can change from solid to liquid and back again. When they melt, they absorb heat, and when they solidify, they release heat. This happens at a specific temperature, known as the melting temperature (T_m), where the material can absorb heat without getting hotter.

The main benefit of storing latent heat is that you can keep energy at a narrow temperature range, close to the phase change temperature. Changes in crystal forms without changing the physical state can also store latent heat, but these changes usually involve less heat than melting and solidifying. Liquid–gas phase changes are not commonly used for heat storage because gases take up a lot of space. So, solid–liquid changes are the main focus for thermal storage. The heat stored during melting, known as the latent heat of melting ($\Delta h_{m,mol}$), is measured in J/kg.

A rough estimate of the latent heat for different PCMs can be figured out using Eq. (1.2) where the melting temperature, T_m, is in Kelvin, and R is the molar gas constant:

$$2.5RT_m < \Delta h_{m,mol} < 5RT_m \tag{1.2}$$

This relationship reveals that a material's latent heat of fusion (ΔH_f) scales positively with its phase change temperature, and low-molar-mass compounds offer advantages due to higher ΔH_f values expressed in J/kg. For semiconductors and similar materials, ΔH_f typically falls within an empirically constrained range determined by T_m and other factors. When heated across substantial temperature intervals, a material's cumulative thermal energy integrates both latent heat (phase change) and sensible heat contributions from solid and liquid phases. Crucially, ΔH_f generally exceeds sensible heat when evaluated over narrow temperature ranges. Since T_m is material-specific, it serves as a critical parameter for storage medium selection.

1.3.2.1 Storage Materials

Phase Change Materials are characterized primarily by their phase transition temperature and enthalpy, enabling diverse applications. For low-temperature uses (e.g., building cooling), water/ice and water-based slurries (brine or glycol solutions) are common. In the medium range (up to 150 °C), paraffins and salt hydrates serve in transport containers, electronics thermal management, wearable heaters, textiles, and HVAC systems. Above 120 °C, organic PCMs encounter challenges like thermal degradation, oxygen reactivity, and high vapor pressure, necessitating alternative materials.

Anhydrous salts/mixtures are viable for phase changes > 130 °C. Typical formulations combine alkali/alkaline earth cations (Li^+, Na^+, K^+, Ca^{2+}, Mg^{2+}) with anions like nitrates, nitrites, hydroxides, halides, carbonates, or sulfates. Extensive anhydrous salt miscibility enables diverse single/ternary/binary systems. Figure 1.5 catalogs high-enthalpy PCMs (> 300 °C melting point). Though often low-cost, salts exhibit poor thermal conductivity ($\leq$ 1 W/m K), demanding enhanced heat transfer strategies.

Metallic PCM candidates include silicon, magnesium, aluminum, and their alloys. Lithium hydride (LiH) achieves an exceptional specific enthalpy due to its minimal molar mass. Metals/alloys generally offer superior thermal conductivity versus salts, benefiting performance-critical applications.

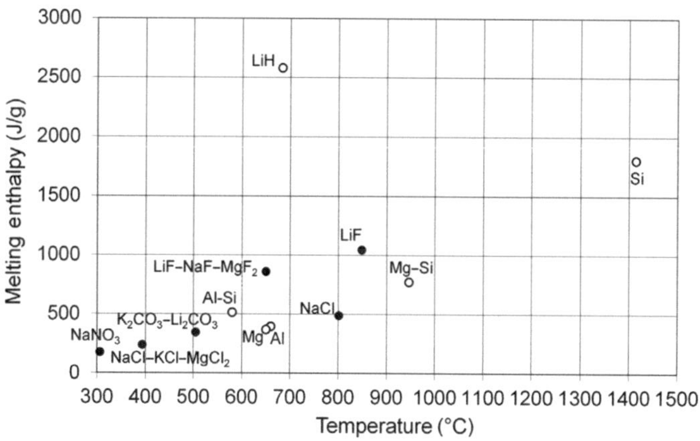

Fig. 1.5 Properties of selected inorganic high-temperature PCMs: salts and salt mixtures (full circles) and metals and other alloys (empty circles) [7]

1.3.2.2 Storage Components and Heat Transfer Concepts

Designing a latent heat storage system starts with PCM selection according to the application's required phase change temperature. Effective PCM storage necessitates optimized heat transfer strategies to minimize the ΔT between the PCM's melting point and HTF temperatures, particularly for low-thermal-conductivity PCMs. The PCM's thermal conductivity critically governs system power density and heat exchanger design. Charge/discharge power depends directly on PCM melting/solidification durations. Figure 1.6 surveys enhanced-heat-transfer PCM storage configurations for HTF-material interfaces.

Analytical solutions for nonlinear phase change problems originated in the nineteenth century but remain limited to simple geometries (e.g., plane walls, tubes). Complex geometries and phenomena like free convection require numerical resolution [20].

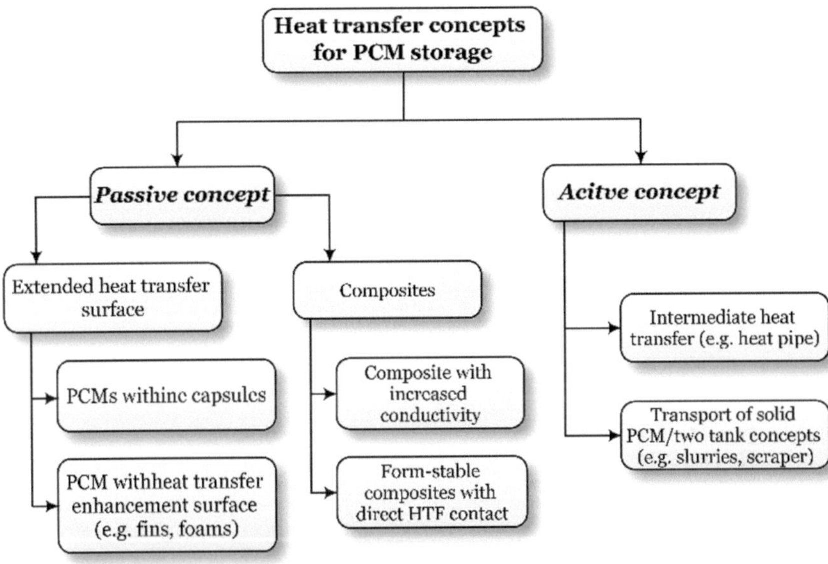

Fig. 1.6 Heat transfer enhancement concepts for latent heat storage systems [7]

1.3.3 *Passive Phase Change Material Heat Transfer Concepts*

1.3.3.1 Extended Heat Transfer Surface Concepts

In the realm of passive phase change material heat transfer, two primary concepts stand out for extending heat transfer surfaces: encapsulation and the use of additional conductive structures.

Encapsulation of PCM

Encapsulation involves enclosing the PCM within capsules to prevent leakage during its liquid phase. This can be done on a micro or macro scale, known respectively as microencapsulation and macroencapsulation. Microcapsules, often embedded in a solid matrix or suspended in a heat transfer fluid (creating a microencapsulated PCM slurry), offer a large surface area, enhancing heat transfer rates. Macrocapsules alternatively achieve comparable thermal performance by immersion in HTF (e.g., molten salts or metals). Critical design parameters include:

- Wall thickness (flexible/stiff configurations)
- Internal/external pressure differentials
- Material compatibility between capsule, PCM, and HTF
- Expansion allowance volume for PCM melting
- Thermomechanical resilience and lifecycle stability.

Encapsulation is generally applicable below 120 °C using polymers. Above 120 °C, metals/graphite/ceramics are preferred. Capsule designs can be categorized into two main types based on their structural characteristics. The first type is the thin-walled, or flexible capsule, which is designed to maintain equal internal and external pressure. In contrast, the second type is the thick-walled, or stiff capsule. This design accommodates pressure differentials and includes a gas buffer to compensate for the expansion of phase change materials. It's important to note that stiff designs tend to limit the fill volume of the PCM; for instance, when using nitrates, the fill volume can be approximately 60% due to the need for expansion compensation [18]. Other critical aspects include ensuring leak tightness and managing the material volume required for the capsules.

Conductive Structures for Enhanced Heat Transfer

Another approach to extending heat transfer surfaces involves incorporating conductive structures within the PCM. These structures, such as fins, foams, meshes, wires, and fibers, aim to reduce the distance for heat transfer in the low-conductive PCM. Dispersed conductive particles, while less effective due to low interconnection, can still enhance heat transfer if designed appropriately. Design criteria for these structures include:

- Compatibility with the PCM
- Minimizing contact resistance between the conductive structures and the heat carrier (e.g., tubes)

- Managing thermomechanical stress due to PCM density changes and ensuring long-term stability
- Optimizing the geometry to minimize material use and ensure effective interconnection.

A crucial factor is the volume fraction of the conductive structure relative to the total volume (PCM plus conductive structure). This ratio helps balance the cost-effectiveness of the conductive materials, which typically include aluminum, graphite, steel, and copper.

Finned Tube Design

Finned tube configurations offer distinct advantages for high-temperature applications, corrosive PCMs, and pressurized HTF. They expand the tube's effective heat transfer surface, enabling substitution of costly pressure-rated piping with more economical ambient-pressure conductive structures. Fin geometry (longitudinal/axial), spacing, and thickness can be tailored to specific power demands. Mastery of these principles allows engineers to substantially boost PCM-based heat transfer efficiency, enabling advanced thermal management solutions with enhanced reliability [18].

1.3.3.2 Conductive Composites

Conductive composites substantially enhance heat transfer by merging PCMs' high latent heat with additives' superior thermal conductivity. Critical to their performance is maintaining composite stability, ensuring no phase separation occurs across repeated melting/solidification cycles. Such PCM composites are fabricated via techniques like compression and infiltration.

Form-Stable Composites

Structurally stable composites maintaining direct HTF contact are highly advantageous. When engineered with microscale dimensions or open HTF channels, these composites achieve exceptional heat transfer rates. For high-temperature applications, salt ceramic composites have emerged, combining anhydrous salt PCMs with ceramic matrices for stability. Systems using PCMs with 700–900 °C phase change temperatures and air HTF in direct contact [19] demonstrate viability.

These material/design innovations enable significant thermal management performance gains, facilitating novel implementations in extreme-temperature environments.

1.3.4 *Active Phase Change Material Heat Transfer Concepts*

1.3.4.1 Active Heat Transfer Enhancement Methods

To actively intensify heat transfer, several advanced techniques can be utilized. One strategy employs an intermediate HTF to shuttle thermal energy between the primary HTF and PCM. Heat pipes operating via intermediate HTF evaporation/condensation cycles are particularly effective. Here, both intermediate HTF and PCM undergo phase transitions, minimizing ΔT while maximizing exergy efficiency during charge/discharge cycles. Where direct PCM-intermediate HTF contact is feasible, primary heat exchangers can be substantially smaller than PCM-embedded units. Critical intermediate HTF requirements include alignment between its boiling/condensation temperatures and the PCM's melting point [1].

An alternative approach addressing heat transfer constraints from PCM solidification involves solid PCM transport. This enables segregated liquid/solid PCM reservoirs, permitting smaller heat exchangers versus PCM-embedded designs. Ice slurries implement this concept in cooling applications. At elevated temperatures, granular PCM transport systems utilize screw heat exchangers, scrapers, or surface pre-heating [17].

1.3.5 *Storage Integration and Applications*

Operational prototypes for high-pressure steam storage (capacity: hundreds of kWh) have been realized and thoroughly analyzed. Candidate PCM melting temperatures correspond to water/steam phase change conditions, spanning 120 °C (2 bar) to 340 °C (150 bar). These latent heat systems employ finned heat exchanger pipes embedded within PCM to boost thermal transfer. The operational cycle of the system involves two key processes: charging and discharging. During the charging phase, steam condensation inside the pipes leads to the melting of PCM. This absorbed energy prepares the system for its next function. In the discharging phase, the PCM solidifies, releasing stored heat which in turn evaporates water to generate steam. This process is crucial for applications such as direct steam generation in concentrated solar power plants and in high-pressure steam backup systems [11]. Other potential applications and PCM storage concepts include:

- Storing steam for industries that need a lot of hot water or steam.
- Using different types of PCMs together to cover a wider range of temperatures.
- Creating PCM systems that can work over a temperature range instead of just one specific temperature.
- Building thermocline storage with sections of PCM to keep the temperature steady in certain applications.
- Storing electrical energy in steam power systems and converting it into heat and light.

- Using high-temperature PCMs in special solar power systems like Dish-Stirling and Fresnel technology.

By combining these advanced storage and heat transfer ideas, engineers can create better and more useful systems for managing heat in many different areas.

1.3.6 Thermochemical Storage

Thermochemical energy storage systems utilize reversible reactions' enthalpy changes for energy storage. These systems offer superior energy density versus other TES types [22], with key advantages:

- *Ambient temperature storage*: Reactants/products remain storable at room temperature, reducing thermal losses—ideal for seasonal/long-term storage.
- *Variable temperature levels*: Unlike sensible/latent TES, thermochemical systems enable different charge/discharge temperatures through gas-phase pressure adjustments. This facilitates heat upgrading/downgrading ("heat transformers" "chemical heat pumps") for optimal temperature delivery.
- *Simple discharge mechanism*: Thermochemical systems can be discharged easily, even from room temperature, by simply bringing the reactants together, such as by opening a valve for gas supply.

Energy storage involves generating chemical compounds A and B via endothermic reactions. Recovery occurs through exothermic recombination, forming AB (Eq. 1.3):

$$AB \overset{\Delta H_r}{\Leftrightarrow} A + B \tag{1.3}$$

The thermal energy stored/released equals the reaction enthalpy ΔH_r, typically exceeding latent heat transition enthalpies or sensible heat capacities across practical temperature intervals. Consequently, thermochemical materials achieve higher storage densities than latent/sensible alternatives.

While many thermochemical TES concepts remain in earlier development phases versus mature sensible/latent systems, low-temperature applications (< 150 °C) like sorption technologies are already commercially deployed in niche markets. Thermochemical storage's potential was identified during CSP technology's formative development. Multiple research groups actively investigate solar-driven chemical processes for producing solar fuels and marketable commodities that simultaneously function as energy storage systems.

1.3.6.1 Classification of Thermochemical Storage Systems

Thermochemical storage systems can be broadly classified into two main categories: open-type and closed-type systems.

Open-Type Systems

Open-type systems engage in gas exchange with the surrounding environment. Charging involves releasing gases into the environment, whereas discharging uses an environmental gas. This configuration removes the necessity for compressing and storing gas, thereby reducing system complexity. Relevant gases are O_2, N_2, H_2O, and potentially CO_2. However, open designs can admit undesirable environmental contaminants like dust, SO_2, CO_2, and organic compounds, potentially degrading system performance. Utilizing filters can alleviate these concerns.

Closed-Type Systems

In closed-type systems, storing the unpressurized gas phase is usually impractical due to the large volume required. Instead, these systems typically compress or condense the gas, allowing for convenient storage of the pressurized gas or liquid. Alternatively, the gas may be reabsorbed by a secondary chemical reaction. This approach highlights the importance of considering the overall system design rather than focusing solely on a single chemical reaction.

Sorption Processes

Sorption processes can be viewed as chemical reaction systems exhibiting weaker chemical interactions compared to the strong covalent bonds typical of other chemical systems. Within a sorption heat storage context, the sorbent is heated during charging, causing vapor desorption from the material. During discharging, lower-temperature vapor undergoes adsorption (for solid sorbents) or absorption (for liquid sorbents), releasing thermal energy at a higher temperature.

Adsorption processes using H_2O as the working fluid alongside materials like silica gel and zeolites are commonly used for low-temperature applications. Alternatively, absorption mechanisms can be employed, using compounds such as $MgSO_4$, LiCl, LiBr, $CaCl_2$, $MgCl_2$, KOH, and NaOH [21, 23]. Crucially, sorption systems are generally incompatible with high-temperature applications, therefore, these will not be explored further here.

Reversible Reactions

Figure 1.7 classifies reversible reactions into three physical phases: solid–gas, liquid–gas, and gas–gas reactions. For temperatures above 300 °C, solid–gas reactions are primarily relevant, with some consideration given to liquid–gas reactions. By understanding these classifications and the specific requirements of each system type, engineers can design more efficient and effective thermochemical storage solutions tailored to various applications.

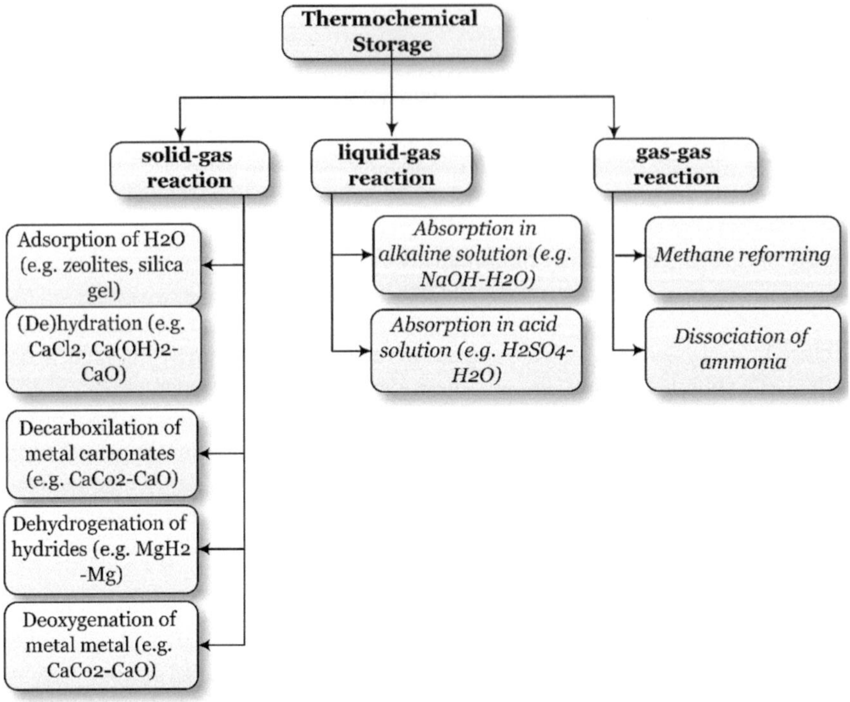

Fig. 1.7 Classification of thermochemical energy storage by the reaction type

References

1. Adinberg R, Epstein M (2007) Conception and design of a thermal energy storage system. In: 5th IASME/WSEAS international conference on heat transfer, Athens, Greece
2. Arce P, Medrano M, Gil A, Oró E, Cabeza LF (2011) Overview of thermal energy storage (TES) potential energy savings and climate change mitigation in Spain and Europe. Appl Energy 88:2764–2774
3. Barde A, Jin K, Shinn M, Nithyanandam K, Wirz RE (2018) Demonstration of a low cost, high temperature elemental sulfur thermal battery. Appl Therm Eng 137:259–267
4. Bauer T, Pfleger N, Laing D, Steinmann W-D, Eck M, Kaesche S (2013) High-temperature molten salts for solar power application. In: Lantelme F, Groult H (eds) Molten salts chemistry. Elsevier, Oxford, pp 415–438
5. Cabeza LF (2012) Thermal energy storage. In: Sayigh A (ed) Comprehensive renewable energy. Elsevier, Oxford, pp 211–253
6. Cverna F (2002) Introduction to thermal properties, chap 1. In: Thermal properties of metals, pp 1–8
7. Datas A (2020) Ultra-high temperature thermal energy storage, transfer and conversion. In: Woodhead publishing series in energy. Woodhead Publishing
8. Esence T, Bruch A, Molina S, Stutz B, Fourmigué J-F (2017) A review on experience feedback and numerical modeling of packed-bed thermal energy storage systems. Sol Energy 153:628–654
9. Forsberg CW, Peterson PF, Zhao H (2006) High-temperature liquid-fluoride-salt closed-Brayton-cycle solar power towers. J. Sol. Energy Eng. 129:141–146

10. Fritsch A, Flesch J, Geza V, Singer C, Uhlig R, Hoffschmidt B (2015) Conceptual study of central receiver systems with liquid metals as efficient heat transfer fluids. Energy Procedia 69:644–653
11. Johnson M, Vogel J, Hempel M, Hachmann B, Dengel A (2017) Design of high temperature thermal energy storage for high power levels. Sustain Cities Soc 35:758–763
12. Laing D, Bahl C, Bauer T, Fiss M, Breidenbach N, Hempel M (2012) High-temperature solid-media thermal energy storage for solar thermal power plants. Proc IEEE 100:516–524
13. Martin C, Bonk A, Braun M, Odenthal C, Bauer T (2018) Investigation of the long-term stability of quartzite and basalt for a potential use as filler materials for a molten-salt based thermocline storage concept. Sol Energy 171:827–840
14. Mosaffa AH, Infante Ferreira CA, Rosen MA, Talati F (2013) Optimization of free cooling systems using multiple PCM latent heat thermal storage unit. In: 4th IIR conference on thermophysical properties and transfer processes of refrigerants, Delft, The Netherlands, p 19
15. Niedermeier K, Marocco L, Flesch J, Mohan G, Coventry J, Wetzel T (2018) Performance of molten sodium vs. molten salts in a packed bed thermal energy storage. Appl Therm Eng 141:368–377
16. Pacheco J, Showalter S, Kolb W (2002) Development of a molten-salt thermocline thermal storage system for parabolic trough plants. J Sol Energy Eng 124:153
17. Pointner H, Steinmann W-D (2016) Experimental demonstration of an active latent heat storage concept. Appl Energy 168:661–671
18. Steinmann W-D, Tamme R (2007) Latent heat storage for solar steam systems. J Sol Energy Eng 130:011004
19. Tamme R, Taut U, Streuber C, Kalfa H (1991) Energy storage development for solar thermal processes. Sol Energy Mater 24:386–396
20. Tien CL (1999) Annual review of heat transfer. Begell House
21. Weber R, Dorer V (2008) Long-term heat storage with NaOH. Vacuum 82:708–716
22. Wentworth WE, Chen E (1976) Simple thermal decomposition reactions for storage of solar thermal energy. Sol Energy 18:205–214
23. Zsembinszki G, Solé A, Barreneche C, Prieto C, Fernández AI, Cabeza LF (2018) Review of reactors with potential use in thermochemical energy storage in concentrated solar power plants. Energies 11:2358

Chapter 2
Phase Change Materials

2.1 Introduction

Phase change materials are substances that experience a solid–liquid phase transition, typically termed the melting-solidification cycle, within the temperature range appropriate for a specific thermal energy application. While transitioning from solid to liquid, these materials take in energy from their environment while sustaining a constant or nearly constant temperature. This assimilated energy elevates the vibrational state of their constituent atoms or molecules. At the melting point, atomic bonds weaken, permitting the material to change from a solid to a liquid state.

The reverse process, solidification, occurs when the material releases energy to its surroundings. As the material cools, the molecules lose energy and reorganize into a solid phase. This cyclical process of melting and solidification enables PCMs to effectively store and release thermal energy, making them valuable for various thermal management applications.

2.2 The Marvel of Phase Change Materials

Within thermal management applications, specific natural materials distinguish themselves through their exceptional capacity to experience phase transitions. These substances, designated as phase change materials, exhibit high inherent heat storage capacity, termed latent heat capacity. This distinctive characteristic stems from the significant thermal energy necessary to overcome intermolecular forces during solid-to-liquid transitions and vice versa. Thus, melting is strongly endothermic, absorbing thermal energy, while solidification is exothermic, releasing heat.

PCMs are particularly valuable in the field of electronics, where effective thermal management is crucial. During the melting phase, PCMs absorb a significant amount

A. Mosaffa, *Phase Changing Materials*, Materials Horizons: From Nature to Nanomaterials, https://doi.org/10.1007/978-981-95-3742-6_2

of heat, which is later released into the surrounding environment during solidification. This cyclical process helps maintain a stable temperature, protecting sensitive electronic components from overheating.

One of the defining characteristics of PCMs is their ability to melt and solidify within specific temperature ranges (0–100 °C) without consuming excessive energy. This makes them ideal for capturing and storing substantial thermal energy under both transient and continuous power loads. As the temperature stabilizes, the stored heat is gradually released, ensuring the protection of electrical circuits.

Organic PCMs provide multiple benefits, including long-term chemical stability, high latent heat capacity, non-corrosive properties, efficient thermal energy diffusion, and cost-effectiveness. These properties have facilitated pervasive implementation across diverse applications, including lithium-ion battery packs, heating and air conditioning (HVAC) systems, semiconductor chip thermal management, solar thermal power systems, textiles, heat pumps, refrigeration systems, and building thermal comfort systems. Nevertheless, their limited thermal conductivity impedes efficacy in heat transfer applications.

Despite this drawback, PCMs exhibit exceptional performance as latent heat storage media. They assimilate surplus peak energy loads during one phase of the power cycle and liberate stored thermal energy during the subsequent phase, thereby preventing thermal degradation of circuit elements. PCM-based thermal management systems constitute a novel thermal regulation strategy, diminishing thermal resistance and guaranteeing the operational integrity of electronic apparatus.

Forecasting alternating melting-solidification heat transfer cycles is crucial for optimizing latent heat energy storage systems (Fig. 2.1). Nevertheless, PCMs' low thermal conductivity presents difficulties for electronic cooling applications. To overcome this, various technologies have been devised to improve PCM thermal conductivity, including porous structures, pin–fin heat sinks, and nanoparticles in PCM-based heat sinks. These advancements markedly enhance PCM efficiency in thermal management applications.

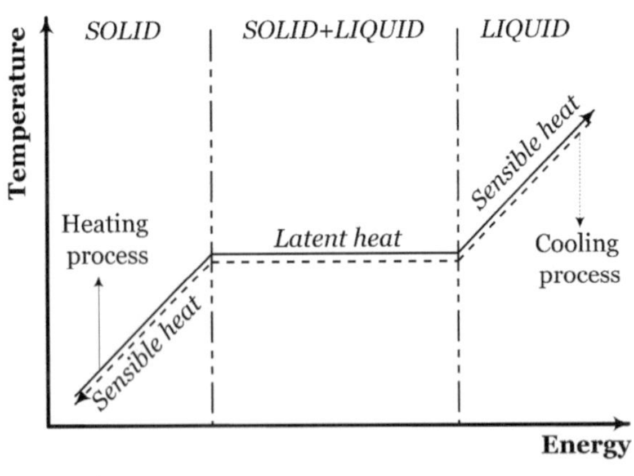

Fig. 2.1 Temperature variations in PCM during the melting/solidification process

2.3 Characteristics of Phase Change Materials

Scientists and researchers have been attracted to PCM owing to their outstanding heat-retaining properties, enabling absorption and release of substantial energy quantities. PCMs consistently store and discharge energy during their phase transition process. Absent this transition, materials cannot accumulate significant amounts. Thermal latent energy storage systems surpass sensible heat storage approaches, providing 5–14 times greater energy storage density due to PCMs' fundamental characteristics. When selecting PCMs, critical attributes/parameters are subject to rigorous assessment, encompassing physical characteristics, kinetic behavior, chemical stability, material thickness, thermal properties, economic viability, reliability metrics, commercial availability, operational longevity, safety protocols, and ecological ramifications; compatibility constraints; and application-specific requirements, as schematically represented in Fig. 2.2.

Choosing a PCM is very important because it affects how much heat the material can hold and release. Here are some key points to think about when selecting a PCM:

- *Chemical stability*: The PCM should stay stable and not break down over time. This is important so it can keep doing its job. It should also not react badly with other materials it comes into contact with.
- *Density*: Density refers to how much mass the material has in a given volume. A higher density means the PCM can store more thermal energy in a smaller space, which is useful if there isn't much room available.

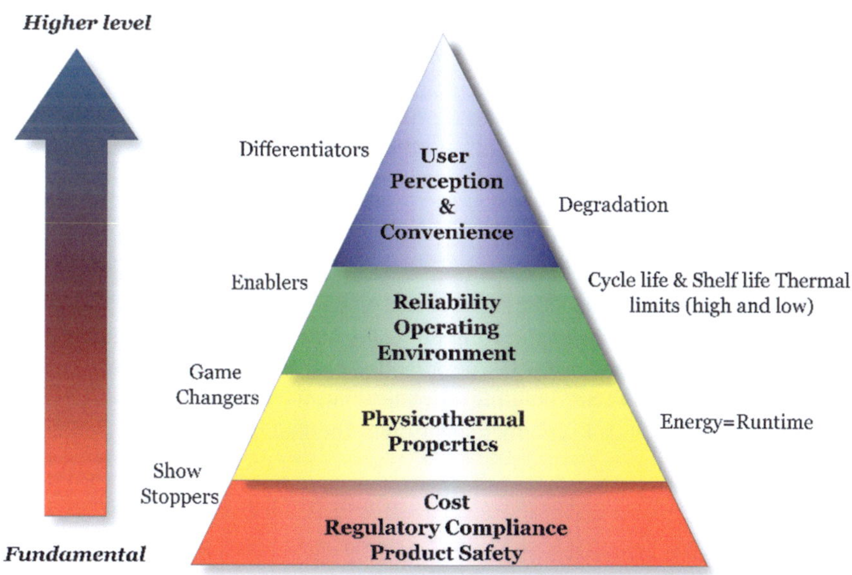

Fig. 2.2 PCM selection criteria [6]

- *Melting temperature*: The temperature at which the PCM melts should be suitable for its intended use. For example, if the PCM is to be used for building insulation, it should melt above the highest outdoor temperature expected.
- *Heat of fusion*: This is the energy needed to change the PCM from solid to liquid. A PCM with a higher heat of fusion can store more thermal energy during this change, making it more effective.
- *Thermal conductivity*: This refers to how quickly heat can move in and out of the PCM. For situations that need fast changes in temperature, a PCM with high thermal conductivity is better.
- *Specific heat capacity*: This tells us how much energy is needed to raise the temperature of the PCM by a certain amount. A PCM with high specific heat capacity can hold a lot of energy for its weight.
- *Thermal stability*: The PCM should be strong and not break down or change over time when subjected to different temperatures.
- *Compatibility with other materials*: The PCM should work well with materials it will be used with, such as containers or heat transfer fluids.
- *Cost*: Finally, the price of the PCM is very important, especially for large projects. It's essential to find a balance between cost and the performance needed from the PCM.

Considering these factors will help guarantee that the chosen PCM functions effectively for its intended application. Ultimately, selecting the PCM type should be underpinned by meticulous consideration of these parameters and any application-specific requirements.

Equation (2.1) illustrates the volume of stored energy in a latent heat storage system:

$$Q = \int_{T_\mathrm{i}}^{T_\mathrm{m}} mc_\mathrm{p} \cdot \mathrm{d}T \bigg|_\mathrm{solid} + m \cdot h_{\mathrm{liquid}\rightarrow\mathrm{solid}} + \int_{T_\mathrm{m}}^{T_\mathrm{f}} mc_\mathrm{p} \cdot \mathrm{d}T \bigg|_\mathrm{liquid} \tag{2.1}$$

Critical selection parameters for PCMs encompass thermal stability, phase change temperature, chemical stability, economic feasibility, latent heat capacity, and minimal volume expansion. These properties consequently establish PCM applicability across multifarious engineering domains, including electrochemical energy storage systems, refrigeration cycles, desalination processes, advanced textiles, solar thermal energy conversion, building climate control systems, food preservation packaging, thermal diodes, and microelectronics thermal management.

2.4 Drawbacks of Phase Change Materials

Commercialization of latent heat thermal storage (LHTS) units is substantially impeded by problematic thermophysical properties of phase change materials. Primary factors impeding optimal LHTS functionality encompass diminished thermal conductivity, temporal variations in thermophysical properties during extended thermal cycling, operational phase segregation, subcooling phenomena, incongruent melting behavior, volumetric instabilities, and prohibitive economic constraints.

A primary limitation of LHTSs is PCMs' inferior thermal conductivity and thermal stability. Typically, thermal conductivity ranges between 0.15–0.3 W/m K for organic materials and 0.4–0.7 W/m K for salt hydrates. This low conductivity proves especially problematic during energy retrieval, causing significant temperature drops. Consequently, phase change process rates (PCM melting/solidification) fail to meet expectations. Although sufficient energy may exist, systems often cannot utilize it at the required rates.

Economic concerns must be taken into account when choosing PCMs for latent thermal energy storage systems, in addition to acceptable thermophysical, kinetic, and chemical characteristics. Different PCM classes have their special traits, uses, benefits, and drawbacks. Finding the best balance between cost and thermal performance is necessary because no single material has all the necessary qualities for thermal storage.

The lifetime and cost of the storage material have a significant impact on the economic viability of PCM deployment. Therefore, a thorough understanding of PCM thermal stability across repeated thermal cycles is essential to guaranteeing the long-term performance and financial sustainability of latent heat storage systems.

2.5 Thermal Stability of Phase Change Materials

Phase change materials may deteriorate after repeated storage cycles, resulting in undesirable alterations to their thermophysical properties over time. A PCM is regarded as reliable if it maintains thermal, chemical, and physical stability across numerous thermal cycles without degradation of its properties, specifically latent heat and melting point. For latent heat storage applications, a PCM is considered thermally stable when it demonstrates negligible changes in melting point and latent heat of fusion following multiple thermal cycles.

The economic viability of PCM implementation is intrinsically linked to the longevity and cost of the storage medium. Therefore, guaranteeing the sustained performance and economic feasibility of latent heat storage systems mandates rigorous characterization of PCM thermal stability under repeated thermal cycling.

As a result, thermal stability tests should be carried out on PCMs before their actual application. While previous literature has extensively addressed the preparation, leakage, thermal conductivity, and thermal storage properties of PCMs, there are relatively few reports focusing on their thermal stability and reliability. To date, only two major review papers have emphasized the importance of thermal stability in the literature.

2.5.1 *Thermal Stability Test*

An LHTS system that uses solar thermal energy goes through a melting and freezing process at least once a day, which we call a normal cycle. A thermostatic chamber, a thermal bath, an electric hot plate, or an oven with a constant temperature are some examples of the several pieces of equipment that can be used to test this process. There are also special machines called thermal cyclers, used mainly in medical research for these tests. When these thermal cycles are done in a controlled lab setting, they are called accelerated thermal cycle tests. After a certain number of cycles, a small sample of the material (PCM) is taken out to check its properties, such as melting point and how much heat it can store.

One common tool used in labs to measure these properties is called a differential scanning calorimeter (DSC). DSC works by comparing the heat needed to raise the temperature of the sample with that of a reference material. It's important to keep the temperatures of both the sample and the reference very close to each other. For this analysis, alumina (Al_2O_3) is often used as the reference material.

During the DSC test, heat is added to both the sample and the reference so that their temperatures stay steady. When the sample changes state, like melting, it will need more or less heat compared to the reference to stay at the same temperature. By measuring the heat that flows to the sample, we can find out how much heat is absorbed or released during these changes. This information is illustrated on a graph called a DSC curve.

From this curve, we can calculate the latent heat of fusion, which is the heat required for the sample to change from solid to liquid. The melting point is found by analyzing the beginning of the peak on the graph. The "mushy zone," when the material first softens and then completely melts, is the region where this melting occurs. We can assess the PCM's stability by tracking changes in the melting point and heat of fusion across a number of cycles.

Figure 2.3 depicts the DSC analysis outcomes for the tested Merck P42-44 paraffin. Analyses were performed across a temperature range of 0–80 °C at a heating rate of 5 °C/min. A 5 mg specimen was hermetically sealed within an aluminum pan for analysis. The sample's melting temperature range was determined by identifying the onset temperature and peak temperature on the DSC thermogram. The onset temperature was ascertained by extending a tangent at the point of maximum slope along the curve's leading edge and extrapolating the baseline proximate to this leading edge [1].

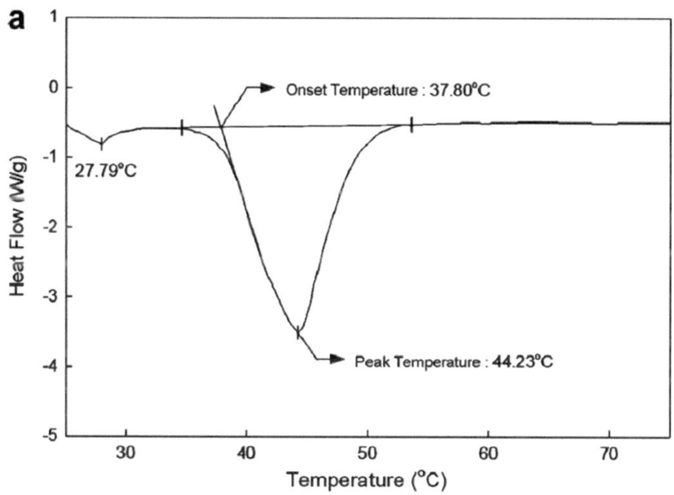

Fig. 2.3 Results of the DSC analysis for Merck P42-44 paraffin considered [1]

2.6 Classification of Phase Change Materials

Phase change materials are classified into three principal categories: organic, inorganic, and eutectic. Organic PCMs are further subdivided into paraffin and non-paraffin classifications. These PCM categorizations are illustrated in Fig. 2.4.

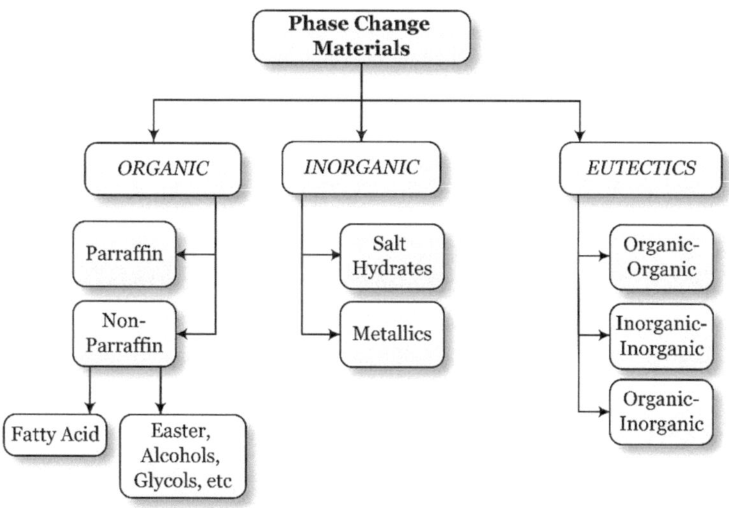

Fig. 2.4 PCMs classification [7]

PCMs undergo four distinct phase transition modalities: solid-to-liquid, solid-to-solid, solid-to-gas, and liquid-to-gas. Solid-to-liquid conversions and their reverse transitions represent the predominantly utilized modality for latent heat storage applications. During heating cycles, these substances absorb significant enthalpic energy, subsequently liberated upon solidification. An exemplary characteristic transition is the water–ice phase change, which occurs isothermally at 0 °C.

Organic PCMs include all types of paraffin and non-paraffin, while inorganic PCMs consist of metallic and salt hydrate composites. Organic PCMs, which are carbon-based compounds, are categorized into paraffin and non-paraffin. Paraffin belongs to a family of hydrocarbons with the chemical formula C_nH_{2n+2} and is widely used in electronic cooling applications due to its beneficial properties, such as non-reactivity, stability, and reusability. They also exhibit minimal volume expansion and subcooling.

Organic PCMs are especially appropriate for building cooling applications owing to their high latent heat of fusion. This enables significant heat absorption or release during phase transitions, effectively regulating indoor temperatures. During daytime, organic PCMs absorb heat to maintain lower indoor temperatures; at night, they release stored heat, supporting cooling and reducing energy consumption.

Organic PCMs typically comprise substances like fatty acids, esters, and alcohols. Examples include:

- *Fatty acids*: Organic compounds featuring carboxylic acid groups bonded to long hydrocarbon chains; examples such as stearic acid, palmitic acid, and lauric acid melt between 20 and 70 °C.
- *Esters*: Formed by alcohol-carboxylic acid reactions, these possess diverse characteristics; examples like butyl stearate, methyl palmitate, and ethyl oleate melt between − 20 and 100 °C, with this broad melting range enabling versatile applications.
- *Alcohols*: Organic compounds containing a hydroxyl group (–OH) attached to a carbon atom; examples, including one-hexadecanol, one-octadecanol, and one-octanol, exhibit melting points from − 10 to 60 °C.

Organic PCMs provide advantages such as affordability, non-corrosivity, and low toxicity. They are widely implemented in thermal energy storage systems, building insulation, and vehicle temperature control. However, disadvantages include poor thermal conductivity and potential flammability, necessitating careful consideration during selection and use.

Inorganic PCMs, conversely, are substances undergoing phase transitions at specific temperatures. These materials typically comprise inorganic compounds, including salt hydrates, eutectic mixtures, metals, and their alloys. Examples are:

- *Salt hydrates*: Inorganic substances with crystalline structures incorporating water molecules; examples such as lithium nitrate trihydrate ($LiNO_3{\cdot}3H_2O$), sodium sulfate decahydrate ($Na_2SO_4{\cdot}10H_2O$), and calcium chloride hexahydrate ($CaCl_2{\cdot}6H_2O$) melt between 0 and 100 °C.

- *Eutectic mixtures*: Inorganic PCMs formed by combining ≥ 2 inorganic substances, exhibiting specific melting points; examples like potassium nitrate-sodium nitrate (KNO_3–$NaNO_3$), calcium chloride-magnesium chloride ($CaCl_2$–$MgCl_2$), and sodium hydroxide-sodium acetate ($NaOH$–CH_3COONa) melt between − 25 and 150 °C.
- *Metals and their alloys*: Inorganic PCMs possessing high melting points and substantial thermal energy storage capacity; examples include aluminum, copper, and their alloys that melt above 100 °C.

Inorganic PCMs provide benefits such as long-term stability, high thermal conductivity, and high energy storage density. They are commonly employed to augment thermal energy storage systems, regulate indoor temperatures via building insulation, and manage vehicle temperatures. However, potential drawbacks include expense, corrosiveness, and hazard risks, requiring careful consideration during selection and implementation.

2.6.1 Advantages and Disadvantages of Organic and Inorganic PCMs

Phase change materials are conventionally categorized into inorganic and organic compounds. Organic PCMs, encompassing paraffin waxes, are noted for their chemical stability and non-corrosive properties. They exhibit negligible supercooling, elevated latent heat capacity, and recyclability, facilitating integration with diverse building materials. Nevertheless, drawbacks comprise low thermal conductivity, substantial volumetric alterations during phase transitions, and pronounced flammability. Consequently, organic PCMs often necessitate encapsulation before deployment in thermal energy storage systems. This encapsulation increases associated costs and reduces system performance attributable to augmented thermal resistance [3].

Conversely, inorganic PCMs, such as salt hydrates and salt composites, provide greater latent heat per unit volume, superior thermal conductivity, reduced costs, recyclability, and non-flammability [2]. Nevertheless, they exhibit corrosiveness toward most metals, resulting in shorter lifespans and higher packaging and maintenance expenses [4]. Additionally, inorganic PCMs may undergo phase separation and supercooling, adversely affecting phase change properties [2, 5]. Metallic materials, distinct from inorganic salts, avoid these drawbacks, representing a potential solution for high-temperature PCMs.

2.7 Summary

This chapter provides a comprehensive analysis of PCMs, substances characterized by their capability to store and release significant quantities of thermal energy during phase transitions. PCMs undergo solid–liquid phase transformations, absorbing thermal energy upon melting and discharging it during solidification while maintaining near-isothermal conditions. This characteristic renders them advantageous for diverse thermal management and energy storage implementations.

The chapter delineates distinctive PCM properties, encompassing high latent heat capacity, specific melting temperature ranges, and their ability to store substantial thermal energy without significant temperature fluctuations. Organic PCMs, exemplified by paraffin and fatty acids, are highlighted for their chemical stability, cost-effectiveness, and compatibility with construction materials. However, inherent limitations include low thermal conductivity and pronounced flammability. Conversely, inorganic PCMs, such as salt hydrates and metallic alloys, offer enhanced thermal conductivity and greater volumetric energy density but may exhibit corrosiveness, phase segregation, and supercooling phenomena.

Applications of PCMs span heating, cooling, solar thermal systems, electronics, and building insulation. The chapter emphasizes their role in regulating indoor temperatures, enhancing energy efficiency, and reducing peak energy loads. Advanced technologies, such as nanoparticles, pin–fin heat sinks, and porous structures, have been developed to improve the thermal conductivity of PCMs, making them more suitable for rapid thermal cycling applications.

Despite their advantages, PCMs face challenges in commercialization due to high costs, thermal instability over cycles, and material-specific drawbacks. The chapter concludes by underscoring the importance of careful PCM selection based on thermal, chemical, and economic parameters, along with continued research to optimize their performance and expand their applicability in sustainable energy systems.

References

1. Akgün M, Aydin O, Kaygusuz K (2008) Thermal energy storage performance of paraffin in a novel tube-in-shell system. Appl Therm Eng 28:405–413
2. Bruno F (2004) Using phase change materials (PCMs) for space heating and cooling in buildings. In: 2004 AIRAH performance enhanced buildings environmentally sustainable design conference, pp 26–31
3. Cabeza LF (2021) Advances in thermal energy storage systems: methods and applications. In: Woodhead publishing series in energy. Woodhead Publishing
4. Chen C, Wang L, Huang Y (2008) Morphology and thermal properties of electrospun fatty acids/polyethylene terephthalate composite fibers as novel form-stable phase change materials. Sol Energy Mater Sol Cells 92:1382–1387
5. Liu H, Awbi HB (2009) Performance of phase change material boards under natural convection. Build Environ 44:1788–1793

6. Nazir H, Batool M, Bolivar Osorio FJ, Isaza-Ruiz M, Xu X, Vignarooban K, Phelan P, Inamuddin, Kannan AM (2019) Recent developments in phase change materials for energy storage applications: a review. Int J Heat Mass Transf 129:491–523
7. Said Z, Pandey AK (2023) Nano enhanced phase change materials: preparation, properties and applications. In: Materials horizons: from nature to nanomaterials. Springer Nature Singapore

Chapter 3
Shortcomings of PCMs in LHTS

3.1 Introduction

While PCMs provide significant benefits over sensible storage materials, they present several hurdles in creating reliable and practical storage systems. Challenges such as incongruent melting, supercooling, low thermal conductivity, and inadequate long-term stability constrain their use. Nevertheless, ongoing R&D is tackling many issues. Although identifying a PCM meeting all Chap. 2 requirements is difficult, various techniques have been developed to circumvent or minimize potential problems. The subsequent sections outline some strategies.

3.2 Incongruent Melting/Phase Separation

Most hydrated salts decompose when melting as temperature increases, producing H_2O and a lower-hydrated salt—termed incongruent melting, shown in Fig. 3.1. The denser, lower-hydrated salt usually sinks, leaving only a thin top layer to recrystallize during freezing. This irreversible process markedly diminishes storage efficiency [21]. Experiments indicate that PCM vertical thickness influences incongruent melting in inorganic salt-water solutions. For example, no incongruent melting occurred with very small PCM thickness (e.g., < 40 mm). However, salt deposition happened at the container bottom when the thickness exceeded ~ 70 mm. A salt-water solution with 10 wt.% salt and 90 wt.% H_2O stays homogeneous above – 4 °C. Below this temperature, H_2O freezes out, raising salt concentration in the remaining solution. Repeated freezing/melting causes separation into original components.

Four approaches prevent incongruent melting: incorporating suspension media/thickening agents, applying the extra water principle, chemical modification, and dynamic melting [57]. Suspension media/thickening agents assist in suspending the lower-hydrated salt, facilitating recrystallization. Nevertheless, these displace

A. Mosaffa, *Phase Changing Materials*, Materials Horizons: From Nature to Nanomaterials, https://doi.org/10.1007/978-981-95-3742-6_3

Fig. 3.1 Incongruent melting in hydrated salt [9]

some PCM, reducing the system's volumetric heat storage capacity and lowering the melting point compared to the pure substance. The extra water principle involves adding extra H_2O to the hydrated salt, forming a saturated salt solution at the melting point. This also diminishes volumetric heat storage capacity and broadens the melting range [45].

Chemical modification, proposed by Carlsson [11], adds a small quantity of isomorphous salt to the hydrated salt to form a solid solution, changing the melting point and hindering lower-hydrated salts. Dynamic melting mixes the PCM in a liquid state to preserve homogeneity [57]. This approach avoids reducing volumetric storage capacity or altering the PCM's melting point but necessitates a mixing mechanism.

3.3 Subcooling

When specific molten phase change materials are cooled, they can exhibit subcooling, solidifying at temperatures below their melting points, as illustrated in Fig. 3.2 during a freezing experiment with a eutectic inorganic salt solution. This subcooling phenomenon arises due to the sluggish rates of nucleation and crystal growth, directly impacting the performance of PCMs and potentially obstructing effective heat recovery if the subcooling is pronounced [19]. During the solidification process, latent heat is not released at the designated phase change temperature. Instead, the material's temperature decreases progressively until crystallization commences. If crystallization is inhibited, the latent heat becomes trapped, resulting in the material only being able to store sensible heat. Therefore, subcooling presents a critical challenge in PCM storage applications [45], as it necessitates lower operational temperatures to initiate freezing, ultimately diminishing the efficiency of the cooling system [61].

To mitigate subcooling, one effective strategy involves the incorporation of nucleating agents possessing crystal structures that are analogous to those of the PCM. These agents, often referred to as "seed crystals," serve as nucleation sites for PCM crystal formation during the freezing process. Alternatively, the cold finger technique can be employed, where a nucleating device is maintained at a temperature lower than the maximum supercooling threshold, facilitating nucleation through crystal formation. Additionally, increasing surface roughness, through methods such as electrolytic polishing or creating porous surfaces, can also reduce instances of subcooling.

It's noteworthy that subcooling is particularly prevalent in encapsulated PCM systems due to the limited volume of PCM within the nodules. Conversely, in bulk PCM applications within larger systems, the occurrence of subcooling tends to be minimized [56].

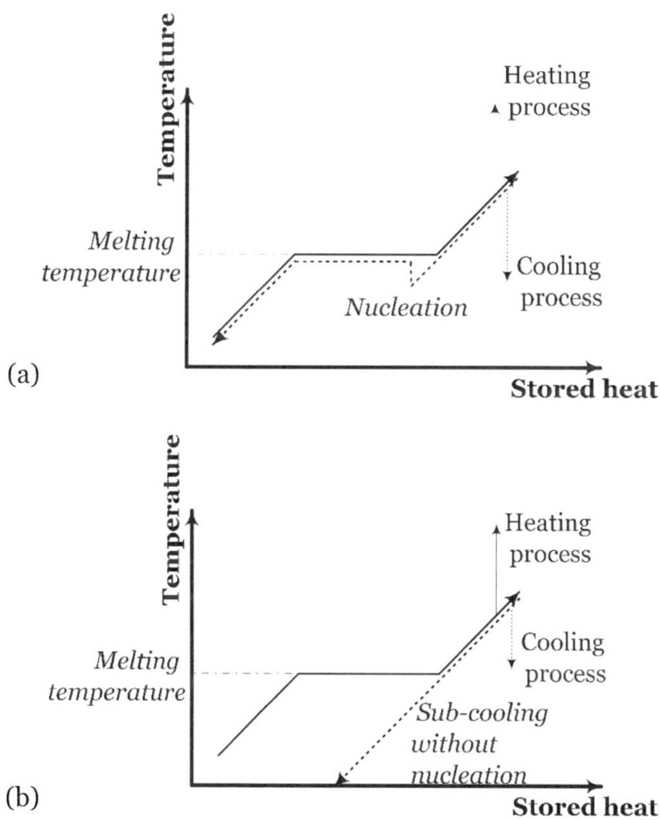

Fig. 3.2 Impact of subcooling on heat storage: **a** minimal subcooling with nucleation, **b** significant subcooling without nucleation [45]

3.4 Low Thermal Conductivity and Heat Transfer Rate

Many high-density PCMs have low thermal conductivity, which means they don't transfer heat well. This makes it necessary to use special techniques to improve heat transfer when storing energy using these materials. During the freezing process, the phase change begins at the surface where heat is being transferred. As the PCM changes from liquid to solid, the boundary between the solid and liquid moves away from the surface. This portion that has changed to solid acts like an insulator, blocking heat from moving to the liquid and making it harder for heat to flow. In solid PCMs, heat transfer happens only through conduction, which is slow because the thermal conductivity is low [56].

Because of this low thermal conductivity, it's not efficient to transfer energy in and out of the PCM. Typically, the melting process (when the PCM changes from solid to liquid) is faster than solidification (when it changes back to solid) because heat can move more easily during melting. To improve heat transfer rates, there are several methods that can be used.

3.4.1 Larger Heat Exchanger Surface

3.4.1.1 Thermal Performance Enhancement Using Pin–Fin Heat Sinks

In thermal energy storage systems employing PCM with HTF, the dominant thermal resistance occurs at the PCM interface. To improve heat transfer efficiency, extended surface areas can be incorporated into the PCM matrix. Experimental studies demonstrate that radial and axial fin configurations on PCM-encased tubes significantly enhance performance, with annular fins proving most effective. Choi et al. [15] observed comparable performance between unfinned tubes and thin-finned variants, though thick-finned designs doubled the heat transfer coefficient.

Ghosh et al. [29] experimentally verified a mathematical model of finned flat walls interacting with paraffin wax during phase change, showing excellent correlation with empirical results. Multiple studies [38] have examined diverse finned tube arrangements, with embedded fins receiving particular attention due to their simple geometry, economic feasibility, and manufacturing convenience.

Metal matrix impregnation presents another thermal enhancement approach. Comparative studies by Agyenim et al. [1] revealed that finned tube systems (both circular and longitudinal) maintained greater radial temperature gradients but smaller axial gradients in liquid phase compared to unfinned systems. While carbon fiber showed better localized heat transfer, carbon brushes delivered superior overall performance.

Experimental research by Castell et al. [12] on vertically oriented finned tubes demonstrated substantial improvements in PCM-to-fluid heat transfer. Their results showed freezing times decreased from 17 min (unfinned) to 13 min (20 mm fins) and 7 min (40 mm fins), representing reductions of 23.53% and 58.82% respectively. Subsequent studies by Fan and Khodadadi [20] comprehensively analyzed PCM thermal conductivity enhancement methods. Sanghvi et al. [53] evaluated various metallic fillers, identifying honeycomb structures as the most effective for thermal conductivity improvement, a finding supported by Zhao et al. [68] who reported an 80% increase in thermal diffusivity using aluminum honeycomb matrices.

De Jong and Hoogendoorn [18] investigated two distinct thermal enhancement approaches: copper finned tubes and metal matrix integration. Their work revealed that while finned tubes substantially shortened solidification periods, metal matrix configurations achieved a sevenfold reduction in solidification duration.

In comparative studies of pin–fin geometries, Ali and Arshad [3] examined n-eicosane PCM systems with 2, 3, and 4 mm diameter circular fins. Their experiments demonstrated superior thermal regulation (40–50 °C range) with 3 mm fins compared to both larger fins and unfinned configurations.

Usman et al. [58] expanded this research by evaluating RT-35, RT-44, and paraffin wax PCMs with different fin arrangements, finding triangular inline fins provided optimal cooling performance at critical thresholds of 45 and 60 °C, particularly with RT-44 PCM.

Avci and Yazici [7] systematically investigated inclination angle effects (0–90°) on flat heat sink performance using advanced imaging techniques. Their data indicated maximum thermal performance occurred at 90° orientation, extending operational duration by 5.5 times compared to horizontal positioning.

Hemati et al. [31] evaluated a power supply system incorporating $CaCl_2{\cdot}6H_2O$ PCM with a fan. Employing 15% fin volume fraction for enhanced thermal conductivity, they tested variable heat fluxes under different fan speeds. Results demonstrated that PCM integration reduced fan power consumption.

Pakrouh et al. [47] numerically optimized RT-44HC-filled four-pin–fin heat sinks using the Taguchi method. Parameters included fin thickness/height, fin count, and base thickness, with genetic algorithms proposed for configuration optimization.

Leong et al. [40] compared paraffin wax and 1-hexadecanol in heat sinks under 0.81 kW/m^2 heat flux. 16-fin paraffin wax configurations outperformed 1-hexadecanol equivalents.

Ali et al. [4] examined circular/rectangular/triangular pin–fin heat sinks with n-eicosane, RT-35HC, SP-31, and RT-44HC PCMs at dual heat inputs. Triangular designs exhibited superior heat transfer without PCM and accelerated integrated circuit response times.

Arshad et al. [5] analyzed circular pin–fin heat sinks with varied cross sections and PCM volumes. 3 mm pin-fins achieved minimal base temperatures and extended operational duration.

Dammak and Hami [17] studied 9%-volume (72-fin) aluminum heat sinks with paraffin wax for electronic temperature stabilization. Cross-fin designs reduced base temperature by 30% using specialized predictive modeling.

Srikanth and Balaji [55] tested 72-pin–fin n-eicosane PCM heat sinks under single versus quad-heater configurations. Thermal performance showed equivalence between single and distributed heating.

Table 3.1 A comprehensive review of published studies on heat sink geometries incorporating PCMs

PCM (melting point (°C))	Heat sink size (mm^3)	Heat sink type	Heat input (kW/ m^2)	References
PCM-HS29P (29) PCM-HS34 (34) PCM-OM37P (37)	50 × 50 × 25	Simple, parallel, and crossed fins	3–5 W	[43]
Plastic paraffin (51.75)	51.5 × 25.5 × 16.5	Unfinned, Cu fins	4.7	[28]
n-Eicosane (35)	42 × 42 × 30	Plate and pin fins	4 and 8	[52]
Paraffin wax (56–58)	114 × 114 × 25	Simple and square pin fins	1.58–3.174	[6]
n-Eicosane (36.5)	80 × 62 × 25	Plate fins	0.99–2.78	[8]

Shang et al. [54] validated reliability of downhole electronics cooling, advocating passive thermal management.

Kalbasi et al. [37] studied finned PCM-based heat sinks using a technique that evaluates the regression model's behavior. Their findings indicated that higher heat density levels resulted in an optimal number of fins. Researchers have examined different shapes, triangular, circular, and rectangular pin–fin heat sinks, filled with PCM at various input power levels. For heat transfer needs at 5 and 8 watts, they recommend using a triangular heat sink. Table 3.1 shows how PCMs are used for cooling electronics efficiently.

3.4.1.2 Thermal Performance Enhancement Using Metallic Foams

Metal foams are prioritized for PCM thermal performance enhancement owing to their exceptional conductivity. Two primary classifications exist: open-cell foams (where fluid permeates the metallic matrix) and closed-cell foams (encased by adjacent cellular structures). This research emphasizes open-cell metallic foams. Porous materials are classified into four types based on pore dimensions (Table 3.2).

Table 3.2 Types of foams and their sizes [22]

Porous support	Porous size
Microporous	< 2
Mesoporous	2–50
Macroporous	> 50
Hierarchical	Ranging from macro to micro

Chow et al. [16] investigated PCM encapsulation geometries and metal/PCM composites, confirming composites substantially increased PCM thermal conductivity. Velraj et al. [60] evaluated three thermal enhancement methods in paraffin-filled aluminum tubes: internal longitudinal aluminum fins with cruciform cross sections, uniformly distributed 1-cm steel Lessing rings; and randomly generated water/vapor bubbles. Lessing rings achieved optimal performance, reducing solid ification time ninefold versus pure paraffin (fins ranked second); bubbles showed negligible improvement. These rings also enhanced paraffin's thermal conductivity tenfold. Lessing rings feature partitioned thin-walled steel cylinders; their non-partitioned counterparts are termed Raschig rings.

Research has also targeted the enhancement of PCM thermal conductivity at the particulate scale. Jegadheeswaran et al. [36] examined solidification rate augmentation in molten salt PCM by dispersing high-conductivity particles, attaining a 17% improvement with 20% particle volume fraction. Substantial work has addressed PCM–graphite composites, leveraging graphite's superior conductivity. Cabeza et al. [10] and Py et al. [49] studied PCM/graphite within water-submerged metal modules, identifying a thermal barrier between modules and water. Yin et al. [65] created a paraffin/expanded graphite composite PCM via liquid paraffin absorption into expanded graphite pores. This achieved 4.676 W/m K thermal conductivity versus pure paraffin's 0.2697 W/m K, with eliminated phase change leakage.

However, Kim and Drzal [39] observed that incorporating expanded graphite reduced the composite's latent heat capacity, consequently, they examined a novel paraffin/exfoliated graphite nanoplatelets (xGnPTM) composite. This exhibited enhanced thermal conductivity with rising xGnP content and no significant difference in latent heat between the paraffin and paraffin/xGnP composite. Chen et al. [13] fabricated a fatty acids/polyethylene terephthalate (PET) ultrafine fibers composite via electrospinning.

The use of heat sinks made with nickel and copper foam has made them work better and last longer. A study by Rehan et al. [51] looked at a special heat sink that used nickel foam and a material called RT-44HC to manage heat in electronic devices when they get hotter. They found that adding PCM and nickel foam inside the heat sink helped transfer heat more effectively.

Another researcher, Ali [2] studied how heat sinks perform when they include porous metal foam and PCM while generating different amounts of heat. Their study compared a regular heat sink with one that had nickel foam and PCM mixed in different amounts (0.6, 0.7, and 0.8). They discovered that using 0.8 amounts of PCM with a nickel foam heat sink made it work four times longer and reduced the temperature by 24%.

Hu and Gong [34] did experiments on a heat sink that used porous materials with different levels of holes (80, 85, 90, and 95%). They found that the heat sink worked best with the material that had the least holes. They also noted that as the power given to the heat sink increased, it ran for less time.

Rehman and Ali [50] looked into how well heat sinks worked using RT-54HC PCM combined with metal foam at different power levels (8, 16, and 24 W). Their study showed that heat sinks with copper foam worked better than those with nickel foam, reducing the base temperature by 26% when both had 0.8 amounts of PCM.

Ferfera and Madani [23] researched heat exchangers that had metal foam and PCM. They performed tests both with computer simulations and in the lab on five different samples that included pure PCM and PCM with foam of different hole sizes. They found that the hole structure was very important for managing heat, and using metal foam helped heat spread much more effectively than just using pure PCM.

Hu et al. [35] looked into how PCM behaves inside low-porosity metal foam with computer models and experiments. Their results were similar in both cases. With foam that had 68% holes, they saw a temperature drop of 83.3% compared to just using paraffin wax.

Mahmoud et al. [43] tested six different heat sink designs with fins and honeycomb materials together with PCM at heat levels of 3–5 W. They concluded that honeycomb materials could replace regular fins. They also found that using PCM with honeycomb material not only made the heat sink last longer but also kept its highest temperature lower.

Zhang et al. [67] worked on improving PCM's poor ability to transfer heat by mixing it with copper foam. They noticed a big difference in heat performance between PCM and copper foam. Veismoradi et al. [59] studied battery cells surrounded by metal foam and PCM to see how the heat and phase change worked. They found that the combination worked very well with strong bursts of heat, making the system seven times more efficient.

Zheng et al. [69] looked at how PCM melted with and without the copper foam that had 95% porosity. They discovered that using copper foam with PCM made it melt 20.5% faster, proving that copper foam helps improve heat movement within PCM.

3.4.1.3 Thermal Performance Enhancement Using Nanoparticles

The integration of nanoparticles into PCMs can markedly enhance their thermal conductivity. Carbon-based NPs, possessing thermal conductivities between 2000 and 6000 W/m K, significantly boost PCM thermal performance relative to metallic oxide NPs due to the latter's lower thermal conductivity. Widely regarded as "the mother of all graphitic forms of carbon," graphene NPs are extensively utilized in PCMs because of their substantial surface area to volume ratio, superior thermal conductivity, facile synthesis, minimized erosion, and high heat transfer capability. Nevertheless, the significant expense of carbon-based NPs represents a major disadvantage. Consequently, incorporating a minimal quantity of carbon NPs into metallic oxide NPs can also efficiently improve the thermal performance of PCMs.

Long-chain fatty acids were utilized, comprising lauric acid, myristic acid, palmitic acid, and stearic acid. This composite demonstrated favorable form-stable morphology and retained outstanding thermal properties following thermal cycle treatment. Pincemin et al. [48] fabricated composites of graphite and inorganic salts for solar applications, attaining a thermal conductivity of 8 W/m K using 40%

natural graphite flakes by weight, however, this substantial graphite content notably diminished the effective thermal storage capacity.

Additional PCM composites have yielded promising results. These encompass polyethylene glycol (PEG)/silica dioxide (SiO_2) composites displaying enhanced conductivity with b-aluminum nitrate; a novel shape-stabilized composite employing PEG and silicon dioxide; oxidized hard Fischer–Tropsch paraffin wax embedded in a UV-cured epoxy resin matrix; and manganese nitrate hexahydrate, alleviating supercooling and heat of fusion [46].

Heat transfer enhancement was also attained by incorporating graphite into PCM slabs. Although PCM/graphite systems effectively improve heat transfer rates, they are expensive. Heat pipes, transferring heat efficiently within compact spaces, reduce storage volume loss. These can transfer heat 500 times more effectively than copper, substantially decreasing the charging and discharging duration of the PCM phase change process. This passive heat transfer device also enables controllable phase change operations, improving system efficiency [61].

Mathematical models integrating heat pipes into latent heat storage systems were introduced by Horbaniuc et al. [32]. Performance enhancements were evaluated through reduced phase change time, increased heat transfer, or effectiveness relative to an ideal scenario. Two thermal conductivity enhancement methods employing carbon fibers were examined in a tube-in-tank system, carbon fibers, possessing higher thermal conductivity than copper, also reduce occupied space within the storage system. The first method dispersed fibers randomly within the PCMs. The second method utilized a fiber brush oriented along the heat flow direction, substantially boosting thermal conductivity in that orientation [26]. Subsequent investigation of carbon-fiber brushes arranged along tubes revealed that transient thermal responses improved with increasing brush length [25].

3.5 Dynamic PCM System

The methods mentioned so far can make systems work better, but they also raise the costs a lot. He and Setterwall [30] first looked into using a special storage system to improve the heat transfer in PCM (Phase Change Materials). This system lets the storage material and the heat transfer liquid come into direct contact, which helps heat move quickly. Martin et al. [44] then tested this idea in experiments using paraffin-based PCM and water. They checked how well the system worked when charging (storing heat) and discharging (releasing heat), as shown in a figure in their study. They found that if they increased the packing factor and the temperature difference, the system could store more heat. However, they also found some problems. For example, when the flow rate increased, the PCM bed expanded, and some liquid PCM got stuck in the frozen parts, which reduced storage capacity. Also, during charging, some frozen PCM shells collapsed, trapping water inside. These issues showed that more research is needed (Fig. 3.3).

Tay et al. [57] also looked into dynamic PCM systems and created a new way to improve heat transfer called dynamic melting. This unit used PCM in a tube-in-tank setup. Their experiments showed that the system's effectiveness increased by 33–89% with a high temperature difference and by 58–82% with a low temperature difference. They found that using a pump to circulate the PCM made the melting process faster and could double the effective thermal conductivity. This was similar to adding fins to the tubes, but without making the system bulkier.

Some problems were found in using PCM for storing energy. These problems included the PCM bed expanding when more fluid flowed through it, liquid PCM getting trapped in the frozen material, which reduced storage ability, and frozen PCM shells collapsing with water inside when charging. This shows that more research is needed.

Tay et al. [57] studied dynamic PCM systems and came up with a new way to improve heat transfer. They used a method called dynamic melting, which involves moving PCM around during the melting process. Their system used PCM in a tube placed inside a tank. Tests revealed that the effectiveness of the system increased by 33–89% when there was a big temperature difference, and by 58–82% with a smaller temperature difference. When they used a pump to circulate the PCM, the time taken for the melting process was shorter, and the ability to conduct heat could even be doubled, similar to using fins on the tubes, without making the system bulkier.

This method helps improve heat transfer when melting PCM in a tube-in-tank system, which significantly boosts energy storage effectiveness. However, more research is needed to explore this method fully. Adding fins and special conductors can greatly strengthen heat transfer during both the melting and freezing of PCM, but there is still a need to optimize it. Using materials like graphite can enhance the process as well, but it can be costly. Dynamic PCM systems offer a good way to improve heat transfer without high costs, and they also help prevent separation in inorganic PCM. More investigation into these systems is essential.

3.5.1 Insufficient Long-Term Stability

Insufficient long-term stability of storage materials and containers poses a challenge restricting the broad adoption of latent heat stores [66]. This issue originates from the inadequate stability of PCMs and corrosion occurring between PCMs and containers. Suitable PCMs need to withstand numerous melting and freezing cycles without degradation. Laboratory tests suggest a minimum of 1000–2000 cycles. Furthermore, compatibility must exist between PCMs and their container materials.

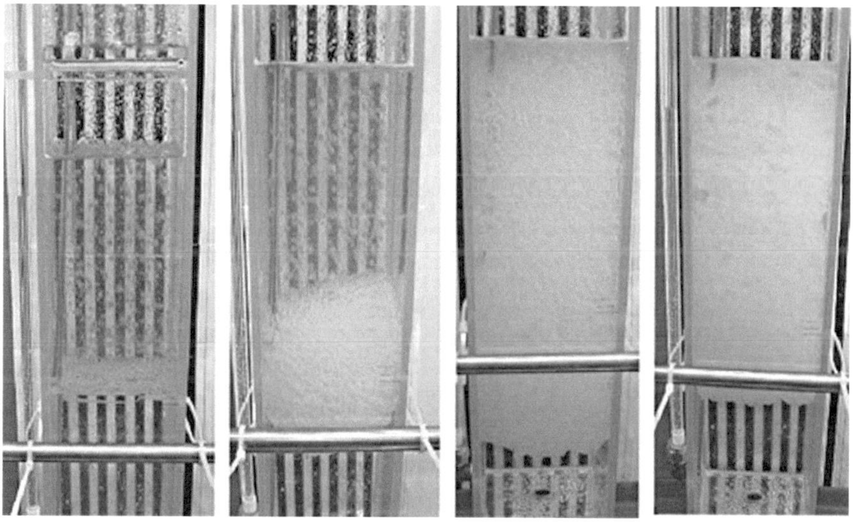

Fig. 3.3 Charging of a direct contact PCM-water cold storage [44]

3.6 Leakage Prevention and Encapsulation Techniques

Phase change materials have rough problems with leakage during solid-to-liquid transitions, particularly in thermal energy storage systems. Leakage occurs when dissipated heat causes the PCM to become molten and escape from the liquid containment, resulting in storage capacity loss, potential peril, and greater expense. This concern worsens during thermal cycling, where rigid container designs tend to be inflexible for volumetric changes accompanying phase transitions. As an example, inorganic salt hydrates as $MgCl_2{\cdot}6H_2O$, melt exhibit phase separation and leakage owing to the density differences of the components, leading to non-uniform heating. Organic PCMs like paraffin also encounter these concerns due to the skeletal stout nature of the organic. Although parasitic liquid can traverse through porous material, over time, this can deteriorate the container's structural integrity [42]. Preventing leakage in PCMs is essential for the long-lasting dependability and effectiveness of these devices.

3.6.1 Microencapsulation Techniques

Microencapsulation refers to sealing PCMs in polymer shells at the micrometer level. This procedure utilizes polymers, silica, or metal oxides to encapsulate cores of PCM. Wang et al. [62] exemplified this by designing microcapsules of polymethyl methacrylate (PMMA) and SiO_2 as shells for n-octadecane that resulted in 62% encapsulation efficiency along with latent heat preservation of 122.31 J/g. The shell structure kept PCM in a molten state while ensuring effective thermal management performance over several cycles. Likewise, Wu et al. [63] used polyurea shells to encapsulate n-octadecane, which minimized supercooling and increased overall thermal stability. Besides preventing leakage, microencapsulation enhances the area available for heat transfer, which, when used with conductive additives such as copper nanoparticles, improves thermal conductivity. However, scaling production while minimizing costs presents challenges, especially for high-temperature applications.

3.6.2 Macroencapsulation Techniques

Macroencapsulation envisions enclosing PCM in bulk containers like metal capsules or polymer bags. This approach supports industrial TES system-level bulk storage with a high thermal storage capacity. Ge et al. [27] designed Cu-Si alloy macrocapsules with Al_2O_3 shells aimed at high-temperature application (750–850 °C) with 816 MJ/m^3 heat storage for over 100 cycles. These capsules also integrated conductive matrices, which improved thermal conductivity. Chen et al. [14] embedded paraffin in graphene aerogel and carbon nanotube (CNT) frameworks, resulting in 0.46 W/m K, which is 1.7 times greater than pure paraffin. Regardless of those advantages, macroencapsulation faces severe challenges of shell fractures from thermal stress, or PCM and container materials compatibility issues.

3.6.3 Composite Matrices for Leakage Prevention

Composite matrices integrate PCMs into porous or fibrous supports to achieve shape stabilization and leakage prevention. Polymer-based matrices, such as styrene-ethylene-propylene-styrene (SEBS), provide flexible frameworks that confine PCM during phase transitions. Lin et al. [42] demonstrated that paraffin/SEBS composites with expanded graphite (EG) achieved thermal conductivity up to 10 W/m K and latent heat of 211 J/g, effectively preventing leakage even at 30 wt% EG loading. Carbon-based matrices, including graphene oxide (GO) and exfoliated graphite (EG), offer high thermal conductivity and structural stability. Li et al. [41] synthesized PEG/GO composites with 95 wt% PCM loading, retaining 93% solar-to-thermal efficiency after 1000 cycles. Porous carbon derived from biomass, such as loofah sponges, has also been used to encapsulate paraffin, achieving latent heat values of 146 J/g. These composites mitigate leakage while enhancing heat transfer, though energy storage density may decrease slightly due to the inert matrix material.

3.6.4 *Impact on Thermal Performance*

Encapsulation techniques significantly improve PCM thermal performance by addressing leakage and enhancing heat transfer. Microencapsulation reduces supercooling and increases nucleation sites, as seen in sodium acetate trihydrate (SAT) composites with disodium phosphate (DSP) nucleating agents, which lowered supercooling to 3.62 °C [24]. Macroencapsulation and composite matrices improve thermal conductivity, for example, EG-enhanced sodium sulfate decahydrate (SSD) composites achieved 1.96 W/m K, 3.6 times higher than pure SSD [41]. However, the addition of supporting materials can reduce latent heat by 10–20% due to non-PCM components (Zhang et al. [33]). Despite this trade-off, encapsulation extends PCM lifespan and enables integration into diverse applications, from building insulation to battery thermal management.

3.7 Economic Viability of PCM Systems

Phase change material-based thermal energy storage systems are progressively becoming quintessential for enhancing the energy efficiency of a given industry and for renewable energy uses. Their economic feasibility is dependent on the raw material costs, manufacturing costs, and lifecycle expenditures of the system in relative comparison to other systems. This section examines these elements based on recent findings from analyzing the performance and economy of high-performance systems and cycle systems.

3.7.1 *Material Costs: Organic Versus Inorganic PCMs*

The selection of the PCM is crucial to the system and will directly affect the pricing. Organic PCMs like paraffin RT64 are commercially available and used because of their moderate costs, chemical stability, and high latent heat of fusion. While salt hydrates are classed as inorganic PCMs and are synthetically more expensive, they sometimes suffer from phase segregation and supercooling. Moreover, they possess superior thermal conductivity and energy density, making their use favorable. Organic PCMs have uncontested superiority in price-sensitive applications, while underregulated high-temperature processes, thermal performance renders high material costs justifiable, making the use of inorganic PCMs favorable. Xie et al. [64] showed that bio-inspired PCM capsules using organic RT64 paraffin exhibited greater energy utilization rates by 48.4% when compared to conventional systems.

Capsule design requirements also impact material costs. While spherical capsules are more easily manufactured, they have poorer heat transfer due to longer thermal pathways. Surpassing traditional waste reduction approaches, bio-inspired designs like chloroplast fin structures optimize surface-to-volume ratios and reduce materials waste. For example, spherical chloroplast fin capsules reduced melting time by 50.4% over spherical designs at the cost of heightened material complexity [64].

3.7.2 Manufacturing Expenses

Manufacturing costs associated with PCM systems depend heavily on the design and fabrication techniques of the capsule. Spherical capsules possess thermal inefficiencies but remain the cheapest and easiest to manufacture due to their standardized processes. More complex geometrical shapes, like chloroplast fin designs, improve thermal performance, but require additive manufacturing or 3D printing, which heightens upfront costs. For instance, 3D-printed bio-inspired capsules achieved a 36.1% reduction in melting times compared to their spherical counterparts, which offset increased manufacturing costs with operational savings.

The division of complexity and scalability is widely regarded as a significant trade-off, as shown in this study. Packed-bed systems with staggered capsule arrangements show improved convective heat transfer, but alignment must be exact, which increases production costs by about 15–20% [64].

3.7.3 Lifecycle Cost Comparisons with Conventional Systems

Lifecycle costs encompass initial investments, operational savings, and maintenance. High-performance PCM systems exhibit 20% higher initial costs than conventional systems but achieve 33.2% faster payback periods due to reduced operational times and energy losses (see Fig. 3.4). For instance, systems using chloroplast-fin capsules reduced energy storage cycle times by 50.4%, translating to annual operational savings of $72,000 for a mid-sized industrial plant [64].

Maintenance costs are also lower for advanced PCM systems. Bio-inspired designs minimize thermal stress on materials, extending system lifespans by 15–20% compared to conventional setups. Additionally, staggered configurations in packed beds reduce pressure drops, lowering pumping energy requirements by 12%. Over 10 years, high-performance systems demonstrated a 35.8% higher return on investment (ROI) than conventional alternatives (see Fig. 3.5).

The balance of material selection, manufacturing creativity, and lifecycle effectiveness determines the economic viability of PCM systems. Organic PCMs like paraffin offer cost benefits while their inorganic counterparts serve high-performance needs. New thermal management techniques can be costly to implement but result in greater return on investment through improved thermal efficiency in the long run. Lifecycle evaluations verify that high-performance systems surpass traditional options, with quicker payback and better ROI. More work is needed in the area of scalable production and hybrid PCM designs.

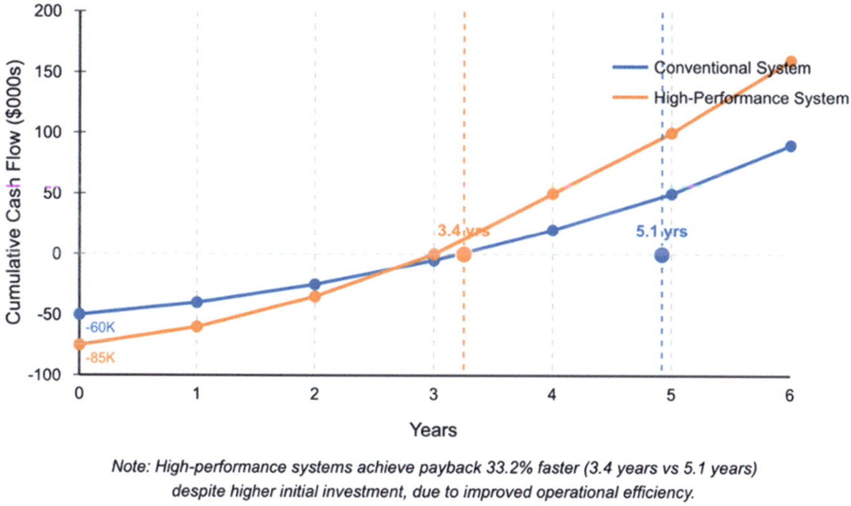

Fig. 3.4 Payback period analysis [64]

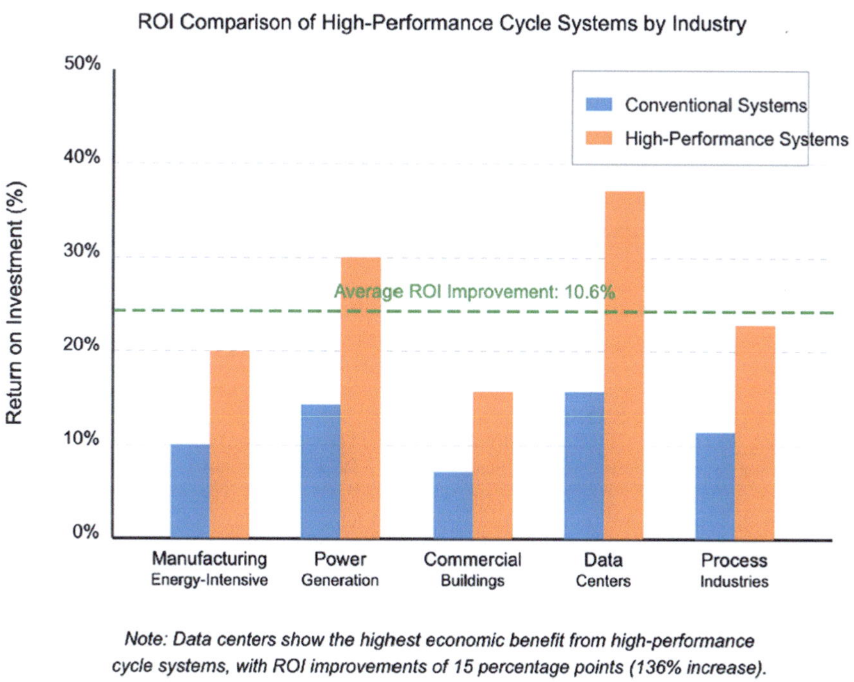

Fig. 3.5 Comparative assessment of return on investment (ROI) for advanced cyclic systems in industrial applications, evaluated by their thermal energy retention performance [64]

3.8 Summary

This chapter examines the limitations of PCMs in LHTS systems despite their widespread use in thermal energy management. While PCMs offer high energy storage density and near-isothermal operation, several significant challenges hinder their broader adoption.

The chapter identifies incongruent melting, subcooling, and low thermal conductivity as primary issues. Incongruent melting leads to phase separation and reduced efficiency, particularly in hydrated salts. Subcooling occurs during solidification, delaying latent heat release and diminishing system performance. Low thermal conductivity impedes heat transfer, slowing melting and solidification processes. Various techniques, such as using fins, metallic foams, nanoparticles, and dynamic systems, are explored to enhance thermal conductivity and improve heat transfer rates.

The chapter also discusses insufficient long-term stability, where PCMs degrade after repeated thermal cycles or corrode storage containers, limiting their reliability. Thermal stability tests are crucial for ensuring PCM longevity in practical applications. While numerous solutions have been proposed, ranging from material modifications to advanced heat exchanger designs, challenges like high costs, material compatibility, and limited scalability remain. The chapter concludes by underscoring the need for continued research and development to address these shortcomings and optimize PCMs for efficient and sustainable energy storage systems.

References

1. Agyenim F, Eames P, Smyth M (2009) A comparison of heat transfer enhancement in a medium temperature thermal energy storage heat exchanger using fins. Sol Energy 83:1509–1520
2. Ali HM (2022) Heat transfer augmentation of porous media (metallic foam) and phase change material based heat sink with variable heat generations: an experimental evaluation. Sustain Energy Technol Assess 52:102218
3. Ali HM, Arshad A (2017) Experimental investigation of n-eicosane based circular pin-fin heat sinks for passive cooling of electronic devices. Int J Heat Mass Transf 112:649–661
4. Ali HM, Ashraf MJ, Giovannelli A, Irfan M, Irshad TB, Hamid HM, Hassan F, Arshad A (2018) Thermal management of electronics: an experimental analysis of triangular, rectangular and circular pin-fin heat sinks for various PCMs. Int J Heat Mass Transf 123:272–284
5. Arshad A, Ali HM, Khushnood S, Jabbal M (2018) Experimental investigation of PCM based round pin-fin heat sinks for thermal management of electronics: effect of pin-fin diameter. Int J Heat Mass Transf 117:861–872
6. Arshad A, Ali HM, Yan W-M, Hussein AK, Ahmadlouydarab M (2018) An experimental study of enhanced heat sinks for thermal management using n-eicosane as phase change material. Appl Therm Eng 132:52–66
7. Avci M, Yazici MY (2018) An experimental study on effect of inclination angle on the performance of a PCM-based flat-type heat sink. Appl Therm Eng 131:806–814
8. Baby R, Balaji C (2013) A neural network-based optimization of thermal performance of phase change material-based finned heat sinks—an experimental study. Exp Heat Transf 26:431–452
9. Cabeza LF (2021) Advances in thermal energy storage systems: methods and applications. In: Woodhead publishing series in energy. Woodhead Publishing

10. Cabeza LF, Mehling H, Hiebler S, Ziegler F (2002) Heat transfer enhancement in water when used as PCM in thermal energy storage. Appl Therm Eng 22:1141–1151
11. Carlsson B (2009) Phase change behaviour of some latent heat storage media based on calcium chloride hexahydrate. Sol Energy 83:485–500
12. Castell A, Solé C, Medrano M, Roca J, Cabeza LF, García D, Castell A, Sole C, Cabeza LF, Garcı D (2008) Natural convection heat transfer coefficients in phase change material (PCM) modules with external vertical fins. Appl Therm Eng 28:1676–1686
13. Chen C, Wang L, Huang Y (2008) Morphology and thermal properties of electrospun fatty acids/polyethylene terephthalate composite fibers as novel form-stable phase change materials. Sol Energy Mater Sol Cells 92:1382–1387
14. Chen G, Su Y, Jiang D, Pan L, Li S (2020) An experimental and numerical investigation on a paraffin wax/graphene oxide/carbon nanotubes composite material for solar thermal storage applications. Appl Energy 264:114786
15. Choi JC, Kim SD, Han GY (1996) Heat transfer characteristics in low-temperature latent heat storage systems using salt-hydrates at heat recovery stage. Sol Energy Mater Sol Cells 40:71–87
16. Chow LC, Zhong JK, Beam JE (1996) Thermal conductivity enhancement for phase change storage media. Int Commun Heat Mass Transf 23:91–100
17. Dammak K, El Hami A (2021) Thermal reliability-based design optimization using Kriging model of PCM based pin fin heat sink. Int J Heat Mass Transf 166:120745
18. De Jong AG, Hoogendoorn CJ (1981) Improvement of heat transport in paraffines for latent heat storage systems. In: den Ouden C (ed) Thermal storage of solar energy. Springer Netherlands, Dordrecht, pp 123–133
19. Dinçer I, Rosen MA (2002) Thermal energy storage systems and applications. Wiley
20. Fan L, Khodadadi JM (2011) Thermal conductivity enhancement of phase change materials for thermal energy storage: a review. Renew Sustain Energy Rev 15:24–46
21. Farid MM, Khudhair AM, Razack SAK, Al-Hallaj S (2004) A review on phase change energy storage: materials and applications. Energy Convers Manag 45:1597–1615
22. Feng D, Feng Y, Qiu L, Li P, Zang Y, Zou H, Yu Z, Zhang X (2019) Review on nanoporous composite phase change materials: fabrication, characterization, enhancement and molecular simulation. Renew Sustain Energy Rev 109:578–605
23. Ferfera RS, Madani B (2020) Thermal characterization of a heat exchanger equipped with a combined material of phase change material and metallic foams. Int J Heat Mass Transf 148:119162
24. Fu W, Zhou W, Lv G, Zhang R, Peng H, Fang T, Liu J, Chen W, Fang Y (2023) Phase change temperature adjustment of $CH_3COONa{\cdot}3H_2O$ to fabricate composite phase change material for radiant floor heating. Case Stud Therm Eng 42:102773
25. Fukai J, Hamada Y, Morozumi Y, Miyatake O (2002) Effect of carbon-fiber brushes on conductive heat transfer in phase change materials. Int J Heat Mass Transf 45:4781–4792
26. Fukai J, Kanou M, Kodama Y, Miyatake O (2000) Thermal conductivity enhancement of energy storage media using carbon fibers. Energy Convers Manag 41:1543–1556
27. Ge Y, Sheng N, Zhu C (2023) Macroencapsulated CuSi phase change material by in situ alloying formation for high temperature thermal energy storage over 800 °C. J Energy Storage 72:108258
28. Gharbi S, Harmand S, Jabrallah SB (2015) Experimental comparison between different configurations of PCM based heat sinks for cooling electronic components. Appl Therm Eng 87:454–462
29. Ghosh D, Guha C, Ghose J (2019) Numerical investigation of paraffin wax solidification in spherical and rectangular cavity. Heat Mass Transf 55:3547–3559
30. He B, Setterwall F (2002) Technical grade paraffin waxes as phase change materials for cool thermal storage and cool storage systems capital cost estimation. Energy Convers Manag 43:1709–1723
31. Hemati R, Veysi F, Qaderi A (2022) Experimental investigation of the simultaneous effect of using phase change material/plate-fin heat sink on thermal performance of a power supply unit. Sustain Energy Technol Assess 52:102009

32. Horbaniuc B, Dumitrascu G, Popescu A (1999) Mathematical models for the study of solidification within a longitudinally finned heat pipe latent heat thermal storage system. Energy Convers Manag 40:1765–1774
33. https://doi.org/10.1016/j.solener.2014.09.003
34. Hu X, Gong X (2021) Experimental study on the thermal response of PCM-based heat sink using structured porous material fabricated by 3D printing. Case Stud Therm Eng 24:100844
35. Hu X, Zhu F, Gong X (2019) Experimental and numerical study on the thermal behavior of phase change material infiltrated in low porosity metal foam. J Energy Storage 26:101005
36. Jegadheeswaran S, Sundaramahalingam A, Pohekar SD (2019) High-conductivity nanomaterials for enhancing thermal performance of latent heat thermal energy storage systems. J Therm Anal Calorim 138:1137–1166
37. Kalbasi R, Afrand M, Alsarraf J, Tran M-D (2019) Studies on optimum fins number in PCM-based heat sinks. Energy 171:1088–1099
38. Khanmohammadi F, Farhadi M, Ali Rabienataj Darzi A (2019) Numerical investigation of heat transfer and fluid flow characteristics inside tube with internally star fins. Heat Mass Transf 55:1901–1911
39. Kim S, Drzal LT (2009) High latent heat storage and high thermal conductive phase change materials using exfoliated graphite nanoplatelets. Sol Energy Mater Sol Cells 93:136–142
40. Leong KY, Chew SP, Gurunathan BA, Ku Ahmad KZ, Ong HC (2019) An experimental approach to investigate thermal performance of paraffin wax and 1-hexadecanol based heat sinks for cooling of electronic system. Int Commun Heat Mass Transf 109:104365
41. Li C, Zhang B, Xie B, Zhao X, Chen J (2020) Tailored phase change behavior of $Na_2SO_4 \cdot 10H_2O$/expanded graphite composite for thermal energy storage. Energy Convers Manag 208:112586
42. Lin X, Zhang X, Liu L, Liang J, Liu W (2022) Polymer/expanded graphite-based flexible phase change material with high thermal conductivity for battery thermal management. J Clean Prod 331:130014
43. Mahmoud S, Tang A, Toh C, Al-Dadah R, Soo SL (2013) Experimental investigation of inserts configurations and PCM type on the thermal performance of PCM based heat sinks. Appl Energy 112:1349–1356
44. Martin V, He B, Setterwall F (2010) Direct contact PCM–water cold storage. Appl Energy 87:2652–2659
45. Mehling H, Cabeza LF (2008) Integration of active storages into systems, pp 181–189
46. Nagano K, Mochida T, Takeda S, Domański R, Rebow M (2003) Thermal characteristics of manganese (II) nitrate hexahydrate as a phase change material for cooling systems. Appl Therm Eng 23:229–241
47. Pakrouh R, Hosseini MJ, Ranjbar AA, Bahrampoury R (2015) A numerical method for PCM-based pin fin heat sinks optimization. Energy Convers Manag 103:542–552
48. Pincemin S, Olives R, Py X, Christ M (2008) Highly conductive composites made of phase change materials and graphite for thermal storage. Sol Energy Mater Sol Cells 92:603–613
49. Py X, Olives R, Mauran S (2001) Paraffin/porous-graphite-matrix composite as a high and constant power thermal storage material. Int J Heat Mass Transf 44:2727–2737
50. Rehman T, Ali HM (2020) Thermal performance analysis of metallic foam-based heat sinks embedded with RT-54HC paraffin: an experimental investigation for electronic cooling. J Therm Anal Calorim 140:979–990
51. Rehman T, Ambreen T, Niyas H, Kanti P, Ali HM, Park C-W (2022) Experimental investigation on the performance of RT-44HC-nickel foam-based heat sinks for thermal management of electronic gadgets. Int J Heat Mass Transf 188:122591
52. Saha SK, Srinivasan K, Dutta P, Asme M, Saha SK, Srinivasan K, Dutta P (2008) Studies on optimum distribution of fins in heat sinks filled with phase change materials. J Heat Transfer 130:034505
53. Sanghvi MR, Tambare OH, More AP (2022) Performance of various fillers in adhesives applications: a review. Polym Bull 79:10491–10553

54. Shang B, Ma Y, Hu R, Yuan C, Hu J, Luo X (2017) Passive thermal management system for downhole electronics in harsh thermal environments. Appl Therm Eng 118:593–599
55. Srikanth R, Balaji C (2017) Experimental investigation on the heat transfer performance of a PCM based pin fin heat sink with discrete heating. Int J Therm Sci 111:188–203
56. Tay NHS, Bruno F, Belusko M (2012) Experimental validation of a CFD and an ε-NTU model for a large tube-in-tank PCM system. Int J Heat Mass Transf 55:5931–5940
57. Tay NHS, Bruno F, Belusko M (2013) Experimental investigation of dynamic melting in a tube-in-tank PCM system. Appl Energy 104:137–148
58. Usman H, Ali HM, Arshad A, Ashraf MJ, Khushnood S, Janjua MM, Kazi SN (2018) An experimental study of PCM based finned and un-finned heat sinks for passive cooling of electronics. Heat Mass Transf 54:3587–3598
59. Veismoradi A, Modir A, Ghalambaz M, Chamkha A (2020) A phase change/metal foam heatsink for thermal management of battery packs. Int J Therm Sci 157:106514
60. Velraj R, Seeniraj RV, Hafner B, Faber C, Schwarzer K (1999) Heat transfer enhancement in a latent heat storage system. Paper presented at the ISES Solar World Congress, Taejon, South Korea, 24–29 August 1997. Sol Energy 65:171–180
61. Wang F, Maidment G, Missenden J, Tozer R (2002) A review of research concerning the use of PCMS in air conditioning and refrigeration engineering. In: Anson M, Ko JM, Lam ESS (eds). Elsevier, Oxford, pp 1273–1280
62. Wang H, Zhao L, Chen L, Song G, Tang G (2017) Facile and low energy consumption synthesis of microencapsulated phase change materials with hybrid shell for thermal energy storage. J Phys Chem Solids 111:207–213
63. Wu Q, Zhao D, Jiao X, Zhang Y, Shea KJ, Lu X, Qiu G (2015) Preparation, properties, and supercooling prevention of phase change material n-octadecane microcapsules with peppermint fragrance scent. Ind Eng Chem Res 54:8130–8136
64. Xie R, Deng X, Yin Y, Fayziyeva D, Eyvazov E, Liu F (2025) Economic viability of high-performance cycle systems: energy and cost efficiency insights. Energy Strateg Rev 59:101675
65. Yin H, Gao X, Ding J, Zhang Z (2008) Experimental research on heat transfer mechanism of heat sink with composite phase change materials. Energy Convers Manag 49:1740–1746
66. Zalba B, Marín JM, Cabeza LF, Mehling H (2003) Review on thermal energy storage with phase change: materials, heat transfer analysis and applications. Appl Therm Eng 23:251-283
67. Zhang P, Meng ZN, Zhu H, Wang YL, Peng SP (2017) Melting heat transfer characteristics of a composite phase change material fabricated by paraffin and metal foam. Appl Energy 185:1971–1983
68. Zhao Y, Zhai Z, Drummer D (2018) Thermal conductivity of aluminosilicate- and aluminum oxide-filled thermosets for injection molding: effect of filler content, filler size and filler geometry. Polymers (Basel) 10:457
69. Zheng H, Wang C, Liu Q, Tian Z, Fan X (2018) Thermal performance of copper foam/paraffin composite phase change material. Energy Convers Manag 157:372–381

Chapter 4
Analyses of the Phase Change Problems

4.1 Introduction

Heating a material through the melting process typically involves three distinct stages. Initially, the material's temperature is uniformly increased from its starting point to its melting point through sensible heating. Once the melting point is reached, the material undergoes a phase change through latent heating, where it transitions from solid to liquid without a further increase in temperature. During this phase, the melt front, which marks the boundary between the solid and liquid phases, moves through the material. As the material fully melts, any additional heat raises the temperature of the liquid material through sensible heating.

In practice, this process can be quite complex. The movement of the melt front within the PCM depends on the boundary conditions at its edges. The melt front can accelerate or decelerate based on the rate of heat added or lost through these boundaries. This results in a situation where the exact position of the melt front at any given time is not predetermined and must be calculated. The interaction of the boundaries causes the temperature distribution within the PCM to become three-dimensional, leading to a non-uniform movement of the melt front, which can progress diagonally rather than uniformly. The onset of natural convection within the liquid phase adds further complexity to the melt front's motion and the associated mathematical modeling.

When using PCMs that are blended materials or those with a melting range rather than a sharp melting point, a broad "mushy zone" develops where the melt front's motion is less distinct. In these materials, melting occurs gradually over a wider temperature range, resulting in a slow softening of the material and the presence of both liquid and solid phases interspersed within the PCM. Due to these complexities, developing precise analytical solutions for the melting and solidification processes is challenging. These problems, known as moving boundary problems, are typically solved analytically only under simplified physical assumptions.

A. Mosaffa, *Phase Changing Materials*, Materials Horizons: From Nature to Nanomaterials, https://doi.org/10.1007/978-981-95-3742-6_4

In this chapter, we focus on one-dimensional thermal conduction equations for phase change problems (solidification). Exact solutions, the description of common boundary conditions, and the nonlinear nature of these equations are discussed. Approximate solutions to simpler models of phase change problems are also provided.

4.2 Mathematical Formulation of the Phase Change Problem (Stefan Problem)

We begin by describing the mathematical formulation of the classic Stefan problem for phase change phenomena. This model lays the groundwork for analyzing more complex phase change problems discussed in subsequent chapters. A general overview of moving boundary problems is provided, highlighting the challenges of determining both the temperature distribution and the position of the phase change interface (e.g., the solidification front).

By starting with assumptions commonly used to simplify phase change problems found in the literature (e.g., [1]), we arrive at the governing equations and boundary conditions for heat transfer in such systems.

4.2.1 General Assumptions

To construct a sound mathematical model based on fundamental physical principles, it is crucial to have a clear understanding of what happens during phase change phenomena. The simplifying assumptions made in the problem should be explicitly stated and related to the physical factors involved in the process. These assumptions are summarized as follows:

1. *Heat transfer via conduction*: Heat transfer occurs uniformly through conduction only, with other effects such as convection, radiation, elasticity, chemical reactions, gravitational effects, and electromagnetic forces neglected. In cases like small containers and pure material freezing, this assumption is entirely reasonable.
2. *Constant latent heat*: The latent heat of fusion (or solidification) is assumed to remain constant at the melting point. This implies that the energy absorbed or released at a constant temperature is independent of the material's thermal history.
3. *Negligible temperature range for phase change*: The phase change occurs at a nearly constant temperature, often treated as isothermal. In phase change materials (PCM), this is a valid assumption as the process generally occurs at a fixed temperature.

4. *Sharp interface*: The interface between phases is considered to have no thickness, allowing it to be treated as a boundary rather than a transitional zone.
5. *Negligible surface tension effects*: Surface tension at the phase boundary is neglected.
6. *Constant thermophysical properties*: The thermal properties, such as thermal conductivity and heat capacity, are assumed to be independent of temperature in each phase.
7. *Neglect of density changes*: Changes in density during phase transition are ignored, and in the case of liquid phases, fluid motion is also neglected. This simplification is reasonable in small-scale systems.

4.2.2 Heat Transfer Equations

Consider a material undergoing a phase change, represented schematically in Fig. 4.1. The system consists of two distinct phases: a solid and a liquid, separated by a moving interface. Assuming heat conduction as the only mode of heat transfer, the temperature distribution within each phase can be described by the heat conduction equations:

For the solid phase:

$$\frac{\partial^2 T_s}{\partial x^2} = \frac{1}{\alpha_s}\frac{\partial T_s}{\partial t}, \quad 0 < x < X(t) \tag{4.1}$$

For the liquid phase:

$$\frac{\partial^2 T_l}{\partial x^2} = \frac{1}{\alpha_l}\frac{\partial T_l}{\partial t}, \quad X(t) > x \tag{4.2}$$

where

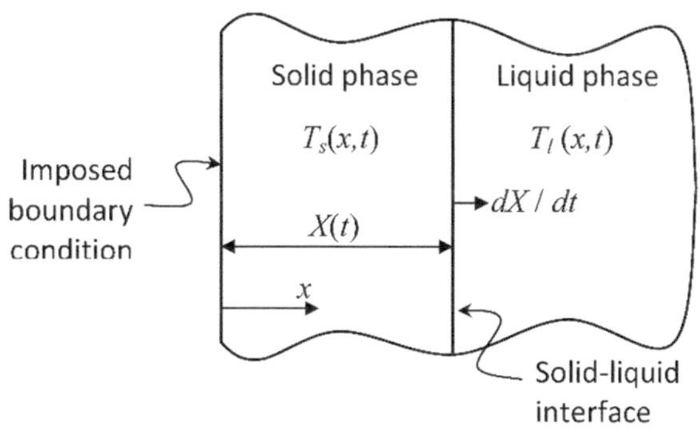

Fig. 4.1 Liquid and solid phases separated by a moving boundary [2]

- T_s and T_l are the temperature distributions in the solid and liquid phases, respectively,
- α_s and α_l are the thermal diffusivities of the solid and liquid phases,
- $X(t)$ represents the position of the moving interface at time t.

4.2.3 Boundary Conditions at the Moving Interface

The moving interface between the solid and liquid phases, $X(t)$, is governed by two boundary conditions derived from physical principles:

1. *Temperature continuity*: The temperature at the interface must remain continuous, as depicted in Fig. 4.2:

$$T_s(X, t) = T_l(X, t) = T_m \tag{4.3}$$

   where T_m is the melting point of the material.
2. *Energy conservation*: The heat absorbed by a pure substance under constant pressure is a thermodynamic property and is referred to as enthalpy. For a pure substance heated under constant pressure with no volume change, enthalpy represents the total energy. The absorbed heat is directly proportional to the temperature increase and follows the equation:

$$dh = c dT \tag{4.4}$$

   where h represents the enthalpy per unit mass (kJ/kg), and cc is the specific heat capacity of the material. The heat associated with a temperature change is called sensible heat, while the heat absorbed or released at constant temperature is referred to as latent heat.

In the phase change process, the melting point temperature T_m is considered the reference temperature, and the sensible heat can be defined as $\int_{T_m}^{T} c\, dT$.

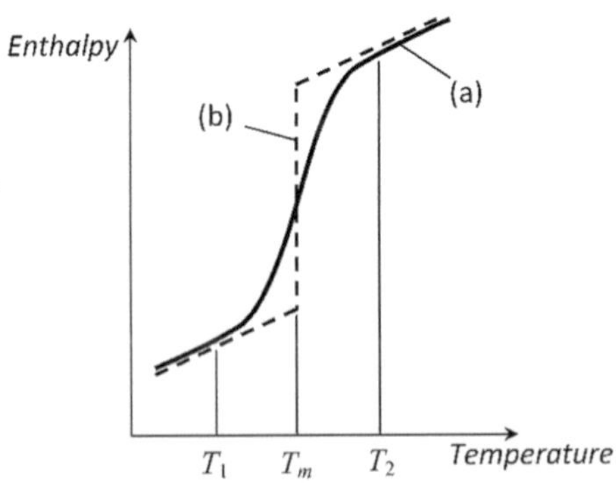

Fig. 4.2 Phase change process, **a** actual curve over a temperature range, and **b** isothermal curve [2]

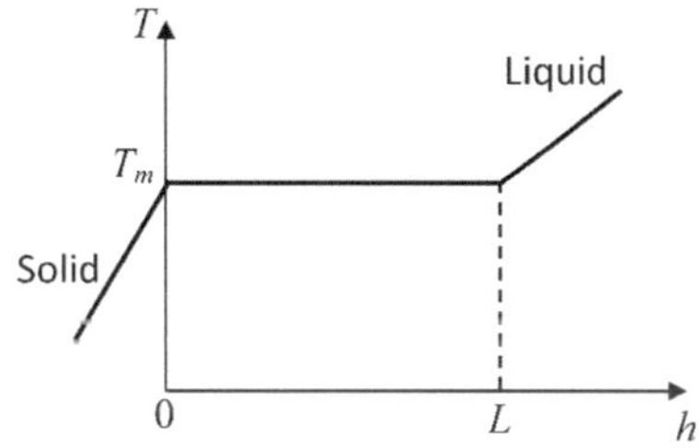

Fig. 4.3 Variations of enthalpy with temperature for constant thermophysical properties [2]

At T_{m}, materials can exist as both liquid and solid. The difference lies in the fact that the liquid phase possesses a latent heat L per unit mass, whereas the solid phase lacks this energy. Therefore, if the enthalpy of the solid phase at T_{m} is considered zero, we have:

$$h = \begin{cases} L + \int_{T_{\mathrm{m}}}^{T} c_{\mathrm{l}} \cdot \mathrm{d}T, & T > T_{\mathrm{m}} \text{ (liquid)} \\ \int_{T_{\mathrm{m}}}^{T} c_{\mathrm{s}} \cdot \mathrm{d}T, & T < T_{\mathrm{m}} \text{ (solid)} \end{cases} \tag{4.5}$$

Thus, at T_{m}, if the enthalpy is zero, the material is in the solid phase; if it equals L, the material is in the liquid phase. If the enthalpy lies between zero and L ($0 < h < L$), the material is undergoing a phase change.

Figure 4.3 illustrates the variations of enthalpy with temperature when c is assumed constant for both the liquid and solid phases.

The law of energy conservation without heat sources or sinks, under constant properties, is expressed as:

$$\rho \frac{\mathrm{d}h}{\mathrm{d}t} + \operatorname{div} \vec{q} = 0 \tag{4.6}$$

Consider a one-dimensional flat wall with a thickness l ($0 < x < l$) undergoing solidification. Heat exits through the boundary at $x = 0$, while the boundary at $x = l$ is insulated. The liquid and solid phases are separated by a moving interface located at $X(t)$ at any given time. Therefore, the total enthalpy is given by:

$$H(t) = A\left\{\int_{0}^{X} \rho c_{\mathrm{s}}[T(x,t) - T_{\mathrm{m}}]\mathrm{d}x + \int_{X}^{l} (\rho c_{\mathrm{l}}[T(x,t) - T_{\mathrm{m}}] + \rho L)\mathrm{d}x\right\} \tag{4.7}$$

where A is the cross-sectional area of the wall. Using Eq. (4.6), the net rate of heat transfer through the wall is:

$$\frac{\mathrm{d}H}{\mathrm{d}t} = A\left\{q''(0,t) - q''(l,t)\right\} \tag{4.8}$$

where $q''(0, t)$ and $-q''(l, t)$ are the heat fluxes exiting the boundaries at $x = 0$ and $x = l$, respectively ($q''(l, t) = 0$). By differentiating Eq. (4.7) and applying Leibniz's integral rule, we have:

$$\frac{1}{A}\frac{\mathrm{d}H}{\mathrm{d}t} = \rho c_s[T(X,t) - T_m]\frac{\mathrm{d}X}{\mathrm{d}t} + \int_0^X \rho c_s \frac{\mathrm{d}T(x,t)}{\mathrm{d}t}\mathrm{d}x - \rho L\frac{\mathrm{d}X}{\mathrm{d}t}$$
$$- \rho c_l[T(X,t) - T_m]\frac{\mathrm{d}X}{\mathrm{d}t} + \int_X^l \rho c_l \frac{\mathrm{d}T(x,t)}{\mathrm{d}t}\mathrm{d}x \tag{4.9}$$

Given that $T(X, t) = T_m$, and using the heat conduction equation $T_t = \alpha T_{xx}$, we obtain:

$$\frac{1}{A}\frac{\mathrm{d}H}{\mathrm{d}t} = k_s T_x(X,t) - k_s T_x(0,t) - \rho L\dot{X} + k_l T_x(l,t) - k_l T_x(X,t) \tag{4.10}$$

Considering that $-k_s T_x(0, t) = q''(0, t)$, $k_l T_x(l, t) = -q''(l, t)$, and using Eq. (4.8), we arrive at:

$$\rho L\dot{X} = k_s T_x(X,t) - k_l T_x(X,t) \tag{4.11}$$

4.2.4 *Stefan Condition*

The energy condition derived in Eq. (4.11) states that the rate of latent energy change is equal to the net heat flux passing through the moving interface between the liquid and solid via conduction. This equation is known as the Stefan condition.

For radial cylindrical and spherical systems at the interface $\Omega(t)$, the Stefan condition can be expressed as:

$$\rho L\dot{\Omega} = k_s T_r(\Omega,t) - k_l T_r(\Omega,t) \tag{4.12}$$

4.3 Nonlinear Nature of the Energy Equation at the Interface

The solution of the conduction problem for the phase change process must ensure that the energy Eq. (4.11) is satisfied. Upon closely examining these equations, we find that they are nonlinear. The nonlinearity arises because the velocity of the moving interface ($\mathrm{d}X/\mathrm{d}t$) depends on the temperature gradient.

To clarify the nonlinear nature of Eq. (4.11), the total derivative of T_s in Eq. (4.3) is computed:

$$\frac{\partial T_s(X,t)}{\partial x}\mathrm{d}x + \frac{\partial T_s(X,t)}{\partial t}\mathrm{d}t = 0 \tag{4.13}$$

Dividing both sides by $\mathrm{d}t$ and noting that $\mathrm{d}x = \mathrm{d}X$, we have:

$$\frac{\mathrm{d}X}{\mathrm{d}t} = -\frac{\partial T_s(X,t)/\partial t}{\partial T_s(X,t)/\partial x} \tag{4.14}$$

Substituting these results into Eq. (4.11), the following equation is obtained:

$$k_s\left[\frac{\partial T_s(X,t)}{\partial x}\right]^2 - k_l\frac{\partial T_l(X,t)}{\partial x}\frac{\partial T_s(X,t)}{\partial x} = -\rho L\frac{\partial T_s(X,t)}{\partial t} \tag{4.15}$$

It is evident that both terms on the left-hand side of Eq. (4.15) are nonlinear.

4.4 Exact Solution to the One-Phase Stefan Problem

One type of phase change problem that has an exact analytical solution is the one-phase Stefan problem with constant thermophysical properties. This assumes that the material undergoing phase change is initially in one phase (either solid or liquid) at a constant temperature, and a constant temperature is applied to one boundary of the system.

The solution to this problem, known as the Neumann Solution, involves the following scenario:

- A one-dimensional rectangular container filled with a phase changing material of length l.
- At $t = 0$, the material is entirely in one phase at the melting point (T_m).
- A constant temperature $T_s < T_m$ is suddenly applied to one end of the container, initiating solidification.

The governing equations, initial conditions, and boundary conditions are:

1. *Governing equation*:

$$T_t = \alpha_s T_{xx}, \quad 0 < x < X, \ t > 0 \text{ (solid)} \tag{4.16}$$

2. *Initial condition*:

$$T(x, 0) = T_m, \quad 0 < x < l \tag{4.17}$$

3. *Boundary conditions*:

$$T(0, t) = T_s < T_m, \quad t \geq 0 \tag{4.18}$$

4. *Stefan condition*:

$$\rho L\dot{X} = k_s T_x(X, t), \quad t > 0 \tag{4.19}$$

4.4.1 Similarity Solution

By defining a similarity parameter χ, the desired solution for the temperature distribution can be written as $T(x, t) = F(\chi)$. It can be shown that the only possible similarity parameter for solving Eq. (4.16) is $\chi = x/\sqrt{t}$. Consequently, the position of the freezing front $X(t)$ is naturally proportional to $\sqrt{t}$. By introducing a coefficient A, this relationship becomes:

$$X(t) = A\sqrt{t} \tag{4.20}$$

Substituting the function $F(\chi)$ into Eq. (4.16) yields the following solution:

$$F(\chi) = B\sqrt{\pi\alpha_s}\,\mathrm{erf}\left(\frac{\chi}{\sqrt{4\alpha_s}}\right) + C \tag{4.21}$$

where erf is the error function, and B and C are integration constants. Using the boundary condition (4.18) and the fact that $T(X, t) = T_m$, we find

$$C = T_s, \quad B = \frac{T_m - T_s}{\sqrt{\pi\alpha_s}\,\mathrm{erf}\left(A/\sqrt{4\alpha_s}\right)} \tag{4.22}$$

where $\lambda = A$.

Defining $\Delta T_s = T_m - T_s$ and substituting Eq. (4.21) into Eq. (4.18), we obtain

$$\lambda \cdot \exp\left(\lambda^2\right) \cdot \mathrm{erf}(\lambda) = \frac{Ste_s}{\sqrt{\pi}} \tag{4.23}$$

where *Ste*, the Stefan number, represents the ratio of sensible heat to latent heat ($Ste = c_s \cdot \Delta T_s$). Sample values of λ are shown in Table 4.1.

Table 4.1 Sample values of κ for various values of Ste_s [3]

Ste_s	λ	Ste_s	λ	Ste_s	λ
0.00	0.00	0.08	0.19801	0.7	0.54193
0.01	0.07055	0.09	0.20948	0.8	0.57420
0.02	0.09952	0.1	0.22030	0.9	0.60404
0.03	0.12159	0.2	0.30638	1.0	0.63191
0.04	0.14006	0.3	0.37034	2.0	0.85426
0.05	0.15826	0.4	0.42257	3.0	1.03670
0.06	0.17260	0.5	0.46725	4.0	1.20490
0.07	0.18577	0.6	0.50658	5.0	1.36270

Thus, the temperature distribution in the solid phase and the position of the freezing front are obtained as functions of λ:

$$T(x,t) = T_s + \Delta T_s \frac{\operatorname{erf}\left(x/\sqrt{4\alpha_s t}\right)}{\operatorname{erf}(\lambda)} \tag{4.24}$$

$$X(t) = \lambda\sqrt{4\alpha_s t} \tag{4.25}$$

where λ is the solution to Eq. (4.23).

It can be easily shown that the function $f(\lambda) = \lambda \cdot \exp(\lambda^2) \cdot \operatorname{erf}(\lambda)$ is strictly increasing for $\lambda \geq 0$. Therefore, any horizontal line intersects $f(\lambda)$ at only one point. In other words, for any $Ste > 0$, Eq. (4.23) has a unique solution.

4.5 Approximate Solutions to Phase Change Problems

Due to the nonlinear nature of the governing equations, exact solutions are limited to specific cases. For most practical problems, approximate methods are used.

4.5.1 *Quasi-Steady Method*

As mentioned earlier, due to the nonlinearity of the energy equation at the moving boundary, analytical solutions for conduction problems in phase change processes are only available for specific conditions. One approximate method that facilitates solving various problems is the quasi-steady method. In this method, the transient heat conduction Eq. (4.1) is replaced with its steady-state form ($T_{xx} = 0$), while the freezing front remains a function of time. Consequently, initial conditions are no longer applicable in this method, which is limited to one-phase problems.

The fundamental assumption for using this method is that the Stefan number (*Ste*) is assumed to be less than one. A small Stefan number implies that sensible heat is negligible compared to latent heat. To better understand the smallness of the Stefan number, consider the extreme case of a material with zero specific heat, where *Ste* = 0.

Such a material would have an infinite thermal diffusivity. This means that thermal effects spread instantaneously, and despite the motion of the interface, the system immediately reaches a steady state. Additionally, one could imagine a material with infinite latent heat (*Ste* = 0) in which the moving boundary remains stationary. Therefore, for small Stefan numbers, the moving boundary moves very slowly, and the temperature distribution at any given moment corresponds to the steady-state solution. In practice, the quasi-steady method is valid for Stefan numbers less than 0.1.

Here, we solve the one-phase freezing problem (4.16)–(4.19) using the quasi-steady method. By removing the transient term on the left-hand side of Eq. (4.16), the steady-state solution is obtained as:

$$T^{\mathrm{qs}} = Ax + B \tag{4.26}$$

where A and B are constants of integration, determined using the boundary conditions. The temperature distribution is then:

$$T^{\mathrm{qs}} = T_{\mathrm{s}} + [T_{\mathrm{m}} - T_{\mathrm{s}}]\frac{x}{X^{\mathrm{qs}}} \tag{4.27}$$

and the position of the freezing front at any time is

$$X^{\mathrm{qs}} = \left[2\frac{k_{\mathrm{s}}}{\rho L}(T_{\mathrm{m}} - T_{\mathrm{s}})t\right]^{1/2} \tag{4.28}$$

4.5.2 Megerlin Method

In this method, the temperature distribution is approximated by a quadratic equation. Consider the one-phase freezing problem (4.16)–(4.19). The temperature distribution is expressed as:

$$T^{\mathrm{Meg}} = T_{\mathrm{m}} + A(x - X) + B(x - X)^2, \quad 0 \le x \le X, \ t > 0 \tag{4.29}$$

where A and B are constants of integration. Using the boundary conditions, the temperature distribution is obtained as

$$T^{\mathrm{Meg}} = T_{\mathrm{m}} + \frac{\rho L}{k_{\mathrm{s}}}\dot{X}(x - X) + \left[\frac{T_{\mathrm{s}} - T_{\mathrm{m}}}{X^2} + \frac{\rho L}{k_{\mathrm{s}}}\frac{\dot{X}}{X}\right](x - X)^2 \tag{4.30}$$

To determine $X(t)$, Megerlin's method ensures that the heat conduction equation is satisfied at the moving boundary. Substituting $X = x$ into the conduction equation at the freezing front yields the following ordinary differential equation:

$$\left(X\dot{X}\right)^2 + 2\alpha_{\mathrm{s}}\left(X\dot{X}\right) - 2\alpha_{\mathrm{s}}^2 \frac{c_{\mathrm{s}}(T_{\mathrm{m}} - T_{\mathrm{s}})}{L} = 0, \quad X(0) = 0 \tag{4.31}$$

By solving the positive root of the above equation, the position of the freezing front at any time can be determined as

$$X^{\mathrm{Meg}} = \left[2\alpha_{\mathrm{s}}\left(\sqrt{1 + 2Ste} - 1\right)t\right]^{1/2} \tag{4.32}$$

4.5.3 Comparison of Approximate Methods

To assess the accuracy of the quasi-steady and Megerlin methods, they are compared with the exact Neumann solution for the temperature distribution and position of the freezing front. For this purpose, consider the isothermal solidification of water at 0 °C under the influence of a fixed wall temperature of – 20 °C. The material properties are:

- Specific heat capacity of solid water (c_s): 2.09 kJ/kg K,
- Thermal conductivity (k_s): 2.24×10^{-3} W/m K,
- Latent heat of fusion (L): 333.4 kJ/kg.

Figure 4.4 compares the freezing front position obtained using the exact, quasi-steady, and Megerlin methods. When $Ste = 0.12$, the maximum difference between the quasi-steady method and the exact solution after 5 h is 1.39 mm (– 2% error). The maximum difference for the Megerlin method is 0.58 mm (0.85% error). As observed, the differences increase over time but remain in good agreement. It appears that the Megerlin method is more accurate in predicting the freezing front position than the quasi-steady method.

Figure 4.5 shows the temperature distribution in the solid phase of water obtained using the three methods. Although both approximate methods agree well with the exact solution, this time, the quasi-steady method shows slightly better agreement. The maximum deviation for the quasi-steady method is 0.4 °C, while for the Megerlin method, it is 0.8 °C.

Unfortunately, exact solutions for other boundary conditions and geometries are not available to evaluate the accuracy of approximate methods. However, since the Stefan number for phase change materials is much smaller than 1 (e.g., the ratio

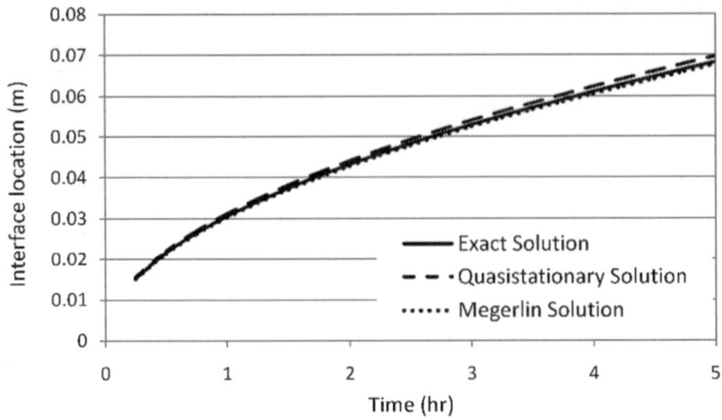

Fig. 4.4 Variations in the freezing front position over time for the exact, quasi-steady, and Megerlin methods [2]

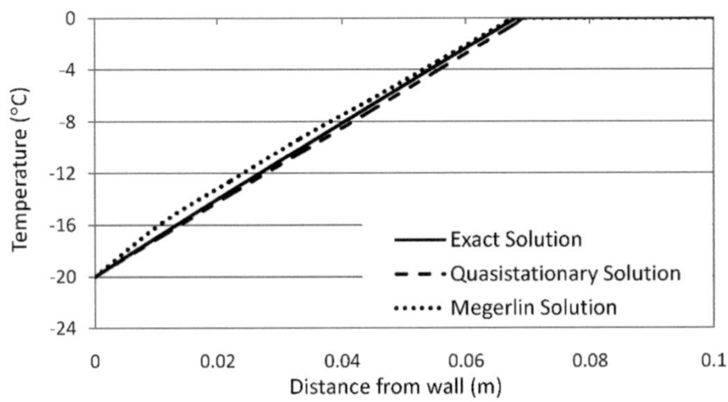

Fig. 4.5 Temperature distribution in the solid phase after 5 h of freezing, using the exact, quasi-steady, and Megerlin methods [2]

of sensible heat to latent heat is approximately 0.01 for every 1 °C temperature difference between the wall and the melting point), these methods provide relatively acceptable results.

In the present semi-analytical study, the quasi-steady and Megerlin methods will be used to determine the freezing front position. In future chapters, semi-analytical solutions for the Stefan problem in finned storage tanks will be presented, employing the assumptions stated in Sect. 4.2.

References

1. Alexiades V, Solomon AD (1992) Mathematical modeling of melting and freezing processes. Taylor & Francis
2. Mosaffa AH (2014) Approximate analytical and numerical investigation on solidification process in the finned thermal energy storage using phase change material. PhD thesis. University of Tabriz
3. Mosaffa AH, Talati F, Rosen MA (2012) Thermal storage using phase change material: solidification processes, chap 12. In: Rosen MA (ed) Energy storage. Nova Science Publishers, pp 367–392

Chapter 5
Semi-analytical Solutions for the Stefan Problem in Finned Rectangular Storage Tanks

5.1 Introduction

One of the most important parameters in designing thermal energy storage systems is determining the tank's geometry and dimensions, which directly impact the system's efficiency. Therefore, to ensure the optimal performance of the system, the tank's geometry and dimensions must be selected appropriately. Extensive research in this field indicates that rectangular and cylindrical geometries are the most commonly used in thermal energy storage systems. Hence, in this chapter and the next, semi-analytical and numerical solutions for such finned storage tanks will be presented.

Most manufacturers of thermal energy storage tanks recommend rectangular storage tanks due to their ease of construction. Additionally, PCMs in these types of storage tanks undergo phase change in a shorter time compared to cylindrical storage tanks with the same volume and heat transfer surface area. Figure 5.1 shows a simple rectangular thermal energy storage tank.

5.2 Problem Definition: Freezing in Finned Rectangular Storage Tanks

As mentioned previously, an exact two-dimensional solution for the phase change problem of a PCM in a finned storage tank is not feasible. Thus, a simplified analytical model will be presented here. Using this semi-analytical model, the location of the freezing front at any time and, consequently, the total freezing time can be determined. A finned rectangular storage tank is shown in Fig. 5.2.

To derive a semi-analytical solution, a symmetric control volume from this storage tank is considered, where thermal boundary conditions are applied symmetrically to its walls. Figure 5.3a illustrates the dimensions of the chosen control volume. To achieve the semi-analytical solution, this symmetric control volume is divided into

A. Mosaffa, *Phase Changing Materials*, Materials Horizons: From Nature to Nanomaterials, https://doi.org/10.1007/978-981-95-3742-6_5

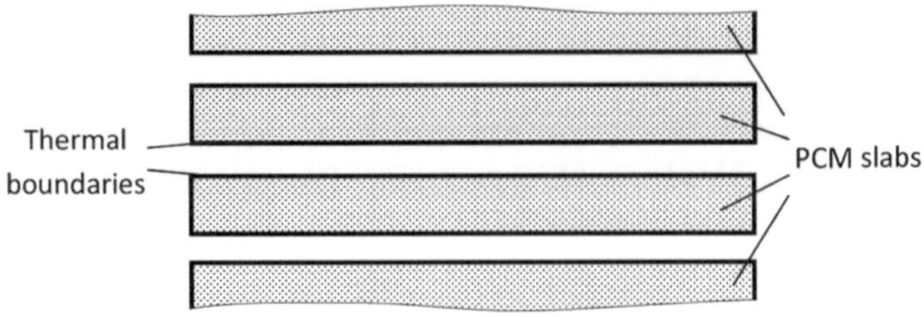

Fig. 5.1 A simple rectangular thermal energy storage tank [1]

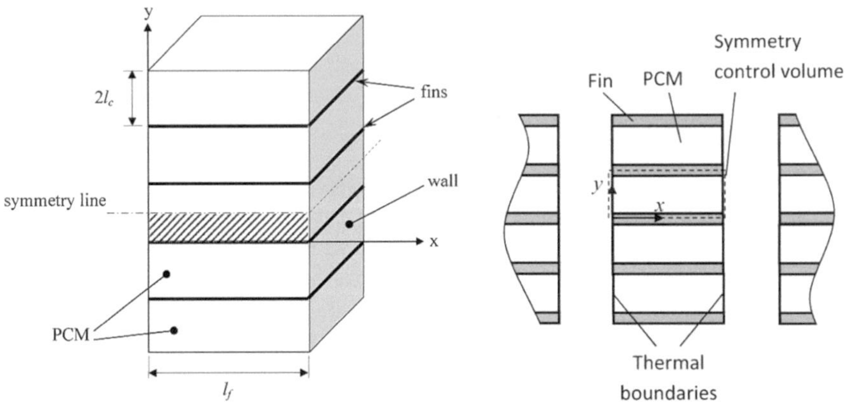

Fig. 5.2 A finned rectangular thermal energy storage tank [2]

two separate regions, as shown in Fig. 5.3b. In the first region, the only source of heat transfer is the boundary conditions applied to the walls. Heat transfer in this region occurs through conduction, is one-dimensional, and takes place in the x-direction. In the second region, the freezing front moves only in the y-direction, and heat is transferred solely through the fins to the surroundings. The approximate nature of this analytical method stems from splitting the two-dimensional problem into two one-dimensional problems.

Due to the non-linearity of the freezing problem, the following assumptions are considered for obtaining the analytical solution:

- Initially, the PCM in the storage tank is in liquid form at the melting temperature T_{m}. The fins' initial temperature is also T_{m}.
- During the freezing process, convective heat transfer in the liquid PCM is negligible compared to conductive heat transfer in the solid PCM.
- The PCM is homogeneous, and its phase change occurs at a constant temperature (isothermal phase change).
- All thermophysical properties are assumed to be constant and independent of temperature.

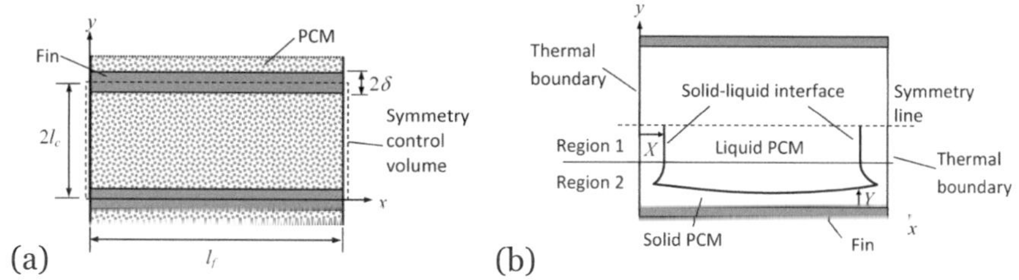

Fig. 5.3 **a** Dimensions of the specified rectangular control volume, **b** division of the two-dimensional control volume into two distinct regions [4]

- Due to the thinness, uniform cross-section, and high thermal conductivity of the fins, conductive heat transfer within the fins is considered one-dimensional and occurs in the x-direction.

The mathematical solution for the freezing problem in finned storage tanks will be analyzed in two separate sections.

5.2.1 Problem Definition in Region#1

In Region#1, where conductive heat transfer occurs only in the x-direction, the Stefan one-dimensional problem with different boundary conditions is defined as follows:

$$\frac{\partial^2 T_s}{\partial x^2} = \frac{1}{\alpha_s}\frac{\partial T_s}{\partial t}, \quad 0 < x < X(t), \quad t > 0 \tag{5.1}$$

$$-\rho_s L \frac{dX(t)}{dt} = -k_s \frac{\partial T_s(X(t), t)}{\partial x} \tag{5.2}$$

$$X(0) = 0 \tag{5.3}$$

$$T_s(x, 0) = T_m, \quad 0 < x < l_f \tag{5.4}$$

$$T_s(X, t) = T_m \tag{5.5}$$

Boundary condition types:

1. *First Type*:

$$T_s(0, t) = T_w, \quad T_w < T_m \tag{5.6a}$$

2. *Second Type*:

$$-k_{\mathrm{s}}\frac{\partial T_{\mathrm{s}}(0,t)}{\partial x}=q''_{\mathrm{w}},\quad q''_{\mathrm{w}}<0 \tag{5.6b}$$

3. *Third Type*:

$$-k_{\mathrm{s}}\frac{\partial T_{\mathrm{s}}(0,t)}{\partial x}+h[T_{\mathrm{s}}(0,t)-T_{\infty}]=0,\quad T_{\infty}<T_{\mathrm{m}} \tag{5.6c}$$

5.2.2 Problem Definition in Region#2

As mentioned earlier, the freezing front in Region#2 moves in the y-direction. By selecting a small element of thickness dx from the fin (Fig. 5.4), the energy balance for the fin can be written as:

$$-q''_{x}+q''_{x+\mathrm{d}x}+q''_{\mathrm{in}}=\dot{E}_{\mathrm{st}} \tag{5.7}$$

where $\dot{E}_{\mathrm{st}}$ is the rate of internal energy change per unit cross-sectional area of the fin. By expanding the heat flux term in a Taylor series and neglecting higher-order terms, we have:

$$-\left[k_{\mathrm{f}}\delta\frac{\partial T_{\mathrm{f}}}{\partial x}\right]+\left[k_{\mathrm{f}}\delta\frac{\partial T_{\mathrm{f}}}{\partial x}+\frac{\partial}{\partial x}\left(k_{\mathrm{f}}\delta\frac{\partial T_{\mathrm{f}}}{\partial x}\right)\mathrm{d}x\right]+\left[-k_{\mathrm{s}}\frac{T_{\mathrm{f}}-T_{\mathrm{m}}}{Y}\mathrm{d}x\right]$$
$$=\rho_{\mathrm{f}}c_{\mathrm{f}}(\delta\,\mathrm{d}x)\frac{\partial T_{\mathrm{f}}}{\partial t} \tag{5.8}$$

In the above equations, $Y(t)$ represents the distance of the freezing front from the fin surface along the y-axis:

$$\frac{\partial^2 T_{\mathrm{f}}}{\partial x^2}-\frac{k_{\mathrm{s}}}{\delta Y(t)k_{\mathrm{f}}}(T_{\mathrm{f}}-T_{\mathrm{m}})=\frac{1}{\alpha_{\mathrm{f}}}\frac{\partial T_{\mathrm{f}}}{\partial t},\quad 0<x<l_{\mathrm{f}},\ t>0 \tag{5.9}$$

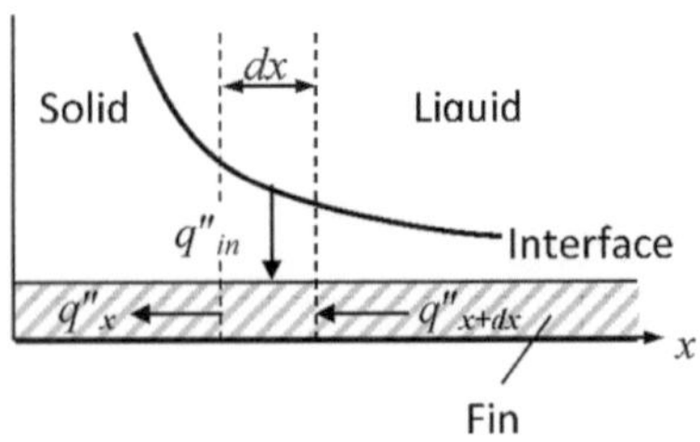

Fig. 5.4 Energy balance for a one-dimensional fin element with a length of dx and a thickness of δ [4]

Boundary condition types:

1. *First Type*:

$$T_f(0, t) = T_f(l_f, t) = T_w, \quad T_w < T_m \tag{5.10a}$$

2. *Second Type*:

$$-k_f \frac{\partial T_f(0, t)}{\partial x} = k_f \frac{\partial T_f(l_f, t)}{\partial x} = q''_w, \quad q''_w < 0 \tag{5.10b}$$

3. *Third Type*:

$$-\frac{\partial T_f(0, t)}{\partial x} + H_f[T_f(0, t) - T_\infty] = \frac{\partial T_s(l_f, t)}{\partial x} + H_f[T_f(l_f, t) - T_\infty] = 0, \quad T_\infty < T_m \tag{5.10c}$$

In the equation above, $H_f = h/k_f$. To solve the above equations, the following dimensionless parameters are defined:

$$\zeta^2 = \frac{l_f^2 k_s}{\delta\, Y\, k_f} \tag{5.11}$$

- **Dimensionless Temperature**

$$\theta = \frac{T_f - T_m}{T_w - T_m} \quad \text{For 1st type boundary condition} \tag{5.12a}$$

$$\theta = \frac{T_f - T_m}{q''_w l_f / k_f} \quad \text{For 2nd type boundary condition} \tag{5.12b}$$

$$\theta = \frac{T_f - T_m}{T_\infty - T_m} \quad \text{For 3rd type boundary condition} \tag{5.12c}$$

- **Dimensionless Location**

$$\eta = \frac{x}{l_f} \tag{5.13}$$

- **Dimensionless Time**

$$\tau = \frac{\alpha_f t}{l_f^2} \tag{5.14}$$

Thus, Eq. (5.9) can be written in dimensionless form as follows:

$$\frac{\partial^2\theta}{\partial\eta^2} - \zeta^2\theta = \frac{\partial\theta}{\partial\tau}, \quad 0 < \eta < 1, \quad \tau > 0 \tag{5.15}$$

5.3 Semi-analytical Solution of the Freezing Problem in Region#1

To solve Eqs. (5.1)–(5.6), the quasi-steady method explained in Sect. 4.5.1 is first used to determine the position of the freezing front. This function is essentially used as an initial guess. Then, using it, the temperature distribution equation in the first region is obtained. Once the temperature distribution function is determined, the exact freezing front position will be obtained using the Stefan condition (5.2).

5.3.1 *Solution of the Freezing Problem in Region#1 for the First-Type Boundary Condition (T_w = constant)*

The steady-state solution of Eq. (5.1) with boundary conditions (5.5) and (5.6a) is as follows:

$$T_{\mathrm{s}}^{\mathrm{qs}} = \frac{T_{\mathrm{m}} - T_{\mathrm{w}}}{X(t)}x + T_{\mathrm{w}} \tag{5.16}$$

By substituting the above equation into the Stefan condition (5.2), the freezing front position can be determined:

$$X(t) = \left[2\frac{k_{\mathrm{s}}}{\rho L}(T_{\mathrm{m}} - T_{\mathrm{s}})t\right]^{1/2} \tag{5.17}$$

The unsteady Eq. (5.1) with the initial condition (5.4) and inhomogeneous boundary conditions (5.5) and (5.6a) can be divided into two separate equations: a steady-state equation with inhomogeneous boundary conditions and an unsteady equation with homogeneous boundary conditions. The solution of the main equation will then be the sum of the solutions of the two mentioned equations [1]:

$$\frac{\mathrm{d}^2 T_{\mathrm{steady}}}{\mathrm{d}x^2} = 0 \tag{5.18a}$$

$$T_{\mathrm{steady}}(0, t) = T_{\mathrm{w}}, \quad T_{\mathrm{steady}}(X, t) = T_{\mathrm{m}} \tag{5.18b}$$

$$\frac{\partial^2 T_{\text{unsteady}}}{\partial x^2} = \frac{1}{\alpha_s}\frac{\partial T_{\text{unsteady}}}{\partial t} \tag{5.19a}$$

$$T_{\text{unsteady}}(x, 0) = T_m \tag{5.19b}$$

$$T_{\text{unsteady}}(0, t) = T_{\text{unsteady}}(X, t) = 0 \tag{5.19c}$$

The solution of Eq. (5.18) will be the same as Eq. (5.16). The solution of Eq. (5.19), using the separation of variables method, is obtained as follows:

$$T_{\text{unsteady}}(x, t) = \frac{2}{\pi}(T_m - T_w)\sum_{n=1}^{\infty}\left[\frac{\exp(-\alpha_s \beta_n^2 t)}{n}\sin(\beta_n x)\right] \tag{5.20}$$

where $\beta_n = n\pi/X$ are the eigenvalues. In this case, the temperature distribution can be written as the sum of Eqs. (5.16) and (5.20):

$$T_s(x, t) = \frac{2}{\pi}(T_m - T_w)\sum_{n=1}^{\infty}\left[\frac{\exp(-\alpha_s \beta_n^2 t)}{n}\sin(\beta_n x)\right] + \left(\frac{T_m - T_w}{X}x + T_w\right) \tag{5.21}$$

5.3.2 *Solving the Freezing Problem in Region#1 for a Second-Type Boundary Condition (q''_w = cte)*

Using the pseudo-steady method for Eq. (5.1) with boundary conditions (5.5) and (5.6b), the position of the freezing front is obtained as follows:

$$X(t) = -\frac{q''_w}{\rho_s L}t, \quad q''_w < 0 \tag{5.22}$$

Using the same method applied to derive Eq. (5.21), the temperature distribution equation can be expressed as [5]:

$$T_s(x, t) = \frac{-8q''_w}{k_s \pi}\sum_{n=1}^{\infty}\left[\frac{\exp(-\alpha_s \beta_n^2 t)}{(2n-1)^2}\cos(\beta_n x)\right] - \frac{q''_w x}{k_s} + \left(\frac{q''_w X}{k_s} + T_m\right) \tag{5.23}$$

where the eigenvalues $\beta_n = \frac{(2n-1)\pi}{2X}$.

5.3.3 Solving the Freezing Problem in Region#1 for a Third-Type Boundary Condition (h, $T_\infty = cte$)

Using the pseudo-steady method for Eq. (5.1) with boundary conditions (5.5) and (5.6c), the position of the freezing front is obtained as follows:

$$X(t) = \left[\frac{2k_s(T_m - T_\infty)}{\rho_s L}t + H_s^{-2}\right]^{1/2} - H_s^{-1} \tag{5.24}$$

where $H_s = h/k_s$. To derive the temperature distribution equation, the same approach as in previous sections is used, yielding [3]:

$$T_s(x,t) = \frac{h(T_m - T_\infty)}{k_s + hX}\left[x + 2\sum_{n=1}^{\infty}\frac{\beta_n X \cdot \cos(\beta_n X) - \sin(\beta_n X)}{H_s + (\beta_n^2 + H_s^2)X} \times (\beta_n^2 + H_s^2) \times \frac{\exp(-\alpha_s \beta_n^2 t)\sin[\beta_n(X - x)]}{\beta_n^2}\right] + \left(\frac{k_s T_m + XhT_\infty}{k_s + hX}\right) \tag{5.25}$$

where the eigenvalues β_n are the roots of $\beta_n \cot(\beta_n X(t)) = -H_s$ which can be solved using the Newton–Raphson method.

5.4 Semi-analytical Solution for the Freezing Problem in Region#2

As is well known, the use of extended surfaces or fins to enhance heat transfer rates plays an undeniable role in modern industries such as air conditioning. In recent years, various methods have been proposed for solving the non-linear heat transfer problem in different types of fins. Here, we will solve Eq. (5.15) [1].

5.4.1 Solving the Freezing Problem in Region#2 for a First-Type Boundary Condition

By considering the variable substitution $\mu = \theta - 1$, the boundary condition becomes homogeneous. Then, using the finite sine Fourier transform, we have

$$\hat{F}[\mu(\eta,\tau)] = \hat{N}(n,\tau) = 2\int_0^1 \mu(\eta,\tau)\sin(n\pi\eta)\cdot d\eta \tag{5.26}$$

By applying integration by parts and substituting these relations into Eq. (5.15) wo obtain

$$\hat{N}_\tau + \left(\zeta^2 + n^2\pi^2\right)\hat{N} = \frac{2\zeta^2}{n\pi}\left[(-1)^n - 1\right] \tag{5.27}$$

The general homogeneous and particular solutions of the above equation result in

$$\hat{N}(n,\tau) = \frac{2\left[(-1)^n - 1\right]n\pi}{\zeta^2 + n^2\pi^2}\exp\left[-\left(\zeta^2 + n^2\pi^2\right)\right] + \frac{2\zeta^2\left[(-1)^n - 1\right]}{n\pi\left(\zeta^2 + n^2\pi^2\right)} \tag{5.28}$$

Finally, using the inverse sine Fourier transform, the solution to Eq. (5.15) with initial condition and boundary condition is obtained as

$$\theta(\eta,\tau) = 1 + \sum_{n=1}^{\infty}\left\{\frac{2\left[(-1)^n - 1\right]n\pi}{\zeta^2 + n^2\pi^2}\exp\left[-\left(\zeta^2 + n^2\pi^2\right)\tau\right] + \frac{2\zeta^2\left[(-1)^n - 1\right]}{n\pi\left(\zeta^2 + n^2\pi^2\right)}\right\}\sin(n\pi\eta) \tag{5.29}$$

After deriving the temperature distribution equation in Region#2, the freezing front position in the y-direction can be determined using the Megerlin method:

$$Y(t) = 2\left[\frac{\sqrt{1 + 2Ste} - 1}{2}\right]^{\frac{1}{2}}\sqrt{\alpha_s t} \tag{5.30}$$

5.4.2 *Solving the Freezing Problem in Region#2 for a Second-Type Boundary Condition*

To homogenize boundary conditions, the following substitution is applied:

$$\mu(\eta,\tau) = \theta - \left(\eta^2 - \eta\right) \tag{5.31}$$

Thus, boundary conditions become

$$\frac{\partial^2\mu}{\partial\eta^2} - \zeta^2\mu - \zeta^2\left(\eta^2 - \eta + \frac{2}{\zeta^2}\right) = \frac{\partial\mu}{\partial\tau} \tag{5.32}$$

For solving the above equation with homogeneous boundary conditions, the cosine Fourier series is used, assuming

$$\mu(\eta, \tau) = \Lambda_0(\tau) + \sum_{n=1}^{\infty} \Lambda_n(\tau)\cos(n\pi\eta) \tag{5.33}$$

Substituting the partial derivatives of $\mu(\eta, \tau)$ into Eq. (5.32) results in

$$\left(\dot{\Lambda}_0 + \zeta^2\Lambda_0\right) + \sum_{n=1}^{\infty}\left[\dot{\Lambda}_n + \left(\zeta^2 + n^2\pi^2\right)\Lambda_n\right]\cos(n\pi\eta) = \zeta^2\left(\eta^2 - \eta + \frac{2}{\zeta^2}\right) \tag{5.34}$$

Integrating the above equation over $0 < \eta < 1$ and considering $\int_0^1 \cos(n\pi\eta)\mathrm{d}\eta = 0$, we have

$$\Lambda_0 = B_1\exp\left(-\zeta^2\tau\right) + \frac{2}{\zeta^2} + \frac{1}{6} \tag{5.35}$$

Multiplying both sides of Eq. (5.34) by $\cos(n\pi\eta)$ and integrating again over $0 < \eta < 1$, we obtain [5]:

$$\Lambda_n = B_2\exp\left[-\left(\zeta^2 + n^2\pi^2\right)\tau\right] + \frac{2\left[(-1)^{n+1} - 1\right]}{n^2\pi^2\left(\zeta^2 + n^2\pi^2\right)} \tag{5.36}$$

The constants $B_1 = -2/\zeta^2$ and $B_2 = \frac{2\left[(-1)^{n+1}-1\right]}{\zeta^2+n^2\pi^2}$ are determined from the initial conditions and Fourier coefficient relations for the cosine series. Thus, the fin temperature distribution equation for second-type boundary conditions is obtained as [5]:

$$\begin{aligned}\theta(\eta, \tau) = {} & \left[-\frac{2}{\zeta}\exp\left(-\zeta^2\tau\right) + \left(\frac{1}{6} + \frac{2}{\zeta^2}\right)\right] + \left(\eta^2 - \eta\right) \\ & + \sum_{n=1}^{\infty}\left\{\frac{2\left[(-1)^{n+1} - 1\right]}{\zeta^2 + n^2\pi^2}\exp\left[-\left(\zeta^2 + n^2\pi^2\right)\tau\right]\right. \\ & \left. + \frac{2\zeta^2\left[(-1)^{n+1} - 1\right]}{n^2\pi^2\left(\zeta^2 + n^2\pi^2\right)}\right\}\cos(n\pi\eta)\end{aligned} \tag{5.37}$$

Finally, the freezing front position in the y-direction can be determined using the Megerlin method:

$$Y(t) = \left[2\sqrt{1 + 2\frac{c_\mathrm{s}(T_\mathrm{m} - T_\mathrm{f})}{L}} - 1\right]^{\frac{1}{2}}\sqrt{\alpha_\mathrm{s}t} \tag{5.38}$$

5.4.3 Solving the Freezing Problem in Region#2 for Third-Type Boundary Conditions

The solution to Eq. (5.15) with the initial conditions and the boundary conditions is considered as the sum of the following two functions:

$$\theta(\eta, \tau) = \psi(\eta, \tau) + \varphi(\eta) \tag{5.39}$$

Using the equation above, Eq. (5.15) is divided into two equations: an ordinary differential equation and a partial differential equation with homogeneous boundary conditions:

$$\frac{\mathrm{d}^2\varphi}{\mathrm{d}\eta^2} - \zeta^2\varphi = 0 \tag{5.40}$$

$$\frac{\partial^2\psi}{\partial\eta^2} - \zeta^2\psi = \frac{\partial\psi}{\partial\tau} \tag{5.41}$$

The boundary and initial conditions are given as follows:

$$-\frac{\mathrm{d}\varphi(0)}{\mathrm{d}\eta} + \kappa\varphi(0) = \kappa \tag{5.42a}$$

$$\frac{\mathrm{d}\varphi(1)}{\mathrm{d}\eta} + \kappa\varphi(1) = \kappa \tag{5.42b}$$

$$-\frac{\partial\psi(0, \tau)}{\partial\eta} + \kappa\psi(0, \tau) = 0 \tag{5.43a}$$

$$\frac{\partial\psi(1, \tau)}{\partial\eta} + \kappa\psi(1, \tau) = 0 \tag{5.43b}$$

$$\psi(\eta, 0) = -\varphi(\eta) \tag{5.43c}$$

The solution to Eq. (5.40) can be obtained using the boundary conditions (5.42) as follows:

$$\begin{aligned}
&\varphi(\eta) = A_1 \exp(\zeta\eta) + A_2 \exp(-\zeta\eta) \\
&A_1 = \frac{\kappa}{\kappa - \zeta} - A_2\frac{\kappa + \zeta}{\kappa - \zeta}, \quad A_2 = \frac{\kappa\zeta\left[1 + \exp(\zeta)\right] - \kappa^2\left[1 - \exp(\zeta)\right]}{2\left(\kappa^2 + \zeta^2\right)\sinh(\zeta) + 4\kappa\zeta\cosh(\zeta)}
\end{aligned} \tag{5.44}$$

The partial differential Eq. (5.41) can be solved using the method of separation of variables. Considering the boundary and initial conditions (5.43), we have [3]:

$$\psi(\eta,\tau) = \sum_{n=1}^{\infty} \frac{c_n}{N(\lambda_n)} \frac{\exp\left[-\left(\lambda_n^2 + \zeta^2\right)\tau\right]}{\lambda_n^2 + \zeta^2} \left(\cos(\lambda_n \eta) + \frac{\kappa}{\lambda_n} \sin(\lambda_n \eta)\right)$$

$$N(\lambda_n) = \frac{\kappa}{2\lambda_n} \sin^2(\lambda_n) + \frac{1}{4}\left(\frac{1}{\lambda_n} - \frac{\kappa^2}{\lambda_n^2}\right) \sin(2\lambda_n) + \frac{1}{2}\left(1 + \frac{\kappa^2}{\lambda_n}\right)$$

$$\begin{aligned} c_n = & -A_1\left\{\left[\lambda_n \sin(\lambda_n) + \zeta \cos(\lambda_n) + \frac{\kappa}{\lambda_n}(\zeta \sin(\lambda_n) - \lambda_n \cos(\lambda_n))\right] \exp(\zeta\eta)\right\} \\ & - A_2\left\{\left[\lambda_n \sin(\lambda_n) - \zeta \cos(\lambda_n) - \frac{\kappa}{\lambda_n}(\zeta \sin(\lambda_n) + \lambda_n \cos(\lambda_n))\right] \exp(-\zeta\eta)\right\} \\ & + (A_1 - A_2)(\zeta + \kappa) \end{aligned} \quad (5.45)$$

The eigenvalues λ_n are determined as the solutions of the $\tan(\lambda_n) = -\frac{2\kappa\lambda_n}{\kappa^2 - \lambda_n^2}$ using the Newton–Raphson method.

The fin temperature distribution equation, which is the sum of $\phi(\eta)$ and $\psi(\eta, \tau)$, can now be obtained. Using the Megerlin method, the position of the freezing front along the y-direction can be determined:

$$Y(t) = \sqrt{\frac{2k_s(T_m - T_f)}{\rho_s L} t} \quad (5.46)$$

5.5 Numerical Solution of the Freezing Problem in a Rectangular Container Based on the Enthalpy Method

Since the phase change problem involves a moving boundary, various numerical methods have been introduced. However, the simplest method, which has attracted the most attention, is the enthalpy method. This is because it does not explicitly require incorporating the behavior of the moving boundary into the problem. In other words, there is no need to track the moving boundary between the liquid and solid phases within the PCM. The formulation for the total energy of the PCM during freezing is given as

$$\frac{\partial H}{\partial t} = \nabla\left(\frac{k_s}{\rho_s} \nabla T\right) \quad (5.47)$$

The total energy H can be divided into two parts: enthalpy (sensible heat) and latent heat:

$$H = h - Lf \quad (5.48)$$

Here, f represents the local freezing fraction, which is the ratio of the frozen PCM mass to the total mass. The enthalpy is given by

$$h = \int_{T_m}^{T} c \cdot \mathrm{d}T \tag{5.49}$$

For isothermal phase change problems, the local freezing fraction is determined as

$$f(T) = \begin{cases} 0 \text{ if } T > T_m \text{ (liquid)} \\ 1 \text{ if } T < T_m \text{ (solid)} \end{cases} \tag{5.50}$$

Substituting Eq. (5.48) into Eq. (5.47) yields

$$\frac{\partial h}{\partial t} = \frac{k_s}{\rho_s}\left(\frac{\partial^2 T}{\partial x^2} + \frac{\partial^2 T}{\partial y^2}\right) + L\frac{\partial f}{\partial t} \tag{5.51}$$

For an interior grid point (i, j), as shown in Fig. 5.5, the discretization of Eq. (5.51) using the backward finite-difference method is:

$$\frac{\partial h}{\partial t} = \frac{k_s}{\rho_s(\Delta x)^2}\left(T_{i-1,j} - 2T_{i,j} + T_{i+1,j}\right) + \frac{k_s}{\rho_s(\Delta y)^2}\left(T_{i,j-1} - 2T_{i,j} + T_{i,j+1}\right) + L\frac{\partial f}{\partial t} \tag{5.52}$$

For isothermal phase change, the temperature within the control volume at grid point (i, j) remains constant and equal to T_m during the phase change process. Thus, from Eq. (5.51), $\partial h/\partial t = 0$. In this case, f lies in the range (0, 1). The discretized equations are solved using the fully implicit method (unconditionally stable), as follows:

$$f_{i,j} = f_{i,j}^{\text{old}} - \frac{k_s \Delta t}{\rho_s L(\Delta x)^2}\left(T_{i-1,j} - 2T_{i,j} + T_{i+1,j}\right) - \frac{k_s \Delta t}{\rho_s L(\Delta y)^2}\left(T_{i,j-1} - 2T_{i,j} + T_{i,j+1}\right) \tag{5.53}$$

If the control volume for grid point (i, j) is fully liquid or fully solidified, f is exactly 0 or 1, and $\partial f/\partial t = 0$. Using the backward finite-difference method for Eq. (5.51), we have:

$$T_{i,j} = T_{i,j}^{\text{old}} + \frac{\alpha_s \Delta t}{(\Delta x)^2}\left(T_{i-1,j} - 2T_{i,j} + T_{i+1,j}\right) + \frac{\alpha_s \Delta t}{(\Delta y)^2}\left(T_{i,j-1} - 2T_{i,j} + T_{i,j+1}\right) \tag{5.54}$$

The above discretized equations have first-order and second-order accuracy for time and space, respectively. By simultaneously solving Eqs. (5.53) and (5.54) for

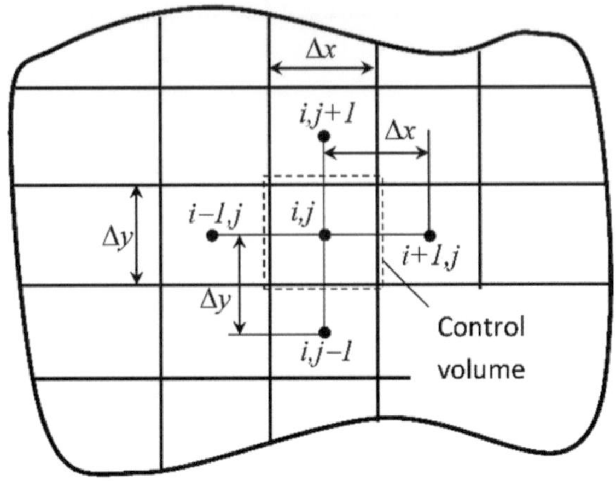

Fig. 5.5 Discretization domain for the two-dimensional phase change problem at an interior grid point [1]

each time step across the entire domain, the temperature and local freezing fraction at all grid points can be determined.

The discretized equation for boundary nodes depends on the boundary conditions and is derived using the energy balance for the control volume around the boundary node (Fig. 5.6).

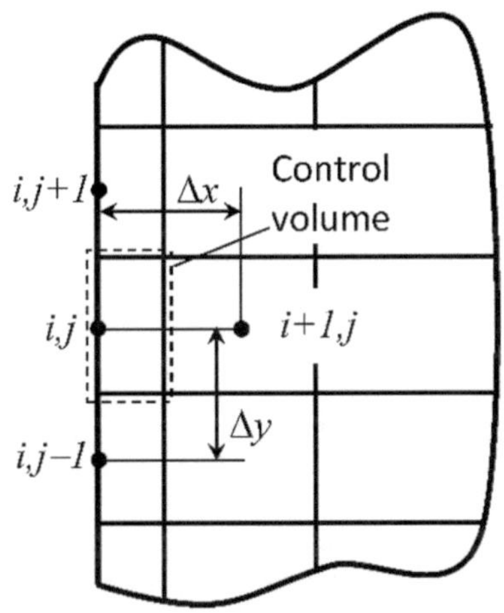

Fig. 5.6 Discretization domain for the two-dimensional phase change problem at a boundary grid point [1]

5.6 Illustrative Examples

5.6.1 *Example 1: Solidification of Climsel C28 in a Rectangular LHTS System with Horizontal Fins*

In this example, we analyze the solidification of the PCM, Climsel C28, a salt hydrate, within a rectangular LHTS unit equipped with horizontal aluminum fins, as illustrated in Fig. 5.3a. Climsel C28 is commonly used in cooling applications, such as laptop thermal management and comfort vests. The material properties of Climsel C28 are as follows:

- *Specific heat capacity*: $c = 3600$ J/kg K
- *Thermal conductivity*: $k = 0.6$ W/m K
- *Latent heat of fusion*: $L = 162{,}000$ J/kg
- *Density*: $\rho = 1480$ kg/m^3
- *Solidification temperature*: $T_{\mathrm{m}} = 28$ °C.

The aluminum fins have the following thermophysical properties:

- *Specific heat capacity*: $c = 875$ J/kg K
- *Thermal conductivity*: $k = 177$ W/m K
- *Density*: $\rho = 2770$ kg/m^3.

The fin thickness is set to 1 cm, with a height of 8 cm and width of 2 cm. The end walls of the storage system are maintained at a constant temperature of $T_{\mathrm{w}} = 10$ °C throughout the process. Given the Stefan number $Ste_{\mathrm{s}} = 0.4$, the transcendental root of Eq. (4.23) is determined to be $\lambda = 0.4226$.

- **Analytical Solution in Region#1**

For Region#1, the Neumann solution provides an exact solution. As an alternative, the quasi-stationary approximation simplifies the governing heat equation as follows:

$$\frac{\partial^2 T_{\mathrm{s}}^{\mathrm{qs}}}{\partial x^2} = 0 \tag{5.55}$$

where $T_{\mathrm{s}}^{\mathrm{qs}}$ represents the approximate quasi-stationary temperature distribution in the solidified PCM. The corresponding boundary conditions are

$$\begin{aligned} T_{\mathrm{s}}^{\mathrm{qs}}(0, t) &= T_{\mathrm{s}}^{\mathrm{qs}}(W, t) = T_{\mathrm{w}} \\ T_{\mathrm{s}}^{\mathrm{qs}}(X^{\mathrm{qs}}(t), t) &= T_{\mathrm{m}} \end{aligned} \tag{5.56}$$

Solving Eq. (5.55) under these boundary conditions gives the quasi-stationary temperature distribution:

$$T_{\mathrm{s}}^{\mathrm{qs}}(x, t) = T_{\mathrm{w}} + (T_{\mathrm{m}} - T_{\mathrm{w}})\frac{x}{X^{\mathrm{qs}}(t)} \tag{5.57}$$

The location of the solid–liquid interface, $X^{qs}(t)$, can be determined using Stefan condition:

$$X^{qs}(t) = \sqrt{\frac{2k_s(T_m - T_w)t}{\rho L}} = 2\sqrt{\frac{Ste_s}{2}}\sqrt{\alpha_s t} \tag{5.58}$$

- **Comparison of Analytical and Numerical Solutions**

Figure 5.7 presents a comparison of the temperature profiles in Region#1, as obtained by the Neumann solution and the quasi-stationary solution at various time intervals. Additionally, Fig. 5.8 illustrates the temperature history at different locations within Region#1. Figure 5.9 showcases a comparison of the solid–liquid interface in Region#1, derived from both methods, under several imposed temperature conditions.

When the Stefan number approaches zero, the sensible heat contribution becomes negligible compared to the latent heat. Under this condition, the transcendental root λ becomes small, and the error function can be approximated as:

$$\mathrm{erf}(x) = \frac{2}{\sqrt{\pi}}\sum_{m=0}^{\infty}\left(\frac{(-1)^m x^{(2m+1)}}{(2m+1)\times m!}\right) \tag{5.59}$$

For small λ, this reduces to:

$$\mathrm{erf}(\lambda) \approx 2\lambda/\sqrt{\pi} \Rightarrow \lambda \approx \sqrt{\frac{Ste_s}{2}} \tag{5.60}$$

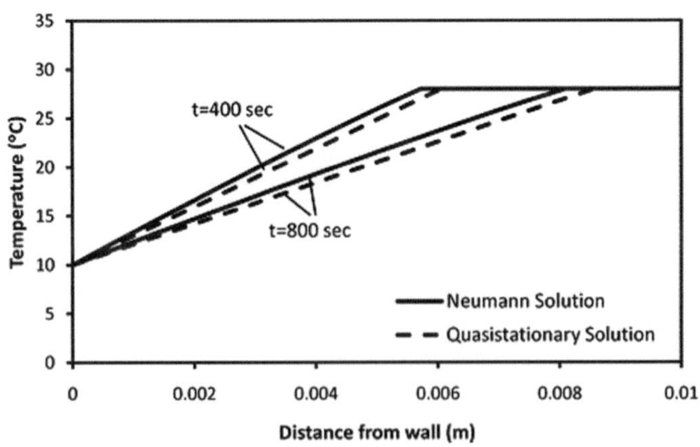

Fig. 5.7 Comparison of the temperature profiles in Region#1 at different time intervals, obtained using the Neumann solution and the quasi-stationary approximation [4]

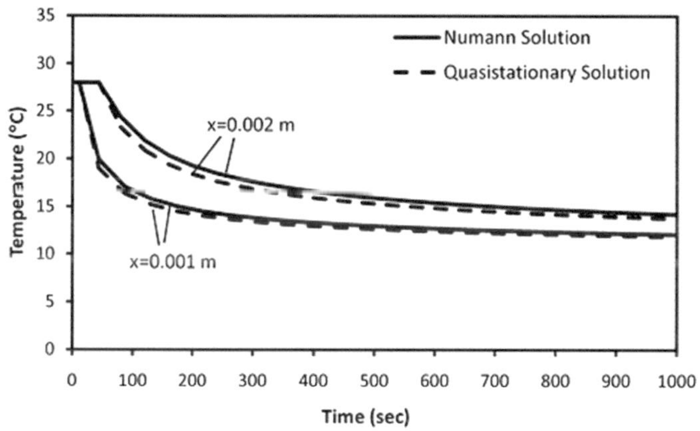

Fig. 5.8 Comparison of the temperature histories at various spatial locations within Region#1 [4]

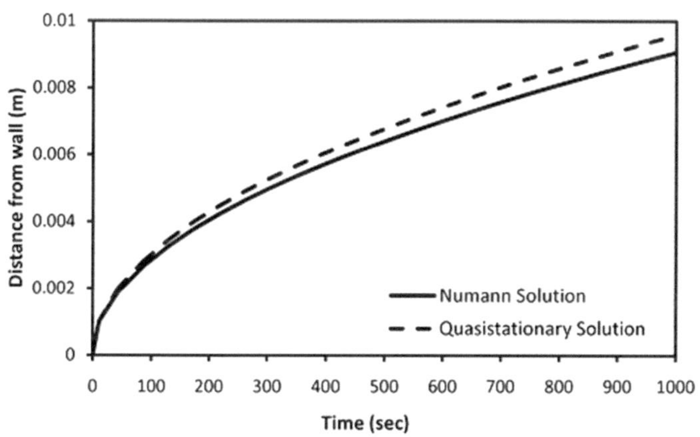

Fig. 5.9 Solid–liquid interface position for different imposed temperatures, calculated using both methods [4]

Thus, when $Ste_s \rightarrow 0$, the difference between the quasi-stationary approximation and the exact Neumann solution is minimal. Table 4.1 demonstrates that the root λ remains slightly smaller than $\sqrt{Ste_s/2}$. Consequently:

- The quasi-stationary method overestimates the thickness of the solidified layer.
- The temperature at specific spatial points in the solidified PCM is underestimated compared to the exact solution.

Figure 5.10 shows the temperature profile of the fin, calculated using both the enthalpy-based numerical method and the approximate analytical solution. Due to the fin's high thermal conductivity, its temperature distribution quickly approaches that of the wall. The difference between the analytical and numerical results is minimal. Figure 5.11 illustrates the history of the solid–liquid interface in Region#2.

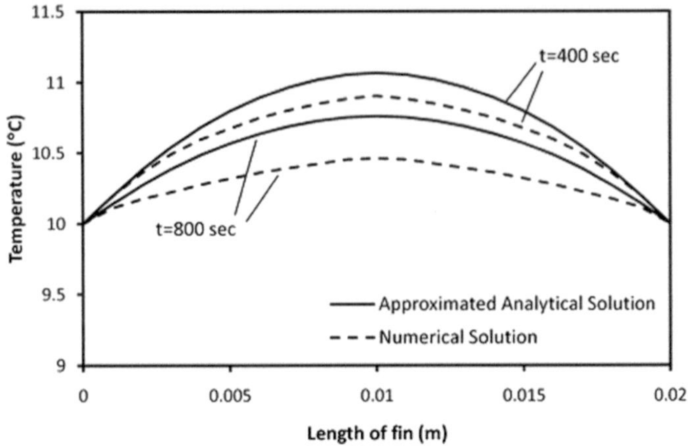

Fig. 5.10 Comparison the fin temperature profiles derived using the enthalpy-based numerical method and the analytical solution [4]

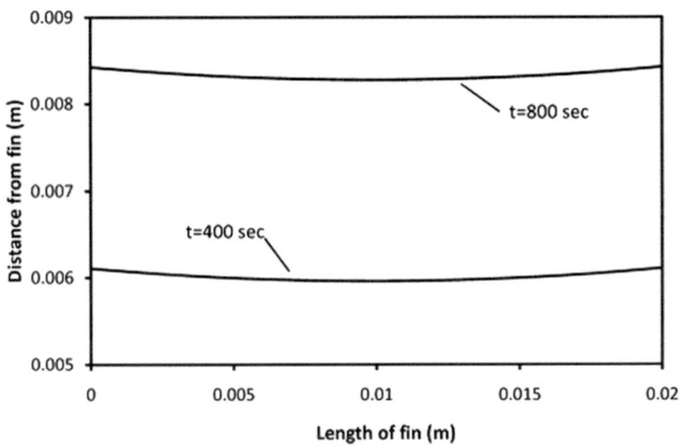

Fig. 5.11 Solid–liquid interface history in Region#2 [4]

Figure 5.12 illustrates the growth rate of the solid fraction of the PCM over time. The solidified PCM fraction indicates the extent of storage solidification at any given moment.

5.6.2 Example 2: Solidification with Imposed Heat Flux

This example considers the solidification of Climsel C28 under a constant heat flux condition ($q''_w = -100$ W/m^2) applied at the storage unit's end wall. Similar to the

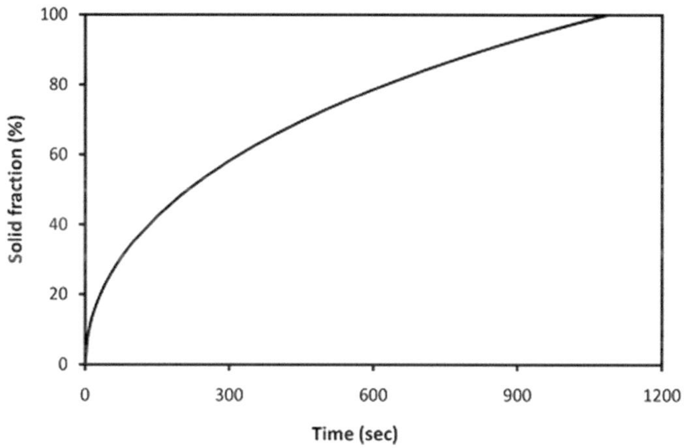

Fig. 5.12 Solid fraction growth over time, providing insights into the rate of PCM solidification [4]

previous case, the quasi-stationary approximation is used, simplifying the governing equation to:

$$\frac{\partial^2 T_s^{qs}}{\partial x^2} = 0 \tag{5.55}$$

The boundary conditions for this case are:

$$\begin{aligned} k_s \frac{\partial T_s^{qs}(0,t)}{\partial x} &= -k_s \frac{\partial T_s^{qs}(W,t)}{\partial x} = q''_w \\ T_s^{qs}(X^{qs}(t),t) &= T_m \end{aligned} \tag{5.61}$$

The resulting quasi-stationary temperature distribution is:

$$T_s^{qs}(x,t) = T_m - \frac{q''_w}{k_s}\left[X^{qs}(t) - x\right] \tag{5.62}$$

The interface location $X^{qs}(t)$ can be obtained as:

$$X^{qs}(t) = \frac{q''_w t}{\rho L} \tag{5.63}$$

Figures 5.13, 5.14 and 5.15 compare the temperature profile, temperature history, and solid–liquid interface location, respectively, between the quasi-stationary approximation and the analytical solution. The discrepancies are minimal, with a maximum error of 0.01 °C in temperature and 0.0001 m in the interface position. Figure 5.16 depicts the temperature profile of the fin at various time intervals,

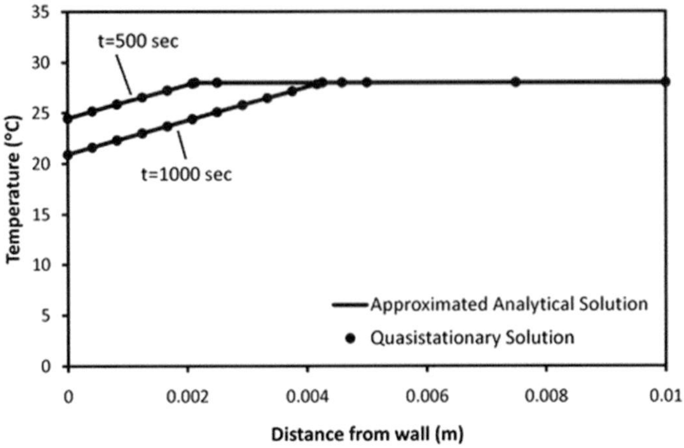

Fig. 5.13 Temperature distribution in Region#1 for $q''_w = -1000$ W/m^2 [4]

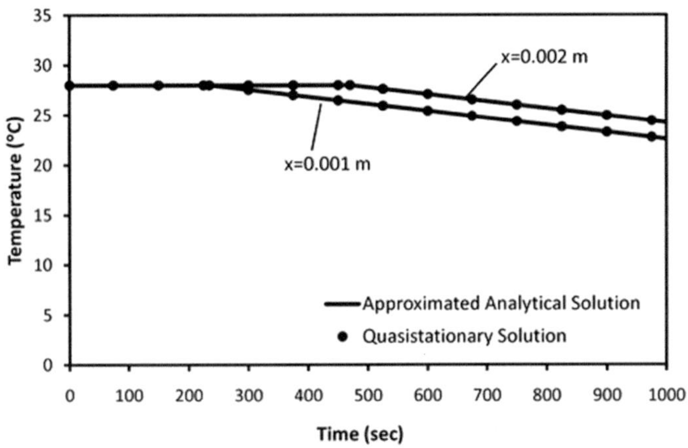

Fig. 5.14 Temperature history in Region#1 for $q''_w = -1000$ W/m^2 after 1800 s [4]

comparing the approximated analytical solution with the enthalpy-based numerical method. The close agreement observed indicates the high accuracy of the analytical method. Figure 5.17 shows the location of the solid–liquid interface in region 2 at different times, as determined by the approximate analytical model. Figure 5.18 illustrates the temporal variation in the fraction of solidified PCM, as predicted by the analytical model.

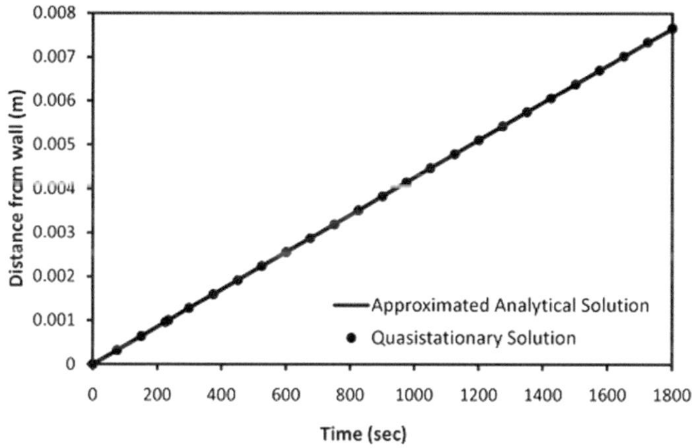

Fig. 5.15 Solid–liquid interface location in Region#1 after 1800 s [4]

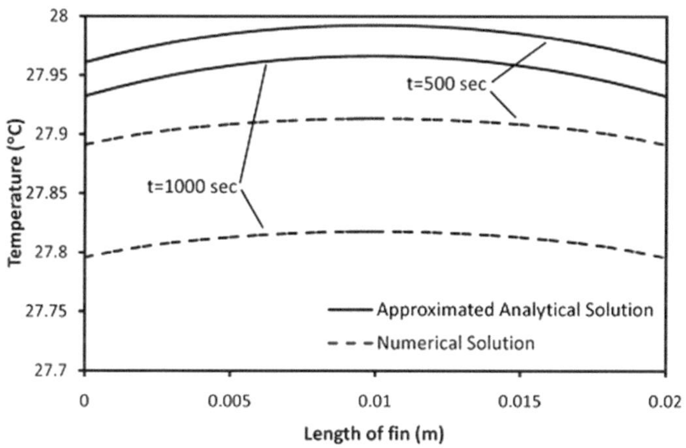

Fig. 5.16 Temperature distribution of fin for $q''_w = -1000$ W/m^2 [4]

5.7 Semi-analytical Solution for 1st Type of Boundary Condition

Following the semi-analytical solution of the freezing problem in finned storage tanks and the presentation of a numerical method based on the enthalpy method, this chapter presents the results. Based on the temperature distribution obtained in each region, the position of the freezing front can be predicted at any given time. Additionally, the results obtained from the analytical and numerical solutions are compared.

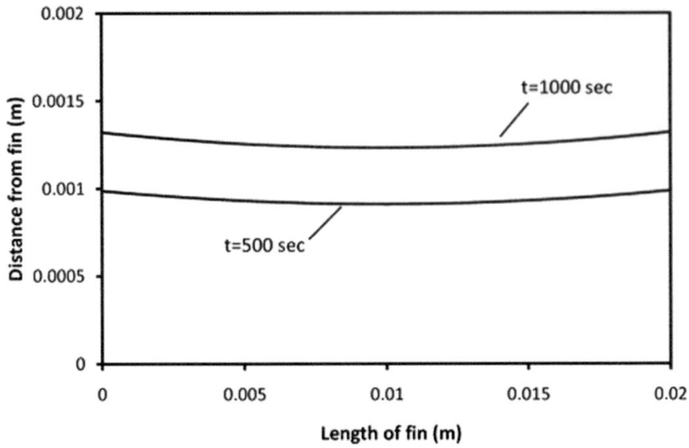

Fig. 5.17 Interface location in Region#2 over time [4]

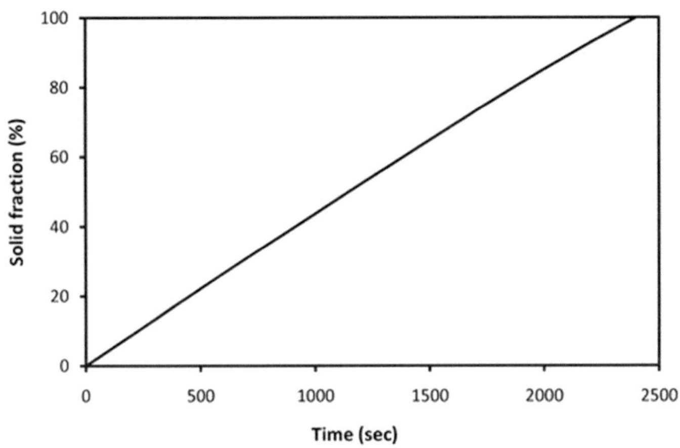

Fig. 5.18 Variation of solid fraction over time in PCM storage, as determined by the approximated analytical method [4]

5.7.1 Material Selection

In this section, phase change materials are selected such that their melting point is within a suitable range for residential applications in buildings. Numerous review articles introduce various types of phase change materials used in air conditioning applications. However, one of the important parameters in design is the chemical compatibility of the phase change material with the container material to prevent corrosion. Studies have been conducted on the corrosive effects of hydrated salts ($Zn(NO_3)_2{\cdot}6H_2O$, $Na_2HPO_4{\cdot}12H_2O$, and $CaCl_2{\cdot}6H_2O$) on common metals such as copper, aluminum, ordinary steel, and stainless steel, and suitable metals have been suggested for each material. Paraffin has been introduced as one of the best materials in terms of compatibility with most metals. Here, four types of phase change materials, including hydrated salt ($CaCl_2{\cdot}6H_2O$), organic material (Paraffin), and commercial materials ClimSel (C23 and C28), were selected. Aluminum was also selected for the fins. Table 5.1 presents the thermophysical properties of the phase change materials and aluminum.

- **Type I Boundary Condition**

As previously mentioned, to demonstrate the influence of fin length and spacing, a parameter 's' was introduced, representing the ratio of fin length to half the fin spacing. To investigate this effect, three values of 's' were considered. The selected dimensions for fin length and spacing are presented in Table 5.2.

The thickness of the aluminum fin is considered to be 1 mm. The PCM chosen for the type I boundary condition is paraffin. The constant temperature applied to the cell walls is $T_w = 20$ °C. As previously mentioned, the PCM is initially in a liquid state at its melting temperature Tm. Therefore, convective heat transfer in the liquid phase is negligible.

Figure 5.19 illustrates the position of the freezing front for the first case, obtained from both analytical and numerical solutions. The largest discrepancy observed in the figure is at the corners, where the curvature in the analytical solution should be slightly greater. This difference arises from dividing the two-dimensional domain into two one-dimensional regions. However, the maximum difference in the first region is 0.1 and 0.13 mm in the second region at 300 s, indicating good agreement between the present semi-analytical method and the numerical method.

Table 5.1 Thermophysical properties of phase change materials in the solid state and aluminum [1]

Property	ClimSel C23	ClimSel C28	$CaCl_2{\cdot}6H_2O$	Paraffin	Al
Heat capacity, c (J/kg K)	2660	3600	1460	1920	875
Thermal conductivity, k (W/m K)	0.6	0.6	1.088	0.514	177
Density, ρ (kg/m^3)	1480	1450	1710	830	2770
Latent heat of fusion, L (J/kg)	148,000	162,000	187,490	251,000	–
Melting temperature, Tm (°C)	23	28	29.7	32	–

Table 5.2 Selected dimensions for fin length and spacing in a rectangular enclosure [1]

Parameter	Case 1	Case 2	Case 3
Length of the fin l_f (m)	0.01	0.015	0.02
Half height of the cell l_c (m)	0.02	0.015	0.01
Cell aspect ratio $\sigma = l_f/l_c$	0.5	1	2

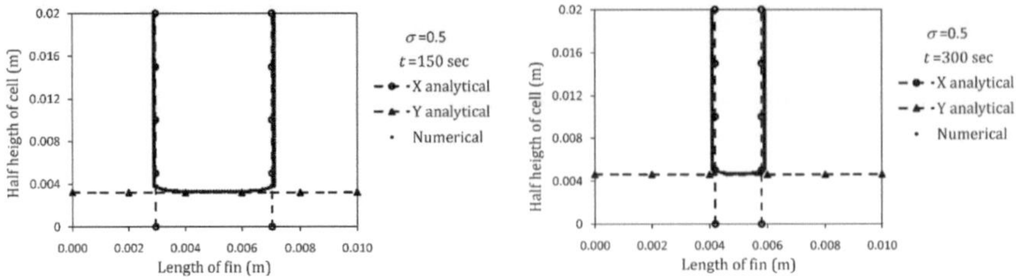

Fig. 5.19 Prediction of the freezing front position within a rectangular PCM (paraffin) storage tank cell using analytical and numerical solutions for $\sigma = 0.5$ and first-type boundary condition ($T_w = 20$ °C) at different times [1]

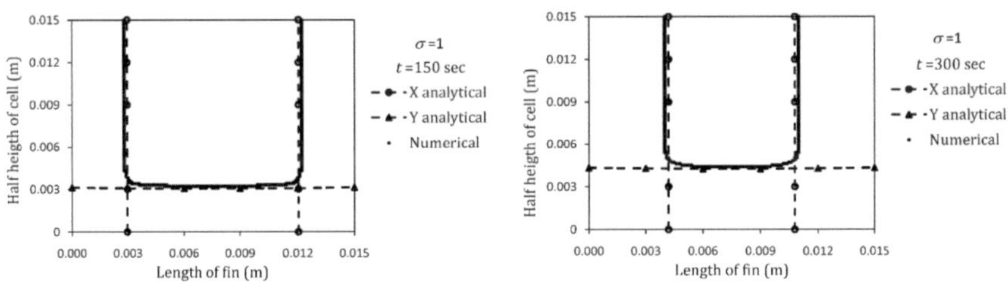

Fig. 5.20 Prediction of the freezing front position within a rectangular PCM (paraffin) storage tank cell using analytical and numerical solutions for $\sigma = 1$ and first-type boundary condition ($T_w = 20$ °C) at different times [1]

Figure 5.20 shows the position of the freezing front for the second case. In this case, the maximum difference is 0.2 mm in the first region and 0.16 mm in the second region at 300 s.

Figure 5.21 shows the position of the freezing front for the third case. In this case as well, the maximum difference is 0.1 mm in the first region and 0.12 mm in the second region at 300 s. As the results show, the greatest difference between the analytical and numerical solutions occurs in the second case, when $\sigma = 1$. It is evident that in this case, the influence of the wall and fin in extracting heat from the paraffin is equal, and as a result, considering one-dimensional heat transfer in two regions increases the difference compared to the two-dimensional case in one region. However, it should be noted that this difference is negligible.

The dimensionless temperature distribution of the fin for three values of s, obtained using analytical and numerical methods, is shown in Fig. 5.22. When the fin length is small ($\sigma < 1$), the fin temperature reaches the wall temperature, T_w, more quickly. As observed in all three cases, the difference between the analytical and numerical solutions decreases with increasing dimensionless time (Fourier number). Also, the least difference is obtained in the first case. This is because in this case, the wall plays a more significant role in extracting heat from the paraffin. Therefore, assuming one-dimensional heat transfer in the second region would be appropriate. This difference is 0.38% for $Fo = 110$. With increasing fin length, the assumption of one-dimensional

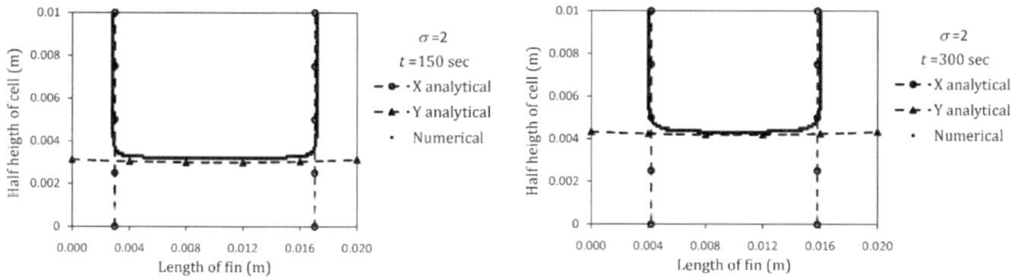

Fig. 5.21 Prediction of the freezing front position within a rectangular PCM (paraffin) storage tank cell using analytical and numerical solutions for $\sigma = 2$ and first-type boundary condition ($T_w = 20$ °C) at different times [1]

heat transfer results in a larger error. Although this difference is still negligible even at the maximum value ($Fo = 55$) (0.92%).

To obtain the total freezing time in each cell, a parameter is defined as the ratio of solid PCM as follows:

$$\omega(t) = \frac{\text{Volume of solid PCM}}{\text{Volume of cell}} = \frac{2X(t)[l_c - \delta - Y(t)] + Y(t)l_f}{(l_c - \delta)l_f} \tag{5.64}$$

where δ is the half height of the fin.

Using this parameter, it can be determined how much PCM has solidified in the cell at any given time. Figure 5.23 shows the variation of the solid PCM ratio with time. As seen in the figure, the complete freezing time of the paraffin in the cell with $s = 0.5$ is less than the other cases. This indicates that the length of the cell wall or the fin spacing plays a more significant role in extracting heat, or in other words, in freezing the PCM, compared to the fin length. In the initial times, the freezing rate is high. However, this rate gradually decreases with increasing thermal resistance due to the solidified PCM. Considering the diagram for the case without fins, it is observed that the ratio of length to fin spacing has an optimal value, and increasing the fin length excessively may reduce its effect.

5.8 Semi-analytical Solution for 2nd Type of Boundary Condition

This section presents a detailed analysis of the thermal behavior of a PCM-based energy storage system, where the PCM under investigation is the salt hydrate ClimSel C28, and the fin material is aluminum. The physical properties of the PCM and the aluminum fins are summarized in Table 5.1.

In this section, the rate of heat transfer is influenced by several geometric factors, including the number of fins, their length, and the spacing between them. The cell aspect ratio plays a significant role in dictating the directionality of heat transfer.

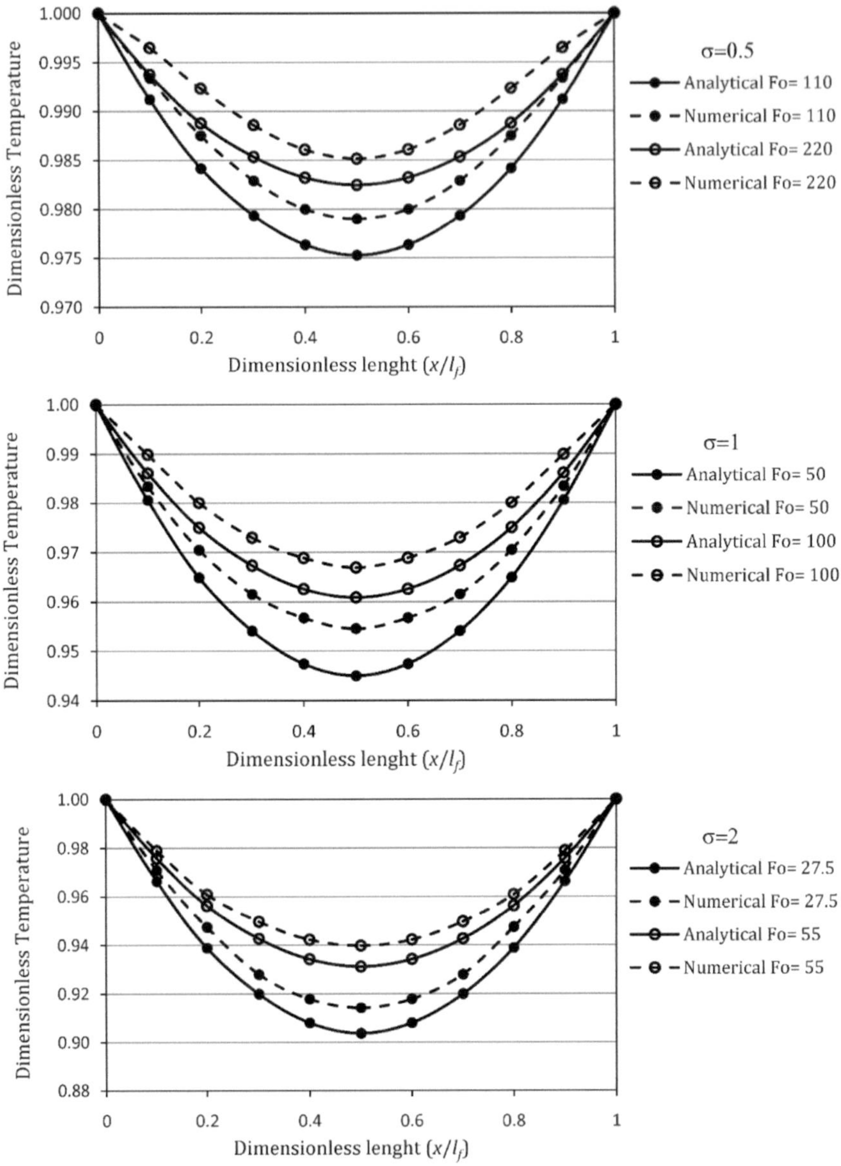

Fig. 5.22 Dimensionless temperature distribution of a rectangular fin obtained from analytical and numerical solutions for different Fourier numbers and first-type boundary condition ($T_{\mathrm{w}} = 20$ °C): $\sigma = 0.5$ (top), $\sigma = 1$ (middle), and $\sigma = 2$ (bottom) [1]

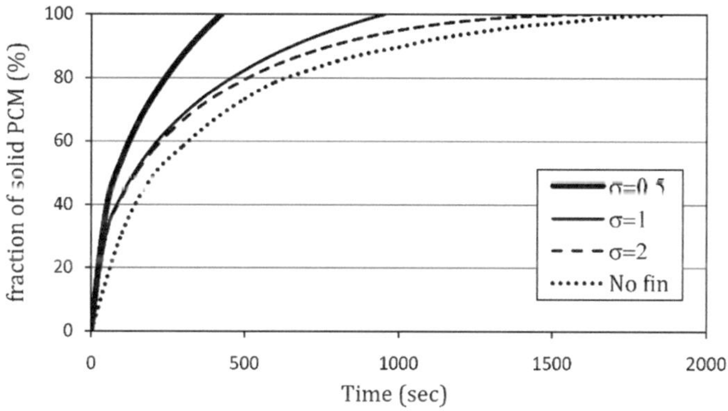

Fig. 5.23 Ratio of the mass of solidified paraffin to the total mass in the rectangular cell, obtained from the analytical solution for different values of *s* and type I boundary condition ($T_w = 20$ °C) [1]

Specifically, when σ is much smaller than one, heat transfer is primarily in the x-direction, while a much larger σ leads to predominant heat transfer in the y-direction. When σ equals one, heat transfer is balanced in both directions.

Three distinct geometries are examined, each corresponding to a different value of σ, specifically 0.5, 1, and 2. In all cases, the half-thickness of the fins, δ, is kept constant at 0.5 mm, and the dimensions used in the simulations are listed in Table 5.2. For all scenarios, the initial temperature of the storage unit, T_m, is set to 28 °C, with the PCM in its liquid phase. Additionally, a constant heat flux of − 1000 W/m^2 is applied throughout the simulations.

- **Solid–Liquid Interface Analysis**

Figure 5.24 display the positions of the solid–liquid interface over time for the different test cases, comparing one-dimensional analytical predictions with two-dimensional numerical results. The sharp corners observed in the solid–liquid interface are due to the high temperature gradients at the boundaries of the system. By comparing the results from both methods, we assess the accuracy of the simplified one-dimensional model.

It is found that the differences between the one-dimensional and two-dimensional predictions are relatively small, particularly for cases where the cell aspect ratio, σ, is close to unity or smaller. The one-dimensional model provides a reasonable approximation of the solid–liquid interface location, especially when heat transfer occurs predominantly in one direction (as in the case of low values of σ). The accuracy of the analytical model is deemed sufficient for cases with a small aspect ratio or where heat transfer is predominantly one-dimensional in the x-direction.

- **Temperature Distribution in the Fin**

The temperature distribution along the fin is calculated using both the one-dimensional analytical model and the two-dimensional numerical method, with

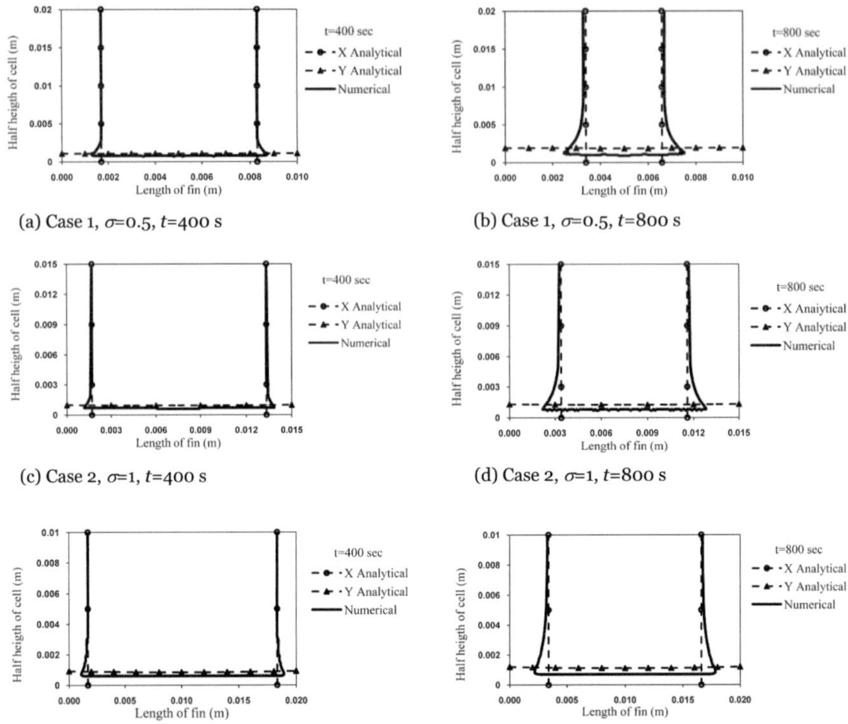

(a) Case 1, σ=0.5, t=400 s

(b) Case 1, σ=0.5, t=800 s

(c) Case 2, σ=1, t=400 s

(d) Case 2, σ=1, t=800 s

(e) Case 3, σ=2, t=400 s

(f) Case 3, σ=2, t=800 s

Fig. 5.24 Position of solid–liquid interface for the one-dimensional analytical model and the two-dimensional numerical method [5]

results presented in Fig. 5.25 for all cases. The aluminum fins, with their high thermal conductivity, rapidly reach thermal equilibrium with the storage unit's end wall, leading to a relatively uniform temperature distribution. When the fin length increases, the temperature difference along the fin also increases, as expected.

For Case 1 ($\sigma = 0.5$), the difference between the analytical and numerical solutions is minimal—approximately 0.07 °C at $t = 400$ s and 0.08 °C at $t = 800$ s. The largest discrepancy is observed in Case 3 ($\sigma = 2$) at $t = 800$ s, with a difference of about 0.12 °C. Notably, when the aspect ratio σ is small, heat transfer is almost entirely one-dimensional in the x-direction, and the one-dimensional model provides a highly accurate prediction of the temperature distribution along the fin. However, for larger σ, the differences between the analytical and numerical solutions remain small but are more noticeable due to the more complex two-dimensional nature of heat transfer.

- **PCM Solidification Behavior**

The fraction of solidified PCM is calculated from the one-dimensional analytical model for each of the test cases, as shown in Fig. 5.26. As expected, larger characteristic lengths (l_c) result in faster solidification of the PCM, as a greater portion

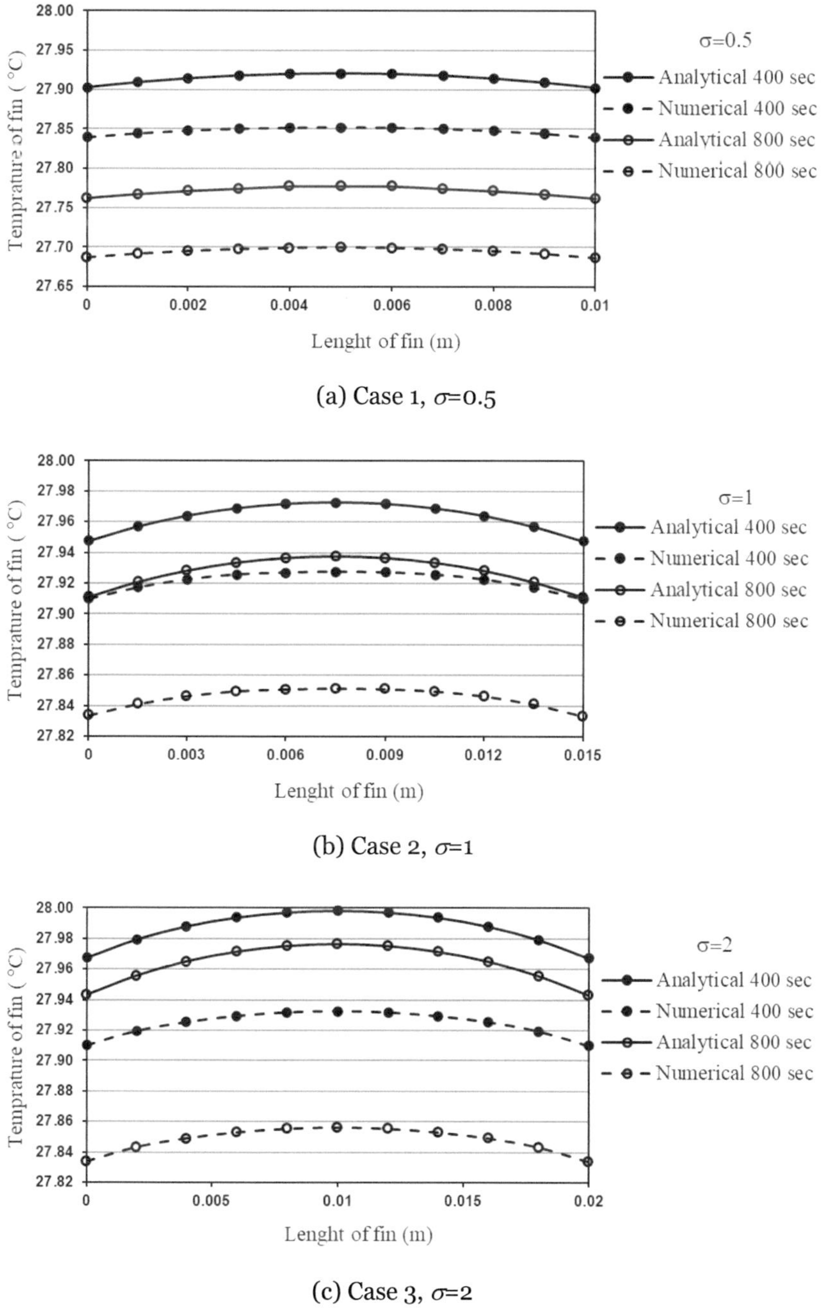

Fig. 5.25 Temperature of the fin for one-dimensional analytical model and two-dimensional numerical method [5]

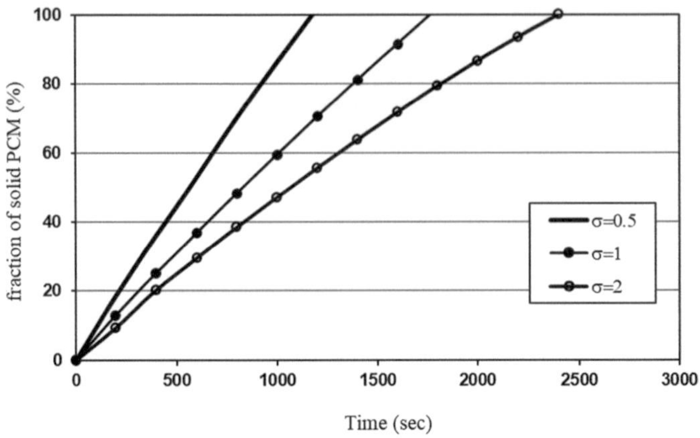

Fig. 5.26 Variation with time of a solid fraction of PCM obtained using the analytical model for all cases [5]

of the heat transfer occurs through the walls of the storage unit. Initially, when the thickness of the solidified PCM is small, thermal resistance is low, allowing for a constant solidification rate and steady movement of the solid–liquid interface. Over time, as the solidified PCM layer thickens, the thermal resistance increases, leading to a decrease in the solidification rate.

5.8.1 Summary and Conclusions

The section demonstrates that the one-dimensional analytical model offers an excellent approximation for the solid–liquid interface location and temperature distribution, particularly when the geometry of the storage unit is close to one-dimensional or when the aspect ratio σ is small. However, as the aspect ratio increases, two-dimensional effects become more significant, and although the differences in predictions remain small, the complexity of heat transfer in such configurations is better captured by numerical simulations. Overall, the analytical and numerical results are in close agreement, validating the use of simplified models for the analysis of PCM-based thermal storage systems with fins.

The thermal performance of the storage system is governed by the geometry of the fins and the PCM's thermal properties. For larger characteristic lengths, the solidification process slows down as the solid PCM layer thickens, indicating that thermal resistance plays a key role in the overall heat transfer efficiency. Further investigations could focus on optimizing the fin geometry and PCM properties to enhance the performance of such energy storage systems.

5.9 Semi-analytical Solution for 3rd Type of Boundary Condition

This section presents an in-depth analysis of the results obtained for the thermal performance of a rectangular finned thermal energy storage system using paraffin as the PCM and aluminum for the fins. Both analytical and two-dimensional numerical solutions were employed to evaluate the heat transfer and phase change behavior, with a focus on varying the cell aspect ratio, denoted as σ.

- **Solid–Liquid Interface Analysis**

Figure 5.27 illustrates the progression of the solid–liquid interface over time, comparing results from the analytical model and the numerical solution for three distinct values of σ. The aspect ratio significantly influences the directionality of heat transfer:

- *For* $\sigma < 1$: Heat transfer predominantly occurs in the x-direction.
- *For* $\sigma > 1$: Heat transfer is primarily in the y-direction.
- *For* $\sigma = 1$: Heat transfer rates are equivalent in both x and y directions.

The analytical model demonstrated good agreement with the numerical solution across all cases, with the largest deviation observed at $\sigma = 2$. At this aspect ratio, the difference in the solid–liquid interface location reached 0.4 mm in the x-direction and 0.31 mm in the y-direction at $t = 500$ s, increasing to 0.67 mm and 0.39 mm, respectively, at $t = 1000$ s. For aspect ratios of 1 and 0.5, the analytical model achieved higher precision, confirming its reliability in estimating the interface location.

- **Temperature Distribution in Fins**

The temperature profiles obtained through both methods for various aspect ratios are depicted in Fig. 5.28. These comparisons validate the accuracy and efficiency of the analytical approach. The high thermal conductivity of aluminum fins minimizes the temperature gradient along the fins, ensuring near-uniform temperature distributions:

- *For* $\sigma < 1$: The heat transfer rate through the fin is relatively lower, and the one-dimensional analytical model delivers reasonable accuracy. The temperature differences between the analytical and numerical solutions were 0.11 °C at $t = 350$ s and 0.18 °C at $t = 700$ s.
- *For* $\sigma > 1$: The analytical model remains accurate in predicting the fin's temperature distribution, although the complexity of heat transfer increases slightly.

Overall, the analytical model provides satisfactory temperature predictions across all scenarios, particularly for shorter fins that quickly achieve equilibrium with the end-wall temperature.

- **Solidification Behavior of PCM**

Figure 5.29 illustrates the predicted solidification time of the PCM for different aspect ratios and varying thermal boundary conditions. The solid fraction, representing the

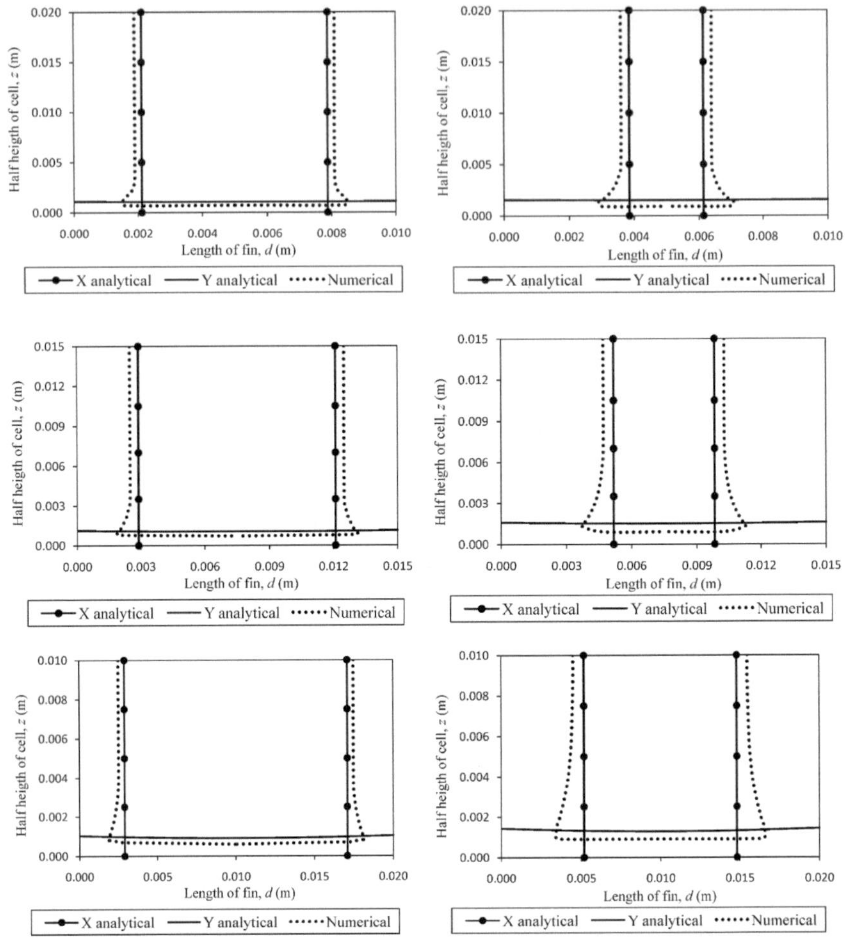

Fig. 5.27 Position of the solid–liquid interface with $h = 65$ W/m^2 K and $T_\infty = 10$ °C for (top-left) $\sigma = 0.5$, $t = 350$ s, (top-right) $\sigma = 0.5$, $t = 700$ s, (middle-left) $\sigma = 1$, $t = 500$ s, (middle-right) $\sigma = 1$, $t = 1000$ s, (bottom-left) $\sigma = 2$, $t = 500$ s, (bottom-right) $\sigma = 2$, $t = 1000$ s [3]

proportion of PCM that has solidified, is a critical metric for evaluating storage efficiency. Key observations include:

- *For small σ values (e.g., $\sigma = 0.5$)*: The PCM solidifies rapidly, with heat transfer predominantly occurring through the walls. At $t = 800$ s, reducing the heat transfer fluid temperature (T_∞) from 15 to 10 °C increases the solid fraction by 17.2%. Conversely, enhancing the convective heat transfer coefficient (h) from 65 to 80 W/m^2 K yields a smaller increase of 10.9%.
- *For larger σ values (e.g., $\sigma = 1$ or 2)*: The impact of T_∞ and h on the solidification rate is less pronounced, though the trends remain consistent.

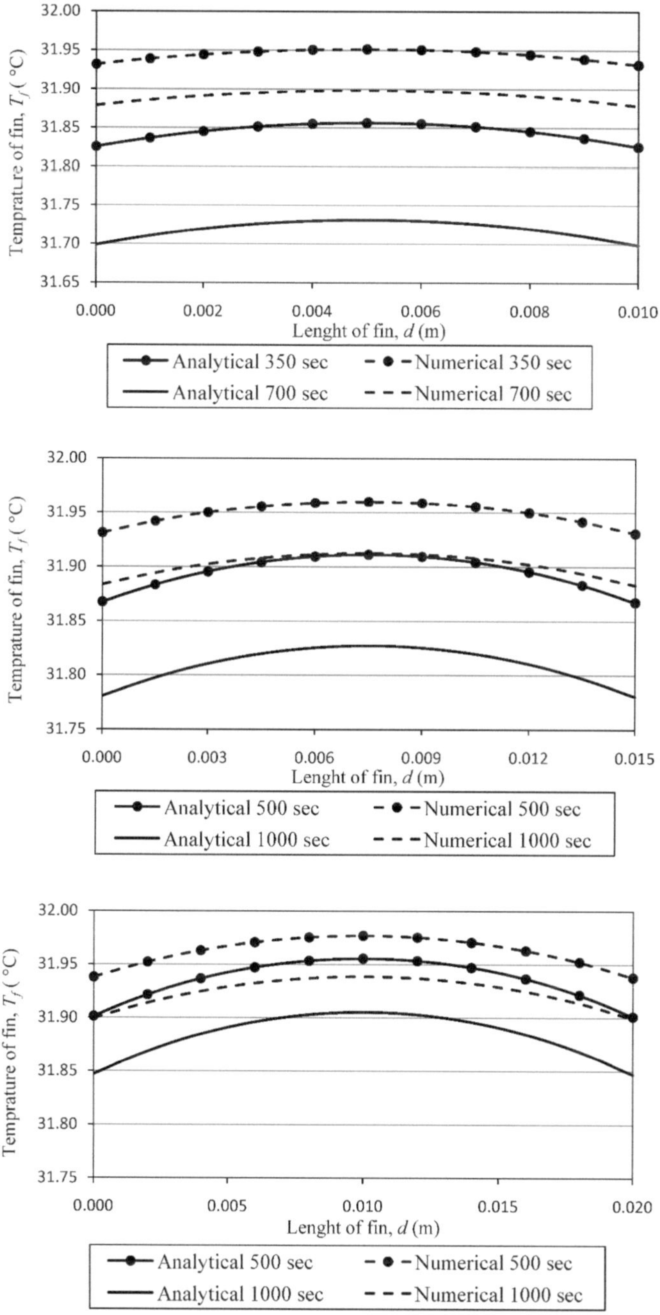

Fig. 5.28 Comparison of fin temperature obtained using the present analytical model and the numerical solution for (top) $\sigma = 0.5$, (middle) $\sigma = 1$, and (bottom-left) $\sigma = 2$ [3]

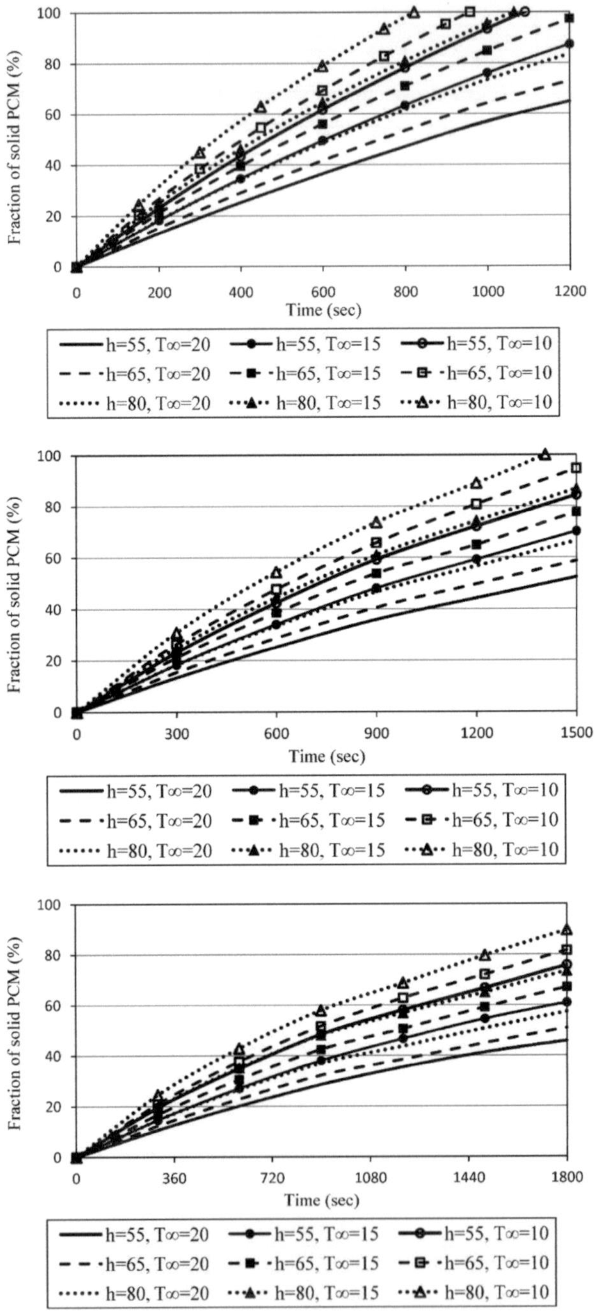

Fig. 5.29 Variation over time of PCM solid fraction predicted using the analytical model for different values of h (W/m^2 K) and T_∞ (°C), for (top) $\sigma = 0.5$, (middle) $\sigma = 1$, and (bottom) $\sigma = 2$ [3]

The results underscore the dominant influence of HTF temperature compared to the convective heat transfer coefficient in accelerating the PCM solidification process.

5.9.1 *Summary of Findings*

The analytical model proves to be a robust tool for predicting heat transfer and phase change phenomena in finned thermal energy storage systems. While numerical methods provide more precise results, the analytical approach offers significant computational efficiency and acceptable accuracy for practical applications. The results demonstrate that the system's thermal performance is strongly influenced by the cell aspect ratio and thermal boundary conditions, providing valuable insights for optimizing PCM-based energy storage designs.

References

1. Mosaffa AH (2014) Approximate analytical and numerical investigation on solidification process in the finned thermal energy storage using phase change material. PhD thesis. University of Tabriz
2. Mosaffa AH, Talati F, Rosen MA (2010) Analytical approximation of solidification time in finned PCM storage with imposed heat flux. In: 10th international conference on clean energy, Famagusta, N. Cyprus
3. Mosaffa AH, Talati F, Rosen MA (2012a) Thermal storage using phase change material: solidification processes, chap 12. In: Rosen MA (ed) Energy storage. Nova Science Publishers, pp 367–392
4. Mosaffa AH, Talati F, Rosen MA, Basirat Tabrizi H (2012) Approximate analytical model for PCM solidification in a rectangular finned container with convective cooling boundaries. Int Commun Heat Mass Transf 39:318–324
5. Talati F, Mosaffa AH, Rosen MA (2011) Analytical approximation for solidification processes in PCM storage with internal fins: imposed heat flux. Heat Mass Transf Stoffuebertragung 47:369–376

Chapter 6
Semi-analytical Solutions for the Stefan Problem in Finned Cylindrical Storage Tanks

6.1 Introduction

One of the most widely used thermal energy storage tanks is the hollow cylindrical storage tank. These types of storage tanks are used in two ways. In one application, the PCM is placed inside the tube, and the boundary conditions are applied to the outer wall of the tube (Fig. 6.1a). In the other type, the PCM is placed in the shell, and the boundary conditions are applied to the inner wall of the shell (Fig. 6.1b). Esen et al. [2] conducted a study on these two types of hollow cylindrical storage tanks. By comparing the effects of thermal and geometric parameters such as cylinder radius, PCM mass, and different temperatures, they concluded that placing the PCM in the shell significantly reduces the phase change time. Another advantage that can be added to this method is that when the heat source is placed inside the tube, heat loss will be less.

This chapter presents a semi-analytical and numerical solution for hollow and finned cylindrical storage tanks.

6.2 Defining the Freezing Problem in Finned Cylindrical Storage Tanks

A finned cylindrical storage tank is shown in Fig. 6.2. To obtain a semi-analytical solution, we consider a symmetrical control volume of this storage tank, to whose inner wall thermal boundary conditions are applied. Figure 6.3a shows the dimensions of the desired control volume. Now, to obtain the semi-analytical solution, like the rectangular storage tank, we divide this symmetrical control volume into two separate regions (Fig. 6.3b).

A. Mosaffa, *Phase Changing Materials*, Materials Horizons: From Nature to Nanomaterials, https://doi.org/10.1007/978-981-95-3742-6_6

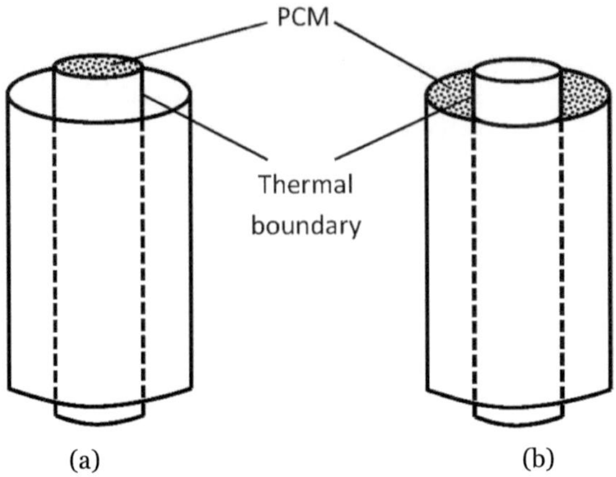

Fig. 6.1 Types of hollow cylindrical energy storage tanks: **a** PCM in the tube, **b** in the shell [4]

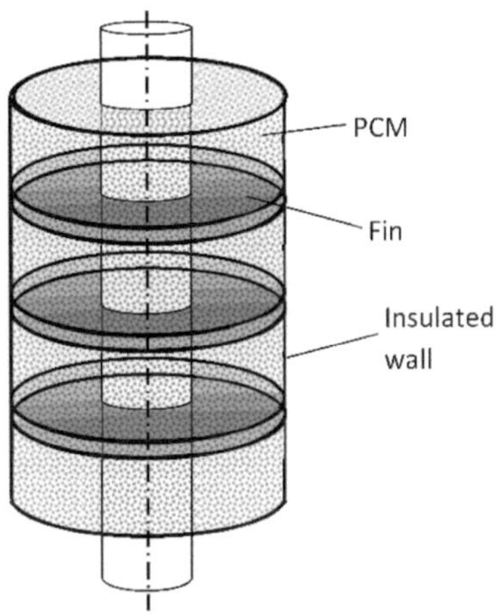

Fig. 6.2 Hollow finned cylindrical thermal energy storage tank [5]

In the first region, the only heat transfer source is the boundary conditions applied to the inner wall. In this region, heat transfer occurs through conduction, one-dimensionally in the r direction. In the second region, the freezing front moves only in the z direction, and heat is transferred to the surrounding environment only through the fins.

Due to the non-linearity of the freezing problem, we reiterate the assumptions made in Sect. 5.2 for this case as well:

- The PCM in the storage tank is initially in liquid form at the melting temperature T_{m}. Also, the initial temperature of the fin is also T_{m}.

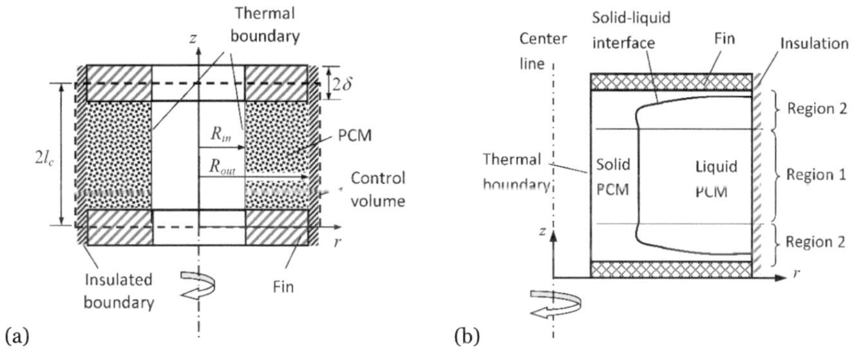

Fig. 6.3 **a** Dimensions of the desired cylindrical shell control volume, **b** dividing the two-dimensional control volume into two different regions [4]

- In the freezing process, convective heat transfer in the liquid PCM is negligible compared to conductive heat transfer in the solid PCM.
- The PCM is homogeneous, and its phase change occurs at a constant temperature (isothermal phase change).
- All thermophysical properties are assumed to be temperature-independent and constant.
- Due to the thinness, constant cross section, and high thermal conductivity of the fin, conductive heat transfer within the fin is assumed to be one-dimensional in the r direction.

6.2.1 Defining the Problem in Region#1

In the first region, where conductive heat transfer occurs only in the r direction, the one-dimensional Stefan problem with different boundary conditions is as follows:

$$\frac{1}{r}\frac{\partial}{\partial r}\left[r\frac{\partial T_s}{\partial r}\right] = \frac{1}{\alpha_s}\frac{\partial T_s}{\partial t}, \quad R_{in} < r < \Omega(t), \quad t > 0 \tag{6.1}$$

$$-\rho_s L\frac{d\Omega(t)}{dt} = -k_s\frac{\partial T_s(\Omega(t), t)}{\partial r} \tag{6.2}$$

$$\Omega(0) = 0 \tag{6.3}$$

$$T_s(r, 0) = T_m, \quad R_{in} < r < R_{out} \tag{6.4}$$

$$T_s(\Omega, t) = T_m \tag{6.5}$$

First type boundary condition:

$$T_s(R_{in}, t) = T_w, \quad T_w < T_m \tag{6.6a}$$

Second type boundary condition:

$$-k_s \frac{\partial T_s(R_{in}, t)}{\partial r} = q''_w, \quad q''_w < 0 \tag{6.6b}$$

Third type boundary condition:

$$-k_s \frac{\partial T_s(R_{in}, t)}{\partial r} + h[T_s(R_{in}, t) - T_\infty] = 0, \quad T_\infty < T_m \tag{6.6c}$$

6.2.2 *Defining the Problem in Region#2*

The movement of the freezing front in the second region is along the z-axis. By selecting a small element with a thickness dr from the annular fin with a rectangular cross-section (Fig. 6.4), the energy balance for the fin can be written as follows:

$$-q''_r + q''_{r+\mathrm{d}r} + q''_{in} = \dot{E}_{st} \tag{6.7}$$

where $\dot{E}_{st}$ is the rate of change of internal energy per unit cross-sectional area of the fin.

The equations for the rate of energy transferred to/from the element in Fig. 6.4 are determined using the heat conduction equation as follows:

- Rate of change of internal energy per unit cross-sectional area of the fin

$$\dot{E}_{st} = \rho_f c_f (2\pi\, r\mathrm{d}r)\delta \frac{\partial T_f}{\partial t} \tag{6.8}$$

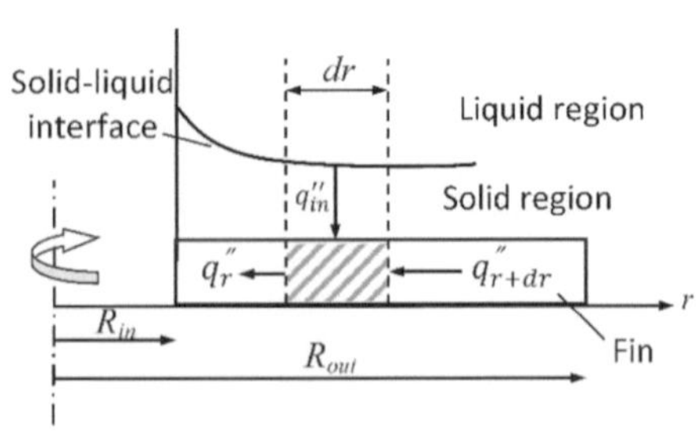

Fig. 6.4 Energy balance for a one-dimensional fin element with thickness dr [4]

- Rate of heat transferred from the freezing front to the fin

$$q''_{\text{in}} = -k_s(2\pi\ r\mathrm{d}r)\frac{T_f - T_m}{Z} \tag{6.9}$$

- Net rate of heat transferred through conduction in the fin

$$\begin{aligned} q''_{r+\mathrm{d}r} - q''_r = &\left[k_f(2\pi\ r\delta)\frac{\partial T_f}{\partial r} + \frac{\partial}{\partial r}\left(k_f(2\pi\ r\delta)\frac{\partial T_f}{\partial r}\right)\mathrm{d}r\right] \\ &- \left[k_f(2\pi\ r\delta)\frac{\partial T_f}{\partial r}\right] \end{aligned} \tag{6.10}$$

In the above equations, $Z(t)$ is the distance of the freezing front from the fin surface in the z-direction. By substituting Eqs. (6.8)–(6.10) into Eq. (6.7), we will have:

$$\frac{1}{r}\frac{\partial}{\partial r}\left(r\frac{\partial T_f}{\partial r}\right) + \frac{k_s}{\delta Z k_f}(T_m - T_f) = \frac{1}{\alpha_f}\frac{\partial T_f}{\partial t}, \quad R_{\text{in}} < r < R_{\text{out}}, \quad t > 0 \tag{6.11}$$

The initial and boundary conditions of the fin are as follows, respectively:

$$T_f(r, 0) = T_m \tag{6.12}$$

$$\frac{\partial T_f(R_{\text{out}}, t)}{\partial r} = 0 \tag{6.13}$$

First-type boundary condition:

$$T_f(R_{\text{in}}, t) = T_w, \quad T_w < T_m \tag{6.14a}$$

Second-type boundary condition:

$$-k_f\frac{\partial T_f(R_{\text{in}}, t)}{\partial r} = q''_w, \quad q''_w < 0 \tag{6.14b}$$

Third-type boundary condition:

$$-\frac{\partial T_f(R_{\text{in}}, t)}{\partial r} + H_f[T_f(R_{\text{in}}, t) - T_\infty] = 0, \quad T_\infty < T_m \tag{6.14c}$$

In the above equation, $H_f = h/k_f$. The position of the freezing front is also obtained using Megerlin's method from the following relationship:

$$Z = \left[2\sqrt{1 + 2\frac{c(T_{\mathrm{m}} - T_{\mathrm{f}})}{L}} - 2 \right]^{1/2} \sqrt{\alpha_{\mathrm{s}} t} \tag{6.15}$$

6.3 Semi-analytical Solution of the Freezing Problem in Region#1

By performing a series of mathematical calculations, it can be shown that the position of the freezing front using the quasi-steady method will be obtained as follows:

First boundary condition:

$$2\Omega^{\mathrm{qs}}(t)^2 \ln\left(\frac{\Omega^{\mathrm{qs}}}{R_{\mathrm{in}}}\right) = \Omega^{\mathrm{qs}}(t)^2 - R_{\mathrm{in}}^2 + \frac{4k_{\mathrm{s}}}{\rho L}(T_{\mathrm{m}} - T_{\mathrm{w}})t \tag{6.16a}$$

Second boundary condition:

$$\Omega^{\mathrm{qs}}(t)^2 - R_{\mathrm{in}}^2 = \frac{2R_{\mathrm{in}}}{\rho L} q_{\mathrm{w}}'' t \tag{6.16b}$$

Third boundary condition:

$$2\Omega^{\mathrm{qs}}(t)^2 \ln\left(\frac{\Omega^{\mathrm{qs}}}{R_{\mathrm{in}}}\right) = \left(1 - \frac{2k_{\mathrm{s}}}{hR_{\mathrm{in}}}\right)\left(\Omega^{\mathrm{qs}}(t)^2 - R_{\mathrm{in}}^2\right) + \frac{4k_{\mathrm{s}}}{\rho L}(T_{\mathrm{m}} - T_{\infty})t \tag{6.16c}$$

To solve Eqs. (6.1)–(6.6) and homogenize the initial condition, we apply the following variable transformations in the equations:

- *For the first and second boundary conditions*:

$$\mu(r, t) = T_{\mathrm{s}} - T_{\mathrm{m}} \tag{6.17a}$$

- *For the third boundary condition*:

$$\mu(r, t) = \frac{T_{\mathrm{s}} - T_{\mathrm{m}}}{T_{\infty} - T_{\mathrm{m}}} \tag{6.17b}$$

Thus, the conductive heat transfer equation and initial and boundary conditions are rewritten as follows:

$$\frac{1}{r}\frac{\partial}{\partial r}\left[r\frac{\partial \mu}{\partial r}\right] = \frac{1}{\alpha_{\mathrm{s}}}\frac{\partial \mu}{\partial t} \tag{6.18}$$

$$-\rho_s L \frac{d\Omega(t)}{dt} = -k_s \frac{\partial \mu(\Omega(t), t)}{\partial r} \tag{6.19}$$

$$\mu(r, 0) = 0, \quad R_{in} < r < R_{out} \tag{6.20}$$

$$\mu(\Omega, t) = 0 \tag{6.21}$$

- **Boundary Conditions**:
  1. *First type*:

$$\mu(R_{in}, t) = T_w - T_m \tag{6.22a}$$

  2. *Second type*:

$$\frac{\partial \mu(R_{in}, t)}{\partial r} = -\frac{q''_w}{k_s} \tag{6.22b}$$

  3. *Third type*:

$$-\frac{\partial \mu(R_{in}, t)}{\partial r} + \frac{h}{k_s}\mu(R_{in}, t) = \frac{h}{k_s} \tag{6.22c}$$

6.4 Solving the Steady-State Part of the Freezing Problem in Region#1

6.4.1 Analytical Solution for Temperature Distribution for 1st and 2nd Type of Boundary Conditions

To analyze the temperature distribution within the annular fin, Eqs. (6.1) through (6.6) are solved using the method of separation of variables. The resulting temperature distribution in region 1 is expressed as [7]

$$T_s(r, t) - T_m = A\left\{ \ln\left(\frac{r}{a}\right) - \sum_{m=1}^{\infty} \frac{\exp(-\alpha_s \beta_m^2 t)}{N(\beta_m)} R^2 [J_0(\beta_m r) Y_0(\beta_m R) - Y_0(\beta_m r) J_0(\beta_m R)] \times \left[Y_0(\beta_m R)\bar{I}_1 - J_0(\beta_m R)\bar{I}_2\right]\right\} \tag{6.23}$$

Here, J_0 and Y_0 are Bessel functions of the first and second kind (zero-order). The constants β_m are determined by the roots of the transcendental equations:

$$\begin{aligned} J_0(\beta_\mathrm{m}a)Y_0(\beta_\mathrm{m}R) - J_0(\beta_\mathrm{m}R)Y_0(\beta_\mathrm{m}a) = 0 \quad \text{for case 1} \\ J_1(\beta_\mathrm{m}a)Y_0(\beta_\mathrm{m}R) - J_0(\beta_\mathrm{m}R)Y_1(\beta_\mathrm{m}a) = 0 \quad \text{for case 2} \end{aligned} \tag{6.24}$$

The constant A is defined as

$$\begin{aligned} A &= \tfrac{T_\mathrm{m}-T_\mathrm{w}}{\ln(R/a)} \quad \text{for case 1} \\ A &= \tfrac{q''_\mathrm{w}a}{k_\mathrm{s}} \quad \text{for case 2} \end{aligned} \tag{6.25}$$

The norm $N(\beta_\mathrm{m})$, along with integrals $\overline{I}_1$ and $\overline{I}_2$, are given by

$$\begin{aligned} N(\beta_\mathrm{m}) &= \tfrac{2}{\pi^2} \times \tfrac{J_0^2(\beta_\mathrm{m}a)-J_0^2(\beta_\mathrm{m}R)}{\beta_\mathrm{m}^2 J_0^2(\beta_\mathrm{m}a)} \quad \text{for case 1} \\ N(\beta_\mathrm{m}) &= \tfrac{2}{\pi^2} \times \tfrac{J_1^2(\beta_\mathrm{m}a)-J_0^2(\beta_\mathrm{m}R)}{\beta_\mathrm{m}^2 J_1^2(\beta_\mathrm{m}a)} \quad \text{for case 2} \end{aligned} \tag{6.26}$$

$$\overline{I}_1 = -\frac{a}{\beta_\mathrm{m}R^2}\ln\left(\frac{R}{a}\right)J_1(\beta_\mathrm{m}a) + \frac{1}{\beta_\mathrm{m}^2R^2}[J_0(\beta_\mathrm{m}R) - J_0(\beta_\mathrm{m}a)] \tag{6.27}$$

$$\overline{I}_2 = -\frac{a}{\beta_\mathrm{m}R^2}\ln\left(\frac{R}{a}\right)Y_1(\beta_\mathrm{m}a) + \frac{1}{\beta_\mathrm{m}^2R^2}[Y_0(\beta_\mathrm{m}R) - Y_0(\beta_\mathrm{m}a)] \tag{6.28}$$

- **Laplace Transform for Small-Time Analysis**

For early time periods or when $R \approx a$, the transcendental equation converges slowly. To address this, a Laplace transform is applied to the time variable. By introducing $\theta(r, t) = T(r, t) - T_\mathrm{m}$, the transformed equations are [7]:

$$\frac{1}{r}\frac{\mathrm{d}}{\mathrm{d}r}\left[r\frac{\partial\overline{\theta}}{\partial r}\right] - \frac{s}{\alpha_\mathrm{s}}\overline{\theta} = 0, \quad a < r < R \tag{6.29}$$

with boundary conditions:

$$\overline{\theta}(R, s) = 0 \tag{6.30a}$$

$$\begin{aligned} \overline{\theta}(a, s) &= \tfrac{T_\mathrm{w}-T_\mathrm{m}}{s} \quad \text{for case 1} \\ \tfrac{\mathrm{d}\overline{\theta}(a,s)}{\mathrm{d}r} &= \tfrac{1}{s}\left(-\tfrac{q''_\mathrm{w}}{k_\mathrm{s}}\right) \quad \text{for case 2} \end{aligned} \tag{6.30b}$$

The solutions for these equations are

$$\frac{\overline{\theta}(r, s)}{T_\mathrm{w} - T_\mathrm{m}} = \frac{1}{s} \times \frac{I_0\left(R\sqrt{s/\alpha_\mathrm{s}}\right)K_0\left(r\sqrt{s/\alpha_\mathrm{s}}\right) - I_0\left(r\sqrt{s/\alpha_\mathrm{s}}\right)K_0\left(R\sqrt{s/\alpha_\mathrm{s}}\right)}{I_0\left(R\sqrt{s/\alpha_\mathrm{s}}\right)K_0\left(a\sqrt{s/\alpha_\mathrm{s}}\right) - I_0\left(a\sqrt{s/\alpha_\mathrm{s}}\right)K_0\left(R\sqrt{s/\alpha_\mathrm{s}}\right)} \quad \text{for case 1} \tag{6.31}$$

$$\frac{\overline{\theta}(r,s)}{q''_{\rm w}\sqrt{\alpha_{\rm s}}/k_{\rm s}} = \frac{1}{s^{3/2}} \times \frac{I_0\left(R\sqrt{s/\alpha_{\rm s}}\right)K_0\left(r\sqrt{s/\alpha_{\rm s}}\right) - I_0\left(r\sqrt{s/\alpha_{\rm s}}\right)K_0\left(R\sqrt{s/\alpha_{\rm s}}\right)}{I_0\left(R\sqrt{s/\alpha_{\rm s}}\right)K_0\left(a\sqrt{s/\alpha_{\rm s}}\right) + I_0\left(a\sqrt{s/\alpha_{\rm s}}\right)K_0\left(R\sqrt{s/\alpha_{\rm s}}\right)} \quad \text{for case 2} \tag{6.32}$$

- **Short-Time Approximation**

For short times after solidification begins, the PCM thickness is small, and the solutions can be approximated using an asymptotic series [7]:

$$\frac{T(r,t) - T_{\rm m}}{T_{\rm w} - T_{\rm m}} = \sqrt{\frac{a}{r}} \sum_{n=0}^{\infty} \left[\mathrm{erfc}(\gamma_n) - \mathrm{erfc}(\omega_n)\right] \quad \text{for case 1} \tag{6.33}$$

$$\frac{T(r,t) - T_{\rm m}}{q''_{\rm w}\sqrt{\alpha_{\rm s}}/k_{\rm s}} = \sqrt{\frac{4ta}{\alpha_{\rm s} r}} \sum_{n=0}^{\infty} \left[\left(\exp\left(-\gamma_n^2\right) - \exp\left(-\omega_n^2\right)\right)\right.$$
$$\left. - \left[\gamma_n \mathrm{erfc}(\gamma_n) - \omega_n \mathrm{erfc}(\omega_n)\right]\right] \quad \text{for case 2} \tag{6.34}$$

where

$$\gamma_n = \frac{2nR - a(1+2n) + r}{\sqrt{4\alpha_{\rm s} t}} \tag{6.35}$$

$$\omega_n = \frac{2R(1+n) - a(1+2n) - r}{\sqrt{4\alpha_{\rm s} t}} \tag{6.36}$$

6.4.2 *Analytical Solution for Temperature Distribution for 3rd Type of Boundary Condition*

The temperature distribution, derived from the solution of Eqs. (6.1) through (6.6), is expressed as [5]

$$\frac{T_{\rm s} - T_{\rm m}}{T_\infty - T_{\rm m}} = \left\{ \ln\left(\frac{r}{a}\right) - \sum_{m=1}^{\infty} \frac{\exp\left(-\alpha_{\rm s}\gamma_{\rm m}^2 t\right)}{N(\gamma_{\rm m})} \Omega^2 \left[J_0(\gamma_{\rm m} r) Y_0(\gamma_{\rm m}\Omega) - Y_0(\gamma_{\rm m} r) J_0(\gamma_{\rm m}\Omega)\right]\right.$$
$$\left. \times \left[Y_0(\gamma_{\rm m}\Omega)\overline{I}_1 - J_0(\gamma_{\rm m}\Omega)\overline{I}_2\right]\right\} \frac{1}{1/(H_{\rm s}R_{\rm in}) + \ln(\Omega/R_{\rm in})} \tag{6.37}$$

where J_0 and Y_0 are Bessel functions of the first and second kinds of zero order, respectively. The parameters $\gamma_{\rm m}$ are determined as the roots of the following transcendental equation:

$$U_0 Y_0(\gamma_{\rm m}\Omega) - W_0 J_0(\gamma_{\rm m}\Omega) = 0 \tag{6.38}$$

where

$$\begin{aligned} U_0 &= -\gamma_\mathrm{m} J_1(\gamma_\mathrm{m} R_\mathrm{in}) - H_\mathrm{s} J_0(\gamma_\mathrm{m} R_\mathrm{in}) \\ W_0 &= -\gamma_\mathrm{m} Y_1(\gamma_\mathrm{m} R_\mathrm{in}) - H_\mathrm{s} Y_0(\gamma_\mathrm{m} R_\mathrm{in}) \end{aligned} \tag{6.39}$$

A systematic method for solving this equation was developed by Carslaw and Jaeger [1]. The normalization factor $N(\gamma_\mathrm{m})$ is given by

$$N(\gamma_\mathrm{m}) = \frac{2}{\pi^2} \times \frac{U_0^2 - (H_\mathrm{s}^2 + \gamma_\mathrm{m}^2) J_0(\gamma_\mathrm{m}\Omega)}{\gamma_\mathrm{m}^2 U_0^2} \tag{6.40}$$

For small time values or when $\Omega \approx R_\mathrm{in}$, the transcendental equation converges slowly. To address this, the Laplace transform is applied to the time variable.

- **Heat Transfer in the Fluid**

The energy balance governing heat transfer in the thermal fluid is expressed as:

$$hA_\mathrm{c}(T_\mathrm{w} - T_\infty) = (\dot{m}c_\mathrm{p})_\mathrm{air} \frac{\partial T_\infty}{\partial z} \tag{6.41}$$

A solution for laminar flow with a fully developed velocity profile is obtained using the correlation provided by Kays et al. [3]:

$$\Lambda(z^*) = \frac{T_\mathrm{w} - T_\infty}{T_\mathrm{w} - T_{\infty,\mathrm{inlet}}} = \exp(-4z^*\overline{\mathrm{Nu}}) \tag{6.42}$$

where $z^* = 2zD/\mathrm{Re}{\cdot}\mathrm{Pr}$, and the heat transfer coefficient in the entry region for fully developed flow is expressed as:

$$\overline{\mathrm{Nu}} = -\frac{1}{2z^*} \ln\left[8 \sum \left(\frac{G_n}{\lambda_n^2}\right) \exp(-\lambda_n^2 z^*)\right] \tag{6.43}$$

where G_n and λ_n are constants and eigenvalues, respectively.

Since the wall temperature varies along the flow direction, Eq. (6.42) cannot be directly applied. Instead, the problem is solved using the superposition method, resulting in [5]:

$$T_{\infty,\mathrm{outlet}} - T_{\infty,\mathrm{inlet}} = \sum_{i=1}^{n} \left[1 - \Lambda(l^* - z_i^*)\right](T_{\mathrm{w},i} - T_{\mathrm{w},i-1}) \tag{6.44}$$

where i represents the cell number and $l^* = 2zD/\mathrm{Re}{\cdot}\mathrm{Pr}$.

6.5 Semi-analytical Solution of the Freezing Problem in Region#2

Here, to obtain the temperature distribution equation of the fin, various solutions can be chosen. The Green's function is introduced as a solution for solving physical equations with non-linear partial derivatives and non-homogeneous and variable boundary conditions in many references, along with its details and physical significance. However, here, due to the fixed boundary conditions, we choose the same method used in rectangular storage tanks.

Figure 6.5 illustrates the geometry of an annular fin with a rectangular profile and a schematic of the thermal energy storage system. The inner radial side of the fin and the outer boundary are assumed to be insulated. To evaluate the performance of the system, a comparison is conducted between analytical and numerical solutions to assess the accuracy of the approximate analytical method.

6.5.1 Mathematical Model Description

A transient analysis is performed for the solidification of PCM on an annular fin under the following assumptions:

1. The thermo-physical properties of both the fin and solidified PCM are constant.
2. Initially, the fin and PCM are at the solidification temperature, T_m.
3. The fin is considered thin with high thermal conductivity, allowing for a one-dimensional temperature distribution along the radial direction (r).

A differential element of the fin with length is illustrated in Fig. 6.6.

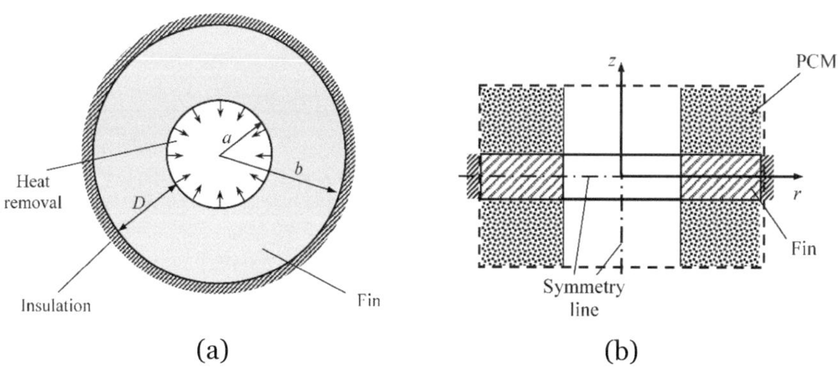

Fig. 6.5 **a** Dimensions and boundary conditions of the annular fin and **b** the schematic arrangement of the cylindrical shell PCM storage [6]

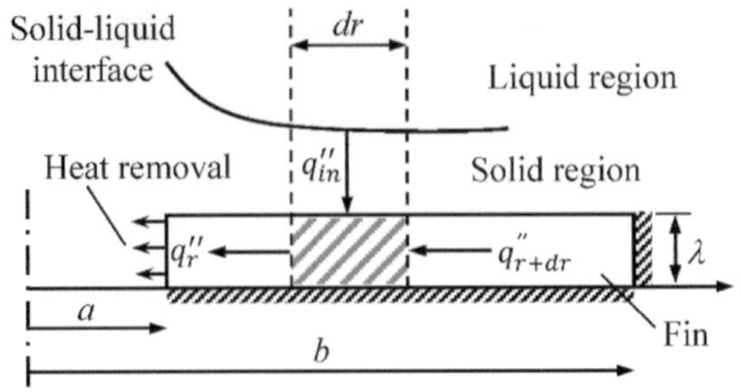

Fig. 6.6 Energy balance in a differential element of the fin [6]

- **Energy Balance of the Fin Element**

Using an energy balance approach, the rate of change of energy stored within the differential element (dr) can be expressed as:

$$\dot{E}_{\mathrm{f}} = -q''_r + q''_{r+\mathrm{d}r} + q''_{\mathrm{in}} \tag{6.45}$$

where natural convection within the liquid PCM is neglected. Substituting relevant expressions for heat fluxes and energy rates yields the following relationships:

- Energy stored in the fin:

$$\dot{E}_{\mathrm{f}} = (\rho\, c)_{\mathrm{f}}(2\pi\, r\mathrm{d}r)\lambda \frac{\partial T_{\mathrm{f}}}{\partial r} \tag{6.46}$$

- Heat flux at position r:

$$q''_r = k_{\mathrm{f}}(2\pi\, r\lambda)\frac{\partial T_{\mathrm{f}}}{\partial r} \tag{6.47}$$

- Heat flux at $r + \mathrm{d}r$:

$$q''_{r+\mathrm{d}r} = k_{\mathrm{f}}(2\pi\, r\lambda)\frac{\partial T_{\mathrm{f}}}{\partial r} + \frac{\partial}{\partial r}\left[k_{\mathrm{f}}(2\pi\, r\lambda)\frac{\partial T_{\mathrm{f}}}{\partial r}\right]\mathrm{d}r \tag{6.48}$$

- Heat flux from PCM interface:

$$q''_{\mathrm{in}} = k_{\mathrm{s}}(2\pi\, r\mathrm{d}r)\frac{T_{\mathrm{m}} - T_{\mathrm{f}}}{S(t)} \tag{6.49}$$

By substituting Eqs. (6.46) through (6.49) into the energy balance Eq. (6.45), the governing equation for the fin is obtained:

$$\frac{1}{r}\frac{\partial}{\partial r}\left(r\frac{\partial T_{\mathrm{f}}}{\partial r}\right)+\frac{k_{\mathrm{s}}}{k_{\mathrm{f}}}\frac{T_{\mathrm{m}}-T_{\mathrm{f}}}{\lambda S}=\frac{1}{\alpha_{\mathrm{f}}}\frac{\partial T_{\mathrm{f}}}{\partial t}, \quad a<r<b, \quad t>0 \tag{6.50}$$

- **Initial and Boundary Conditions**

The following assumptions and boundary conditions are applied:

– *Initial condition*:

$$T_{\mathrm{f}}(r,0)=T_{\mathrm{m}} \tag{6.52}$$

– *Outer boundary condition at* $r=b$:

$$\frac{\partial T_{\mathrm{f}}(b,t)}{\partial r}=0 \tag{6.53}$$

– *Inner boundary conditions at* $r=a$:

1. **Case 1**: Inner wall temperature is constant at T_{w}, where $T_{\mathrm{w}}<T_{\mathrm{m}}$:

$$T_{\mathrm{f}}(a,t)=T_{\mathrm{w}} \tag{6.54a}$$

2. **Case 2**: Inner wall subjected to heat flux q''_{w}:

$$-k_{\mathrm{f}}\frac{\partial T_{\mathrm{f}}(a,t)}{\partial r}=q''_{\mathrm{w}}, \quad q''_{\mathrm{w}}<0 \tag{6.54b}$$

3. **Case 3**: Convective heat transfer at the inner wall with coefficient h and fluid temperature T_{∞}:

$$-k_{\mathrm{f}}\frac{\partial T_{\mathrm{f}}(a,t)}{\partial r}+h[T_{\mathrm{f}}(a,t)-T_{\infty}]=0 \tag{6.54c}$$

6.5.2 Analytical Solution via Green's Function

To solve the governing equation, the following dimensionless variables are introduced [6]:

$$\eta=\frac{r}{D}, \quad \hat{a}=\frac{a}{D}, \quad \hat{b}=\frac{b}{D} \tag{6.55a}$$

$$\tau=\frac{\alpha_{\mathrm{f}}t}{D^2} \tag{6.55b}$$

$$\beta^2 = \frac{k_s D^2}{k_f S \lambda} \tag{6.55c}$$

The solid–liquid interface location is expressed using the Megerlin method:

$$S = \left[2\sqrt{1 + 2\frac{c_s(T_m - T_f)}{L}} - 2\right]^{1/2} \sqrt{\alpha_s t} \tag{6.56}$$

Transforming the equations into dimensionless form, the governing equation becomes

$$\frac{1}{\eta}\frac{\partial}{\partial \eta}\left(\eta \frac{\partial \theta}{\partial \eta}\right) - \beta^2 \theta = \frac{\partial \theta}{\partial \tau}, \quad \hat{a} < \eta < \hat{b}, \quad \tau > 0 \tag{6.57}$$

$$\theta(\eta, 0) = 0 \tag{6.58}$$

$$\frac{\partial \theta(\hat{b}, \tau)}{\partial \eta} = 0 \tag{6.59}$$

$$\theta(\hat{a}, \tau) = T_w - T_m = f_1(\tau) \quad \text{for case 1} \tag{6.60}$$

$$-\frac{\partial \theta(\hat{a}, \tau)}{\partial \eta} = \frac{q''_w D}{k_f} = f_2(\tau) \quad \text{for case 2} \tag{6.61}$$

$$-\frac{\partial \theta(\hat{a}, \tau)}{\partial \eta} + H\theta(\hat{a}, \tau) = H(T_\infty - T_m) = f_3(\tau), \quad H = \frac{hD}{k_f} \quad \text{for case 3} \tag{6.62}$$

- **Green's Function Derivation**

The Green's function satisfying the reciprocity relation is

$$\frac{1}{\eta}\frac{\partial}{\partial \eta}\left(\eta \frac{\partial G}{\partial \eta}\right) - \beta^2 G + \delta(\eta - \eta')(\tau - \tau') = \frac{\partial G}{\partial \tau} \tag{6.63}$$

By solving this equation using separation of variables, the Green's function is expressed as

$$G(\eta, \tau | \eta', \tau') = G(\eta', -\tau' | \eta, -\tau) \tag{6.64}$$

To derive the solution for temperature distribution and the solid–liquid interface movement, we employ Green's function methodology. This approach involves rewriting governing equations, integrating them concerning space and time, and solving for appropriate boundary conditions.

- **Reformulation of Governing Equations**

Rewriting Eq. (6.63) by considering the reciprocity relation from Eq. (6.64) results in

$$\frac{1}{\eta'}\frac{\partial}{\partial \eta'}\left(\eta'\frac{\partial G}{\partial \eta'}\right) - \beta^2 G + \delta(\eta' - \eta)(\tau' - \tau) = -\frac{\partial G}{\partial \tau'} \tag{6.65}$$

Here, the negative sign on the right-hand side originates from replacing τ with $-\tau'$. Equation (6.57) is similarly rewritten by substituting variables η with η' and τ with τ', yielding:

$$\frac{1}{\eta'}\frac{\partial}{\partial \eta'}\left(\eta'\frac{\partial \theta}{\partial \eta'}\right) - \beta^2 \theta = \frac{\partial \theta}{\partial \tau'} \tag{6.66}$$

- **Derivation of General Solution**

By multiplying Eq. (6.65) by θ and Eq. (6.66) by G, then subtracting the resulting equations, we obtain

$$\begin{aligned}&\frac{1}{\eta'}\left[G\frac{\partial}{\partial \eta'}\left(\frac{\partial \theta}{\partial \eta'}\right) - \theta\frac{\partial}{\partial \eta'}\left(\frac{\partial G}{\partial \eta'}\right)\right] - G\beta^2\theta + \theta\beta^2 G\\ &- \theta\delta(\eta' - \eta)(\tau' - \tau) = \frac{\partial}{\partial \tau'}(\theta G)\end{aligned} \tag{6.67}$$

The terms involving β_2 cancel. Integrating with respect to η' and τ' while applying boundary conditions (Eqs. 6.57–6.62), the temperature difference θ is expressed as [6]

$$\begin{aligned}\theta = &\int_{\tau=0}^{t}\left[\eta' G(\eta,\tau|\eta',\tau')\right]_{\eta'=\hat{a}}\frac{f_i(\tau)}{k_f}\mathrm{d}\tau \quad \text{(only for case 2 and 3 where } i = 2, 3)\\ &-\int_{\tau=0}^{t}\left[\eta'\frac{\partial G(\eta,\tau|\eta',\tau')}{\partial \eta'}\right]_{\eta'=\hat{a}} f_i(\tau)\mathrm{d}\tau \quad \text{(only for case 1 where } i = 1)\end{aligned} \tag{6.68}$$

- **Determination of Green's Function**

To obtain the Green's function $G(\eta, \tau \mid \eta', \tau')$, the homogeneous forms of Eqs. (6.57)–(6.62) are solved using separation of variables. The solution is expressed as

$$G(\eta,\tau|\eta',\tau') = \sum_{m=1}^{\infty}\frac{1}{N(\xi_m)}\exp\left[-\left(\beta^2 + \xi_m^2\right)(\tau - \tau')\right]R_0(\xi_m,\eta)R_0(\xi_m,\eta') \tag{6.69}$$

Table 6.1 Transcendental equation of eigenvalues ξ_m in Eq. (6.70) [6]

Case	ξ_m are positive roots of
#1	$J_0(\xi_m\hat{a})Y_1(\xi_m\hat{b}) - J_1(\xi_m\hat{b})Y_0(\xi_m\hat{a}) = 0$
#2	$J_1(\xi_m\hat{a})Y_1(\xi_m\hat{b}) - J_1(\xi_m\hat{b})Y_1(\xi_m\hat{a}) = 0$
#3	$[-\xi_m J_1(\xi_m\hat{a}) - HJ_0(\xi_m\hat{a})]Y_1(\xi_m\hat{b}) - [-\xi_m Y_1(\xi_m\hat{a}) - HY_0(\xi_m\hat{a})]J_1(\xi_m\hat{b}) = 0$

Table 6.2 Normalization integrals, $N(\xi_m)$, in Eq. (6.70) [6]

Case	Norm $N(\xi_m)$
#1	$\frac{2}{(\pi\xi_m)^2}\left(\frac{J_0^2(\xi_m\hat{a})-J_1^2(\xi_m\hat{b})}{J_0^2(\xi_m\hat{a})}\right)$
#2	$\frac{2}{(\pi\xi_m)^2}\left(\frac{J_1^2(\xi_m\hat{a})-J_1^2(\xi_m\hat{b})}{J_1^2(\xi_m\hat{a})}\right)$
#3	$\frac{2}{(\pi\xi_m)^2}\left(\frac{[-\xi_m J_1(\xi_m\hat{a})-HJ_0(\xi_m\hat{a})]^2-J_1^2(\xi_m\hat{b})(H^2+\xi_m^2)}{[-\xi_m J_1(\xi_m\hat{a})-HJ_0(\xi_m\hat{a})]^2}\right)$

where $R_0(\xi_m, \eta)$ is defined as

$$R_0(\xi_m, \eta) = J_0(\xi_m\eta)Y_1(\xi_m\hat{b}) - J_1(\xi_m\hat{b})Y_0(\xi_m\eta) \tag{6.70}$$

Here, J_n and Y_n are Bessel functions of the first and second kind, respectively, of order n. The eigenvalues ξ_m are positive roots of transcendental equations (see Table 6.1) and are solved using the Newton–Raphson method. The normalization integrals $N(\xi_m)$ are provided in Table 6.2.

Using Eq. (6.69) in conjunction with Eq. (6.56), the temperature distribution across the fin and the dynamic movement of the solid–liquid interface are determined. This approach provides a robust solution framework for thermal problems involving cylindrical fin geometries.

6.6 Numerical Solution of the Freezing Problem in a Cylindrical Storage Tank Based on the Enthalpy Method

The formulation for the total energy of the PCM undergoing freezing in a hollow cylindrical storage tank, according to the definition of enthalpy and local freezing fraction, is as follows:

$$\frac{\partial h}{\partial t} = \frac{k_s}{\rho_s}\left(\frac{\partial^2 T}{\partial r^2} + \frac{1}{r}\frac{\partial T}{\partial r} + \frac{\partial^2 T}{\partial z^2}\right) + L\frac{\partial f}{\partial t} \tag{6.71}$$

We discretize Eq. (6.71) for an internal node (i, k) (see Fig. 6.7) using the backward finite difference method as follows:

$$\frac{\partial h}{\partial t} = \frac{k_s}{\rho_s(\Delta r)^2}\left(T_{i-1,k} - 2T_{i,k} + T_{i+1,k}\right) + \frac{1}{i\Delta r}\frac{T_{i+1,k} - T_{i-1,k}}{2\Delta r} + \frac{k_s}{\rho_s(\Delta z)^2}\left(T_{i,k-1} - 2T_{i,k} + T_{i,k+1}\right) + L\frac{\partial f}{\partial t} \quad (6.72)$$

Considering that the phase change occurs isothermally, the temperature inside the control volume corresponding to node (i, k) remains constant and equal to T_m during the phase change. Therefore, the value of becomes 0. In this case, the value of f is in the range (0, 1). The solution method is fully implicit, and we will have

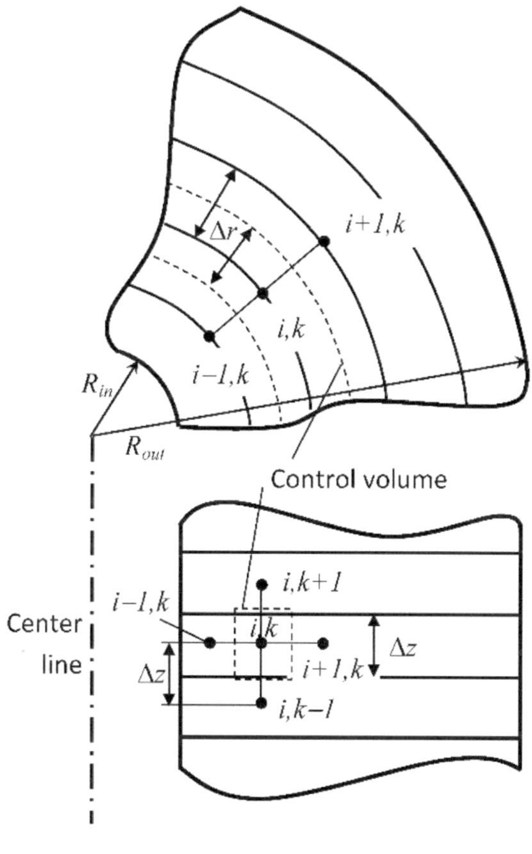

Fig. 6.7 Discretization domain of the two-dimensional phase change problem for an internal node [4]

$$f_{i,k} = f_{i,k}^{\text{old}} - \frac{k_s \Delta t}{\rho_s L (\Delta r)^2}\left(T_{i-1,k} - 2T_{i,k} + T_{i+1,k} + \frac{1}{2i}T_{i+1,k} - \frac{1}{2i}T_{i-1,k}\right) - \frac{k_s \Delta t}{\rho_s L (\Delta z)^2}\left(T_{i,k-1} - 2T_{i,k} + T_{i,k+1}\right) \tag{6.73}$$

However, if the control volume corresponding to node (i, k) is completely liquid or completely frozen, the value of f will be exactly 0 or 1. So this time, will be 0. Using the backward finite difference method for Eq. (6.72), we will have

$$T_{i,k} = T_{i,k}^{\text{old}} + \frac{\alpha_s \Delta t}{(\Delta r)^2}\left[\left(1 + \frac{1}{2i}\right)T_{i+1,k} + 2T_{i,k} + \left(1 - \frac{1}{2i}\right)T_{i-1,k}\right] + \frac{\alpha_s \Delta t}{(\Delta z)^2}\left(T_{i,k-1} - 2T_{i,k} + T_{i,k+1}\right) \tag{6.74}$$

By simultaneously solving Eqs. (6.73) and (6.74) for each time step in the entire solution domain, the value of temperature and local freezing fraction is obtained at all nodes. It should be noted that the discretized equation for the boundary nodes depends on the boundary conditions and is obtained using the energy balance for the control volume around the boundary node.

6.7 Results and Discussion

This chapter presents a detailed analysis of the results obtained through one-dimensional analytical and two-dimensional numerical predictions of the solid–liquid interface. The findings are discussed for various storage configurations, highlighting the impact of geometric factors, thermal resistance, air flow, and inlet temperature on the performance of thermal storage systems.

6.7.1 Numerical Analysis of PCM Solidification in Rectangular and Cylindrical Shell Thermal Storages with Internal Fins

This section discusses a two-dimensional numerical analysis of the PCM ($CaCl_2 \cdot 6H_2O$) solidification process in rectangular and cylindrical shell thermal storages equipped with internal fins. The enthalpy method is employed to model the solid–liquid phase transition and solve the governing equations. The model evaluates key phase change parameters, including the location of the solid–liquid interface and the total solidification time of the PCM under various conditions.

In the simulations, the following assumptions and parameters are applied:

- The inlet temperature of the heat transfer fluid is maintained at a constant value of 10 °C.
- The convection heat transfer coefficient is fixed at 100 W/m^2 K.
- A convergence criterion of 10^{-3} mm is adopted for the solid–liquid interface location.
- Spatial step size: 0.1 mm.
- Time step: 1 s.

- **Comparative Analysis of Rectangular and Cylindrical Shell Containers**

Figure 6.8 compares the temperature–time history of the PCM at three positions in rectangular and cylindrical shell containers. The equivalent distance from the end wall is used to select comparable positions for the cylindrical shell container. For the cylindrical shell, the inner and outer radii are 10 mm and 30 mm, respectively. The fin configuration in the rectangular container matches that of the cylindrical shell container, with $p_f = 20$ mm and $l = 40$ mm.

The results indicate that the solid–liquid interface progresses faster in the cylindrical shell container compared to the rectangular one. This is attributed to the geometric advantages of the cylindrical shell, which facilitates more efficient heat transfer due to its radial symmetry.

- **Total Solidification Time**

Figure 6.9 presents the variation in total solidification time for PCMs packed in rectangular and cylindrical shell containers as a function of PCM mass. The comparison is based on equivalent heat transfer surface areas. While the cylindrical shell container achieves faster total solidification than the rectangular container, the difference is marginal across all PCM masses.

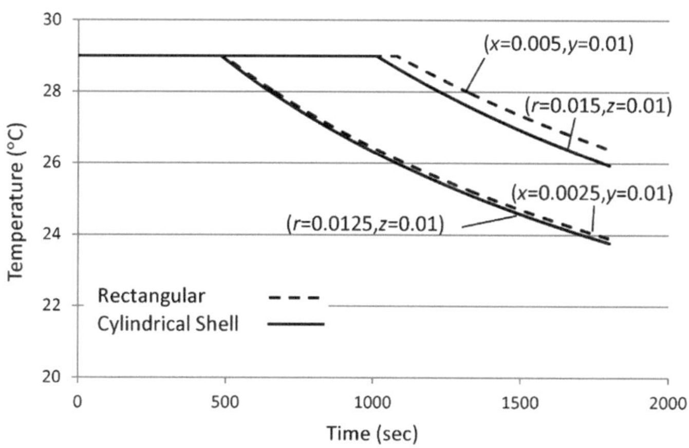

Fig. 6.8 Comparison of the temperature–time history of the PCM at various positions within rectangular and cylindrical shell containers [8]

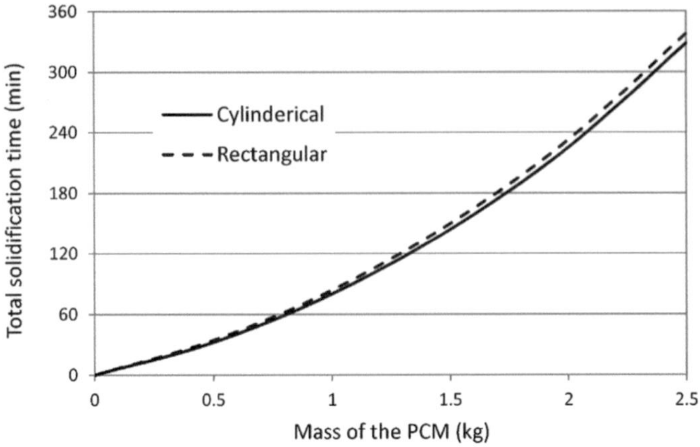

Fig. 6.9 Variation of the total solidification time of the PCM with PCM mass for rectangular and cylindrical shell containers ($p_f = 20$ mm) [8]

- **Effect of HTF Inlet Temperature**

The influence of HTF inlet temperature on the total solidification time is shown in Fig. 6.10. Two key trends emerge from this analysis:

1. *HTF inlet temperature effect*: Lowering the HTF inlet temperature significantly reduces the total solidification time of the PCM. However, the rate of reduction diminishes at lower inlet temperatures. For example, the time difference between HTF temperatures of 20 and 15 °C is more pronounced than the difference between 15 and 10 °C.
2. *PCM mass effect*: The influence of HTF inlet temperature on solidification time becomes more pronounced as the PCM mass increases. This trend highlights the enhanced sensitivity of larger PCM volumes to changes in HTF inlet temperature.

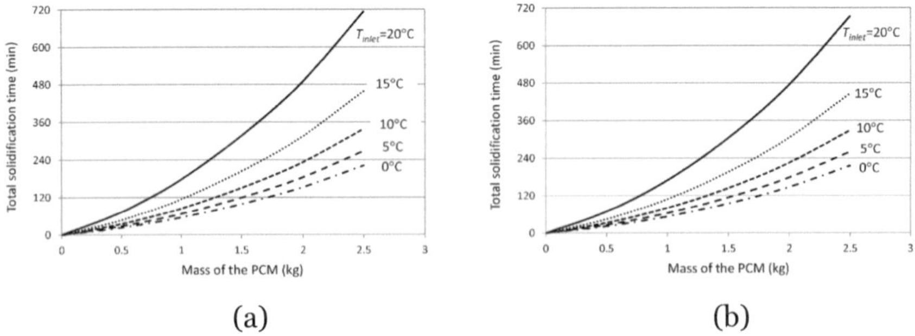

Fig. 6.10 Variation of the total solidification time of the PCM with PCM mass for different storage types and various HTF inlet temperatures ($p_f = 20$ mm): rectangular (**a**) and cylindrical shell containers (**b**) [8]

- **Key Observations**
  - The cylindrical shell geometry accelerates the progression of the solid–liquid interface due to its radial heat transfer mechanism, offering better thermal performance than the rectangular configuration.
  - While cylindrical containers demonstrate faster solidification, the performance difference between the two geometries is relatively small for comparable PCM masses and heat transfer areas.
  - Reducing the HTF inlet temperature not only decreases the solidification time but also allows more effective energy extraction, especially for larger PCM volumes.

6.7.2 Analytical and Numerical Comparison of Fin Temperature Distributions

Here we focus on the thermal performance of an aluminum fin embedded in a PCM, specifically the salt hydrate ClimSel C23. For all calculations, the fin geometry is maintained with a fixed thickness ($\lambda = 1$ mm), an inner radius of $a = 1$ cm, and an outer radius of $b = 3$ cm. These dimensions ensure uniformity and consistency in the analysis.

The initial temperature of the aluminum fin is set to match the solidification temperature of the PCM (T_m). This assumption simplifies the thermal boundary conditions while accurately simulating the solidification process. The combination of high thermal conductivity in the aluminum fin and the latent heat storage capability of the PCM enables efficient heat transfer during phase change, making this configuration an excellent candidate for thermal energy storage and management applications.

The results presented in this section explore the temperature profiles, heat transfer dynamics, and solidification behavior under various thermal conditions. The high thermal conductivity of aluminum ensures efficient heat conduction along the fin, while the PCM undergoes a phase transition, storing thermal energy in the form of latent heat. This interplay between conduction and phase change is analyzed to evaluate system performance and optimize design parameters.

6.7.2.1 Case 1: Temperature Variation with Constant Wall Temperature

Figure 6.11 presents a comparison between the analytical and numerical solutions for the temperature distribution of the fin in Case 1, where the wall temperature is maintained at $T_\mathrm{w} = 7$ °C and $T_\mathrm{w} = 14$ °C. The governing equation for the fin was discretized using the Crank–Nicolson finite difference scheme, and the phase change

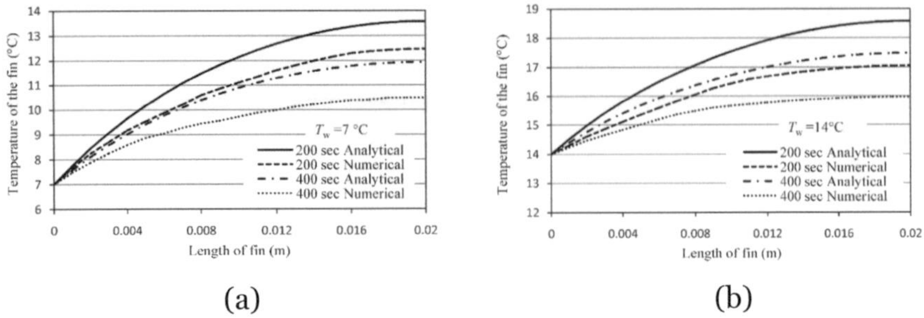

Fig. 6.11 Temperature profile of the fin for Case 1, illustrating the distribution along the fin under specified conditions [6]

problem was solved numerically using the enthalpy method for rectangular finned storage systems.

At lower wall temperatures, the significant temperature difference between the solid–liquid interface and the wall ($T_m - T_w$) drives a rapid advancement of the solidification front. This, in turn, increases the thermal resistance of the solid PCM, prolonging the time required to reach steady-state conditions. The largest deviation between the analytical and numerical results occurs at $T_w = 7$ °C after $t = 400$ s, with an error of approximately 12% (1.5 °C). Despite this, the overall accuracy of the analytical solution is deemed satisfactory, especially given its simplicity relative to the numerical approach.

6.7.2.2 Case 2: Temperature Variation with Constant Heat Flux

Figure 6.12 compares the temperature distribution of the fin for Case 2, where the system is subjected to constant heat fluxes of $q''_w = -500$ W/m^2 and $q''_w = -1000$ W/m^2. The high thermal conductivity of aluminum, combined with the uniform heat flux boundary condition, results in minimal temperature gradients along the fin.

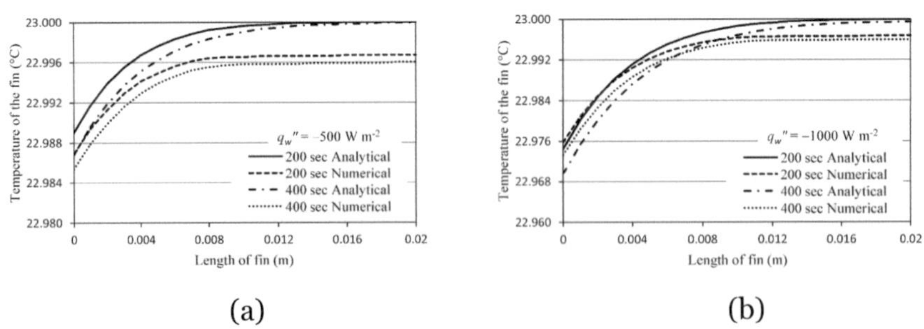

Fig. 6.12 Temperature profile of the fin for Case 2, showing the temperature variation along the fin with constant heat flux boundary conditions [6]

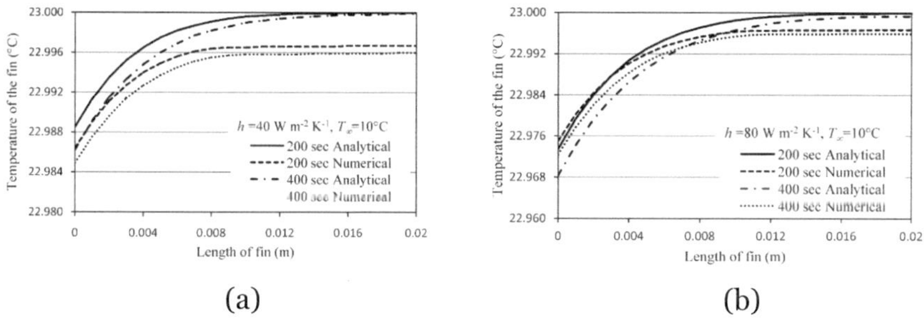

Fig. 6.13 Temperature profile of the fin for Case 3, depicting the impact of convective heat transfer on the temperature distribution [6]

For this case, the analytical solution demonstrates higher accuracy compared to Case 1, with only minor deviations observed between the analytical and numerical results. This improved accuracy is attributed to the reduced temperature variation along the fin, which simplifies the heat transfer dynamics and reduces the influence of thermal resistance effects.

These comparisons highlight the capability of the analytical solution to capture key trends in the temperature distribution of the fin under varying thermal boundary conditions. The results also underscore the effectiveness of the Crank–Nicolson and enthalpy methods in providing accurate numerical benchmarks for validating analytical approaches.

6.7.2.3 Case 3: Temperature Variation with Constant Convective Heat Transfer

Figure 6.13 illustrates a comparison of the temperature distribution along the fin for Case 3, where the fin is subjected to convective boundary conditions. Analytical results are compared to numerical solutions to evaluate the accuracy and performance of the approximate analytical method. The comparison demonstrates a good agreement between the two approaches, validating the reliability of the analytical model in predicting temperature profiles under varying heat transfer conditions.

The results underscore the impact of convection on the thermal behavior of the fin. For instance, increasing the heat transfer coefficient (h) enhances heat dissipation from the fin, thereby altering the temperature distribution and solidification dynamics.

6.7.2.4 Analysis of Solid–Liquid Interface Movement

Figure 6.14 presents the analytically determined position of the solid–liquid interface for all cases, emphasizing the thickness of the solidified PCM along the fin. Key observations include:

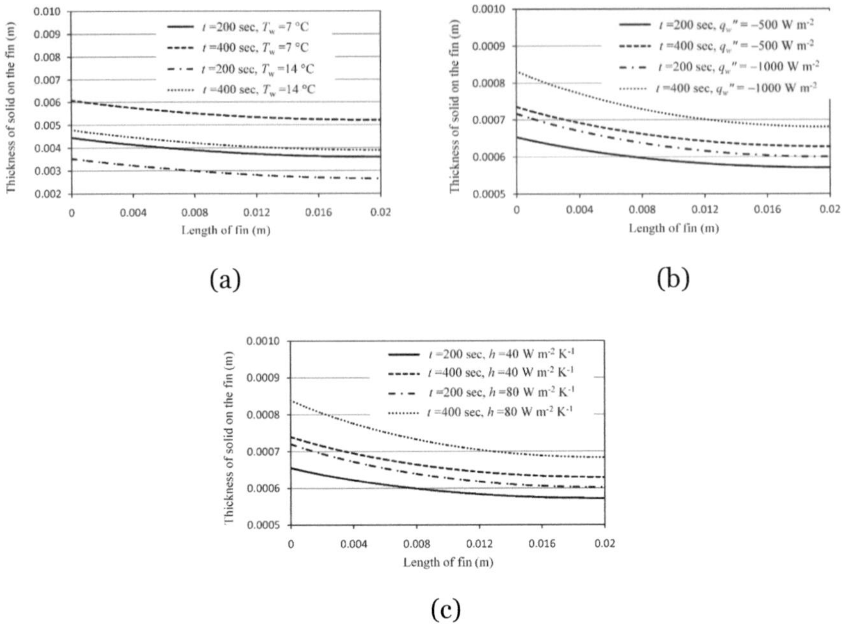

Fig. 6.14 Thickness of the solidified PCM layer along the fin as determined analytically for **a** Case 1, **b** Case 2, and **c** Case 3, with an ambient temperature $T_\infty = 10$ °C [6]

- *Case 1* (Fig. 6.14a): At low wall temperatures, the solid PCM exhibits the greatest thickness. The larger temperature difference ($T_m - T_w$) drives rapid solidification, leading to a steeper increase in the solid layer thickness compared to other cases.
- *Case 2* (Fig. 6.14b): Under constant heat flux conditions, the rate of solidification is relatively moderate. The fixed heat removal rate constrains the solid–liquid interface movement, resulting in slower growth of the solid layer compared to Case 1.
- *Case 3* (Fig. 6.14c): For convective boundary conditions, the solidification process is influenced by the proximity of the fin temperature to the PCM's solidification temperature (T_m). The heat removal rate is nearly constant, leading to a gradual and steady growth of the solid PCM layer. The movement of the solid–liquid interface is slower than in Case 1 but comparable to Case 2.

- **Key Insights**
  - In Case 1, the rapid solidification reflects the strong influence of a low wall temperature, which significantly enhances the thermal resistance of the solid PCM.
  - Cases 2 and 3 exhibit a more uniform solidification rate due to the imposed heat flux and convective cooling, respectively. The steady heat removal in these cases ensures a controlled phase change process.

– The findings suggest that under constant or nearly constant heat removal rates, the solid–liquid interface progresses at a relatively uniform pace, highlighting the importance of boundary conditions in determining solidification dynamics.

These results provide valuable insights into the thermal behavior of the fin-PCM system, aiding in the optimization of heat transfer and phase change processes for energy storage applications.

6.7.3 Solid–Liquid Interface Predictions for Cases 1 and 2

The predictions of the solid–liquid interface location of the ClimSel C23 as PCM for three storage types (see Table 6.3) and Case 1 (1st type of boundary condition) at two different times are illustrated in Fig. 6.15. These results allow a comparative analysis of the accuracy and efficiency of the one-dimensional analytical method against the numerical solution.

For long fins, heat transfer predominantly occurs through the storage medium. In such cases, the one-dimensional analytical model demonstrates a greater error in Region#2 compared to shorter fin configurations. The largest deviation, observed at $t = 200$ s, is approximately 0.4 mm, corresponding to a 10% error. This discrepancy underscores the limitations of the analytical model in scenarios with elongated fins where multidimensional effects become significant.

Figures 6.16 display the interface evolution for storage types 1, 2, and 3, respectively for Case 2 (2nd type of boundary condition). Across all configurations, the solidification rate decreases with time. Initially, the thin solidified layer offers minimal thermal resistance, resulting in a rapid solidification rate. However, as the layer thickens, the thermal resistance increases, and the solidification rate declines.

The fraction of solidified PCM in cylindrical shells and rectangular finned storages is compared in Fig. 6.17 for various storage configurations under Case 1. Here, the height of the storage units is kept constant, while other dimensions are adjusted to ensure equal PCM volume and equivalent heat transfer surface area. The results demonstrate that PCM in cylindrical shell storages solidifies faster than in rectangular storages. This difference arises from the thermal resistance in the solidified PCM, which increases more significantly in rectangular storages. Consequently, the heat transfer is slower, leading to a longer total solidification time.

Figure 6.18 shows the solidification fraction for cylindrical and rectangular storages under Case 2, where a constant heat flux is imposed on the inner wall. The

Table 6.3 Finned cylindrical shell storage geometries [7]

Storage type	Outer radius, b (m)	Fin space, $2z$ (m)
#1	0.030	0.02
#2	0.025	0.03
#3	0.020	0.04

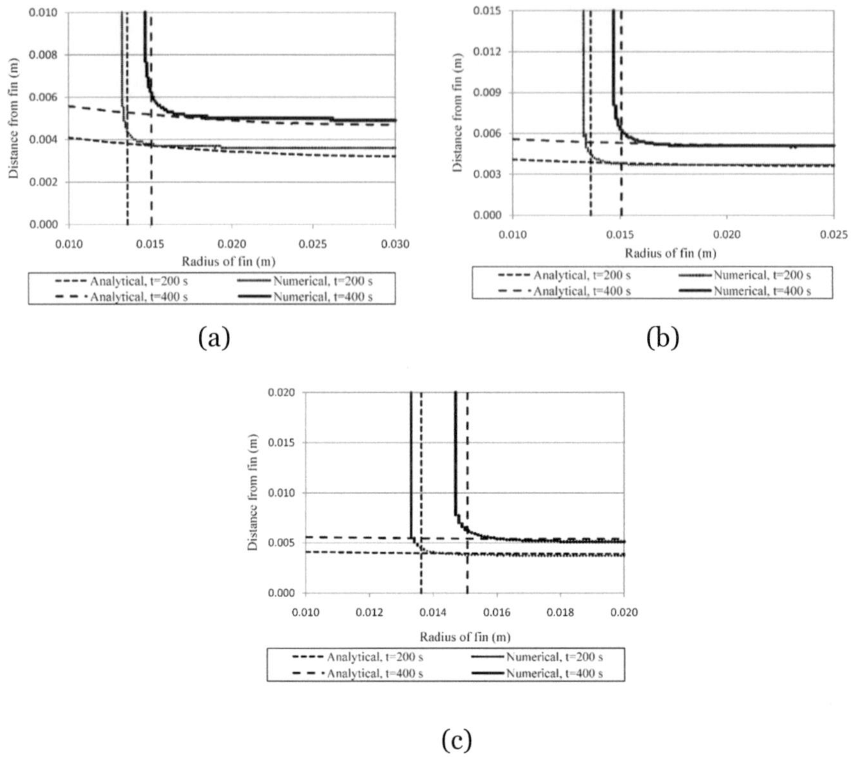

Fig. 6.15 Position of the solid–liquid interface at different time intervals for Case 1: **a** storage type 1, **b** storage type 2, and **c** storage type 3 [7]

findings reveal that the cylindrical shell storage requires a longer time for complete solidification compared to rectangular storage. This outcome is attributed to the heat flux distribution; in cylindrical storage, larger heat flux occurs in Region#1 due to wider fin spacing, leading to faster heat release. As evident in Fig. 6.18, the storage type with the shortest solidification time under both geometric configurations is type 3, due to its optimized design for heat dissipation.

6.7.4 Effect of Geometry on Solidification for Case 3

The storage system specifications for the analysis are given in Table 6.4 while $CaCl_2{\cdot}6H_2O$ is selected as PCM. Figure 6.19 presents a comparison of the solid–liquid interface location obtained using the current analytical model and a two-dimensional numerical method within a single storage cell. The enthalpy method, combined with a finite difference scheme, is employed to simulate the solid–liquid interface location in the cells.

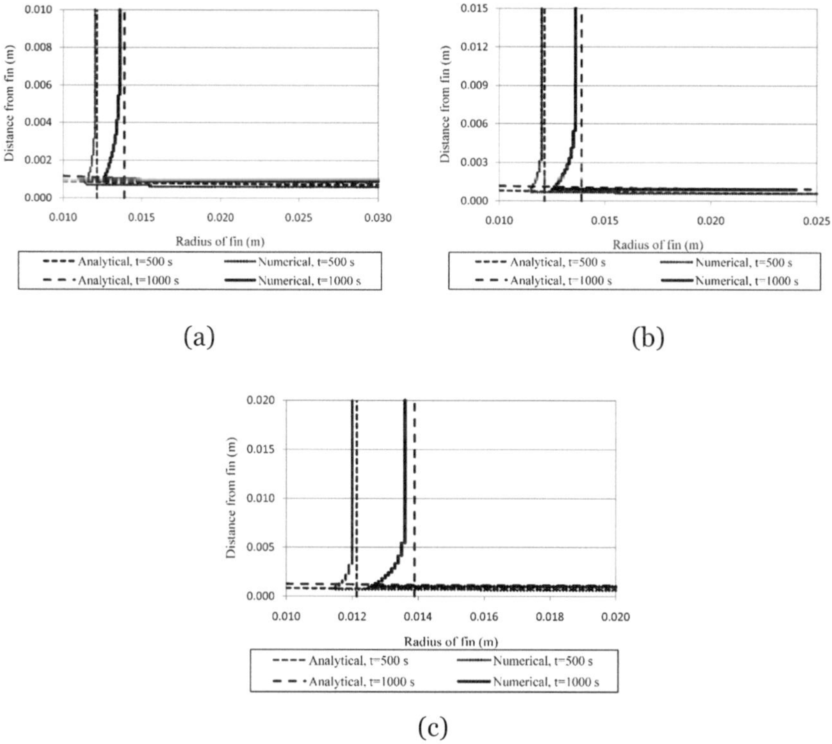

Fig. 6.16 Location of solid–liquid interface at different time intervals for Case 1: **a** storage type 1, **b** storage type 2, and **c** storage type 3 [7]

The role of geometric storage parameters on the solidification process is examined in Fig. 6.20. For various cell aspect ratios (l_c/l_f)—0.5, 1, and 1.6—the heat transfer mechanism shifts based on geometry. At larger aspect ratios, heat predominantly flows through the walls, while at smaller aspect ratios, fins play a more significant role. Lower aspect ratios result in higher solidification rates across all configurations due to enhanced fin-mediated heat transfer.

For cylindrical shell storage with aspect ratios below unity, the analytical model shows a faster solidification rate compared to rectangular storage, as depicted in Fig. 6.21. These findings align with studies by Mosaffa et al. [9].

Figure 6.22 illustrates the impact of air velocity on the variation of the solid fraction of the PCM over time. As anticipated, the solid fraction increases with higher air flow rates. This phenomenon occurs because an increase in air flow rate leads to a higher Reynolds number, which in turn enhances the heat transfer from the PCM to the air. Consequently, the efficiency of the phase change process is improved, resulting in a greater solid fraction of the PCM.

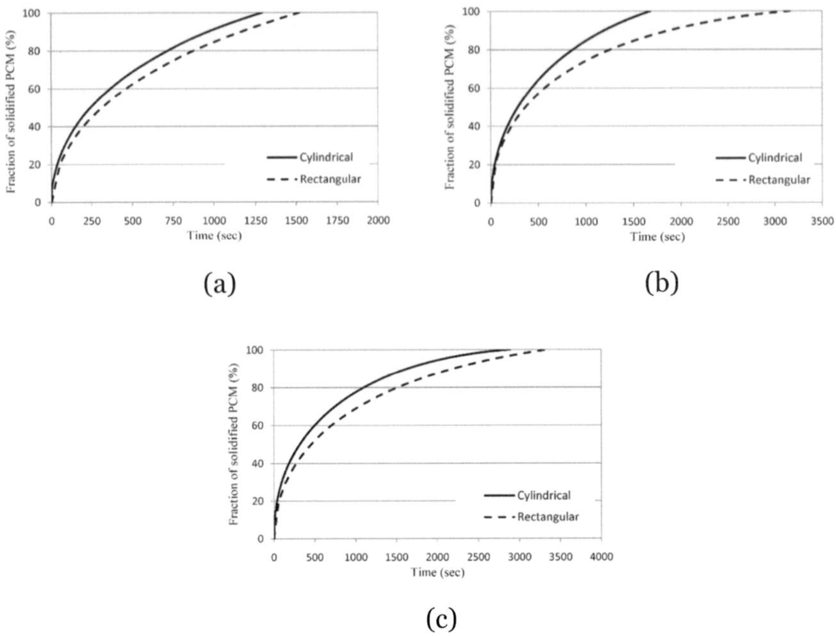

Fig. 6.17 Comparison of PCM solid fraction at different time intervals for Case 1: **a** storage type 1, **b** storage type 2, and **c** storage type 3 [7]

- **Influence of Air Velocity**

The effect of air velocity on PCM solidification is shown in Fig. 6.23. As expected, higher air flow rates lead to increased solidification rates due to the enhanced convective heat transfer and elevated Reynolds numbers.

However, as illustrated in Fig. 6.24 higher air velocities result in lower outlet air temperatures. This is because the increase in mass flow rate outweighs the increase in heat extraction.

The difference in outlet air temperature between successive air velocities diminishes at higher velocities. For instance, at $t = 3600$ s, the temperature difference between velocities 0.5 and 1.0 m/s is 1.38 °C, while between 1.0 and 1.5 m/s, it reduces to 0.58 °C. This suggests that beyond a certain threshold, increasing air velocity has a diminishing effect on thermal storage performance.

- **Influence of Inlet Air Temperature**

The impact of inlet air temperature on PCM solidification is illustrated in Fig. 6.25. Lower inlet air temperatures result in higher solidification rates due to the greater thermal gradient driving heat transfer. Furthermore, Fig. 6.26 highlights the effect of inlet air temperature on outlet air temperature over time. The results indicate that inlet air temperature has a more pronounced influence on thermal storage performance than air velocity.

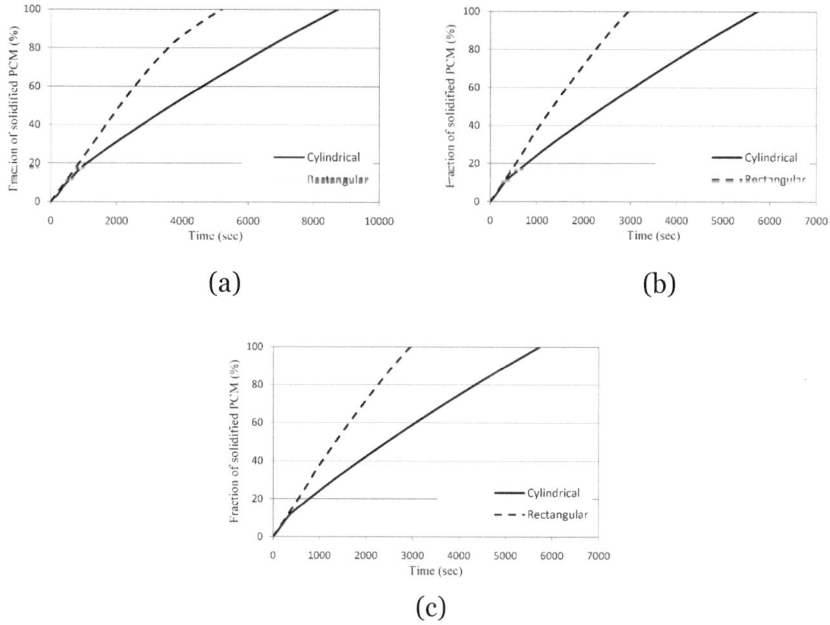

Fig. 6.18 Comparison of PCM solid fraction at different time intervals for Case 2: **a** storage type 1, **b** storage type 2, and **c** storage type 3 [7]

Table 6.4 Thermal storage system specifications used in the analysis [5]

Parameters	Value
Storage height	10 cm
Volume of cells	50.3 cm^3
Inner radius, R_{in}	10 mm
Fin thickness, 2δ	1 mm
Inlet air temperature, $T_{\infty,inlet}$	10 °C
Air velocity, u_∞	1.5 m/s

6.7.5 Summary

The analysis reveals that:

- Geometric configurations significantly influence the solidification rate and thermal performance.
- Cylindrical shell storages generally outperform rectangular storages in terms of total solidification time due to lower thermal resistance.
- Higher air velocities enhance solidification rates but have diminishing effects on outlet temperature at higher values.

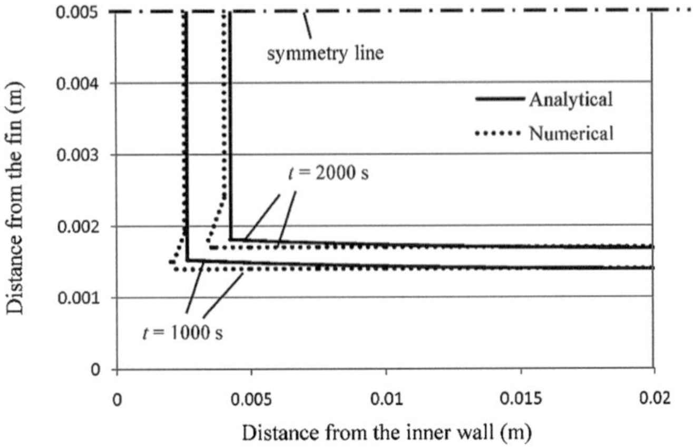

Fig. 6.19 Analytical and numerical results of the solid–liquid interface location in a single cell (l_c = 10 mm, l_f = 20 mm) [5]

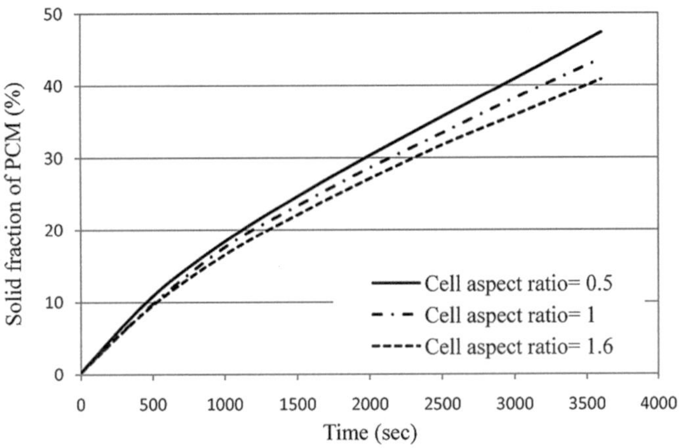

Fig. 6.20 Variation of PCM solid fraction over time for different cell aspect ratios [5]

- Inlet air temperature has a dominant effect on the performance, with lower temperatures yielding better thermal storage efficiency.

These findings offer valuable insights for optimizing PCM thermal storage systems, particularly for applications involving cylindrical shell geometries and variable air flow conditions.

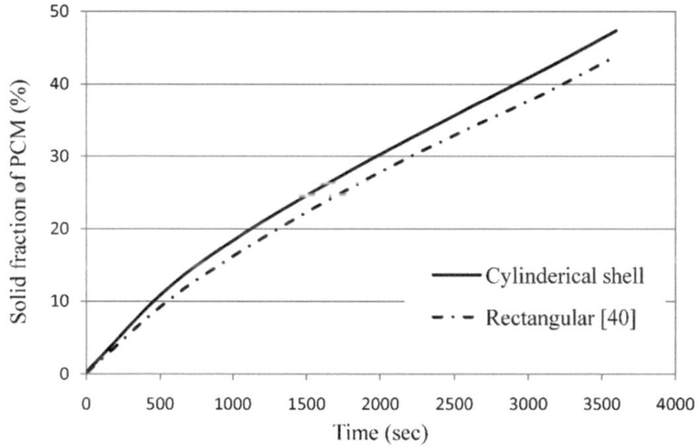

Fig. 6.21 Variation of PCM solid fraction over time for cylindrical shell and rectangular cell configurations [5]

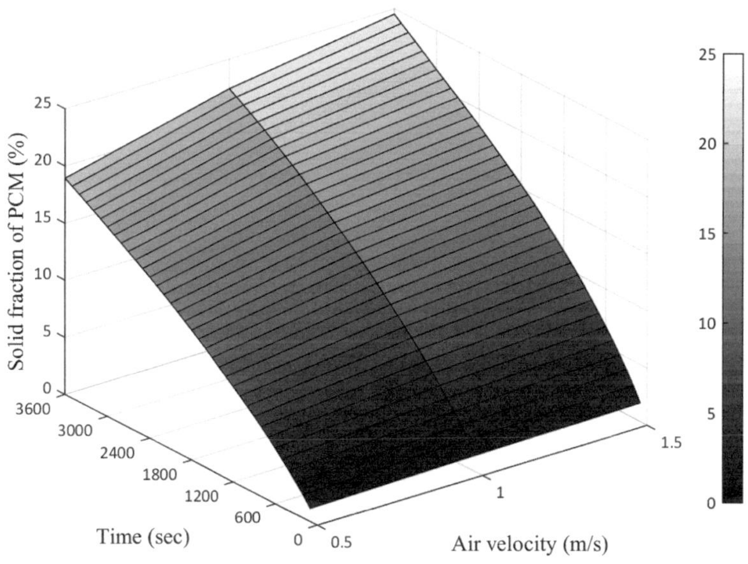

Fig. 6.22 Variation of PCM solid fraction in storage with air velocity and time for a cell aspect ratio of 0.5 [5]

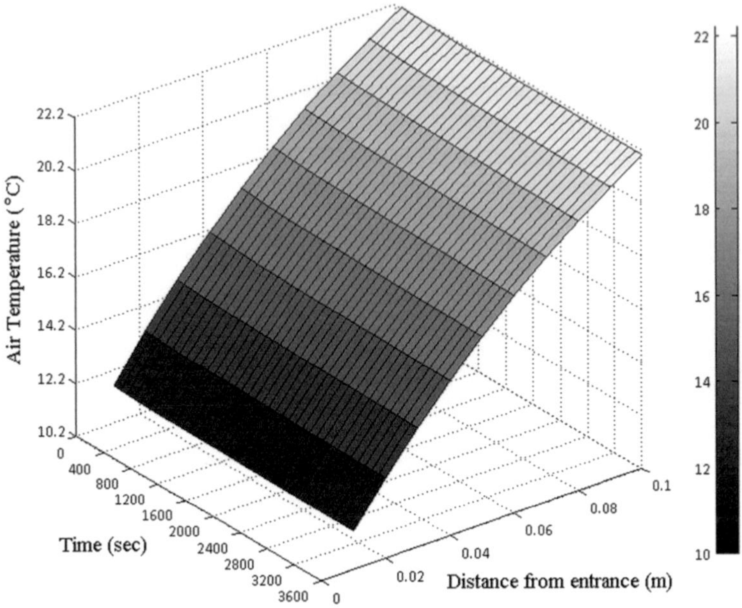

Fig. 6.23 Air temperature distribution along the tube within the thermal storage system for a cell aspect ratio of 0.5 [5]

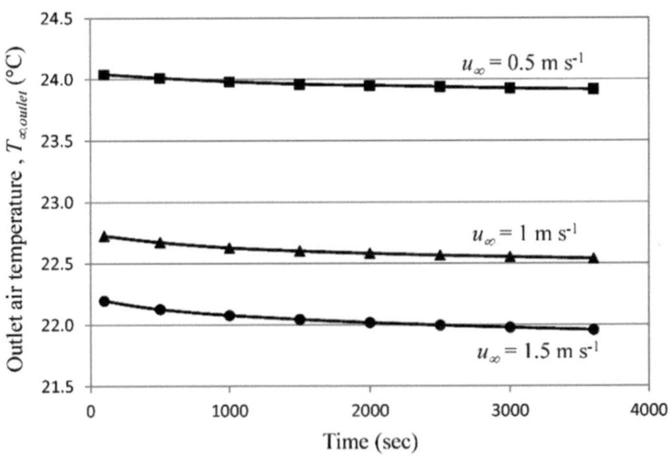

Fig. 6.24 Effect of air velocity on the performance of the thermal storage system for a cell aspect ratio of 0.5 [5]

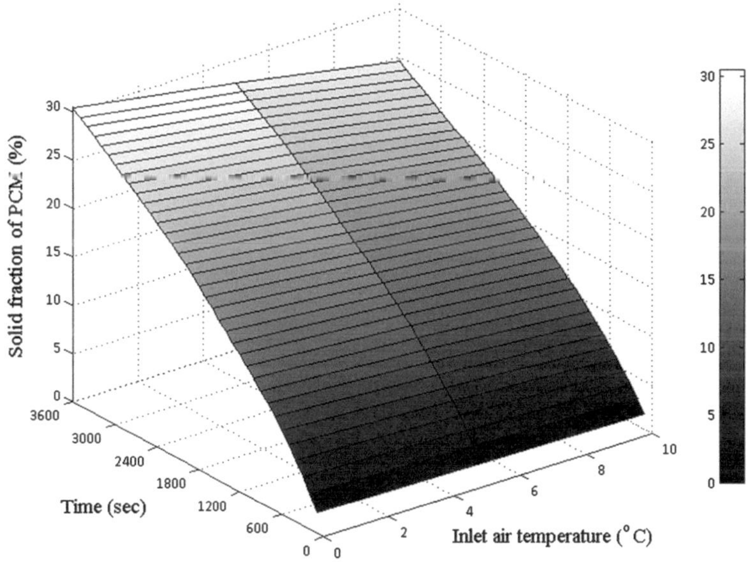

Fig. 6.25 Variation of PCM solid fraction in storage with inlet air temperature and time for a cell aspect ratio of 0.5 [5]

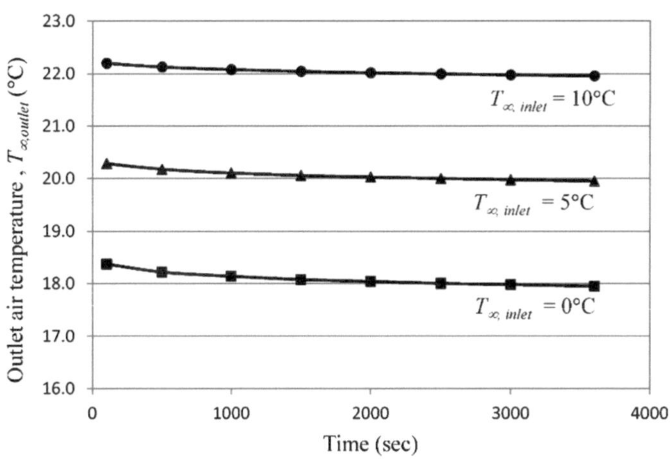

Fig. 6.26 Effect of inlet air temperature on the performance of the thermal storage system for a cell aspect ratio of 0.5 [5]

References

1. Carslaw HS, Jaeger JC (1986) Conduction of heat in solids. Oxford Science Publications, Clarendon Press

2. Esen M, Durmuş A, Durmuş A (1998) Geometric design of solar-aided latent heat store depending on various parameters and phase change materials. Sol Energy 62:19–28
3. Kays W, Crawford M, Weigand B (2005) Convective heat & mass transfer with engineering subscription card. In: McGraw-Hill series in mechanical engineering. McGraw-Hill Companies, Incorporated
4. Mosaffa AH (2014) Approximate analytical and numerical investigation on solidification process in the finned thermal energy storage using phase change material. PhD thesis. University of Tabriz
5. Mosaffa AH, Talati F, Basirat Tabrizi H, Rosen MA (2012a) Analytical modeling of PCM solidification in a shell and tube finned thermal storage for air conditioning systems. Energy Build 49:356–361
6. Mosaffa AH, Talati F, Rosen MA, Basirat Tabrizi H (2012b) Green's function solution for transient heat conduction in annular fin during solidification of phase change material. Appl Math Mech (Engl Ed) 33:1265–1274
7. Mosaffa AH, Talati F, Rosen MA, Basirat Tabrizi H (2013) Phase change material solidification in a finned cylindrical shell thermal energy storage: an approximate analytical approach. Therm Sci 17:407–418
8. Mosaffa AH, Talati F, Rosen MA, Infante Ferreira CA, Basirat Tabrizi H (n.d.) Numerical investigation on PCM solidification in a finned thermal storage based on enthalpy method. In: 10th IIR Gustav Lorentzen conference on natural refrigerants, Delft, The Netherlands
9. Talati F, Mosaffa AH, Rosen MA (2011) Analytical approximation for solidification processes in PCM storage with internal fins: imposed heat flux. Heat Mass Transf Stoffuebertragung 47:369–376

Chapter 7
A Survey of Analytical Solutions for Phase-Change Problems: Theory and Applications

7.1 Introduction

Phase-change materials are essential in energy storage systems and thermal management applications because they can effectively store and release substantial amounts of heat during phase transitions. Analytical modeling of phase-change problems provides important insights into the fundamental thermal behaviors of these materials and serves as a computationally efficient alternative to numerical simulations. This chapter aims to review the development and application of analytical solutions for phase-change problems, emphasizing their theoretical foundations, solution methods, and practical implications.

This chapter examines key analytical approaches to phase-change modeling, including the study of heat transfer in composite PCM systems, latent heat storage units, and enhanced thermal energy storage systems using fins. Each section addresses specific configurations and boundary conditions, highlights critical assumptions, and explores the influence of various thermophysical properties and non-dimensional parameters. By analyzing case studies and comparing analytical results with numerical or experimental findings, this chapter underscores the advantages and limitations of analytical methods in predicting temperature distributions, phase-change front propagation, and overall thermal performance.

The discussion also identifies challenges in current analytical models, such as the neglect of convective effects and the assumption of constant thermophysical properties. Future directions for hybrid analytical–numerical approaches are proposed to bridge the gap between theoretical simplicity and real-world complexity, paving the way for more comprehensive and accurate modeling of phase-change phenomena.

A. Mosaffa, *Phase Changing Materials*, Materials Horizons: From Nature to Nanomaterials, https://doi.org/10.1007/978-981-95-3742-6_7

7.2 Analytical Modeling of Phase Change in a Composite Wall with Two PCMs in Series

This section focuses on the phase-change heat transfer in a composite wall comprising two PCMs arranged in series, heated from one end [6]. The work aims to provide a theoretical framework for predicting temperature distribution and phase-change front propagation, as practical systems often employ multiple PCMs to improve energy storage efficiency. Unlike numerical approaches, the analytical solution offers deeper insights into the fundamental aspects of phase-change processes and identifies key non-dimensional parameters affecting the system's performance.

7.2.1 *Problem Description*

The composite wall consists of two layers of distinct PCMs with different thermophysical properties. The layers have thicknesses L_1 and L_2, and their respective phase-change temperatures are $T_{f,1}$ and $T_{f,2}$. A constant temperature source heats one end of the composite wall, while the opposite end is subject to convective cooling. This setup generates two phase-change fronts that propagate at different rates due to variations in PCM properties (Fig. 7.1). Key assumptions include:

- One-dimensional heat transfer.
- Neglect of convective effects in the liquid phase.
- Constant and uniform thermophysical properties.

The study considers two cases:

1. *Case I*, $T_{f,1} < T_{f,2}$: The first PCM has a lower melting temperature than the second.
2. *Case II*, $T_{f,1} > T_{f,2}$: The first PCM melts at a higher temperature than the second.

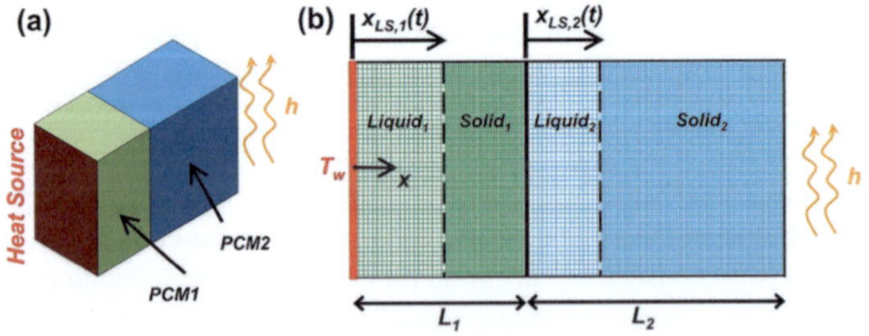

Fig. 7.1 **a** Schematic representation of the problem, illustrating a layered structure of two distinct PCMs heated from one end, leading to phase-change progression within both layers. **b** Detailed depiction of the system geometry, highlighting the liquid and solid regions along with the positions of the phase-change fronts in each layer at a given time [6]

7.2.2 Governing Equations

The analysis divides the phase-change process into multiple stages, each characterized by distinct governing equations. The temperature fields and melting front locations are determined sequentially.

- **Non-Dimensionalization**

$$\xi = \frac{x}{L_1 + L_2}, \quad \tau = \frac{\alpha_1 t}{(L_1 + L_2)^2}, \quad \theta_i = \frac{T_{\mathrm{i}} - T_{\mathrm{f},1}}{T_{\mathrm{w}} - T_{\mathrm{f},1}}, \quad Ste_{\mathrm{i}} = \frac{c_{\mathrm{p,i}}\left(T_{\mathrm{i}} - T_{\mathrm{f},1}\right)}{L_{\mathrm{i}}} \tag{7.1}$$

Case I: $T_{f,1} < T_{f,2}$

The process is split into three stages:

1. Stage A: melting of layer 1
   - Only the first PCM melts initially since its melting temperature is lower.
   - Governing equation:

$$\frac{\partial^2 \theta_1}{\partial \xi^2} = \frac{\partial \theta_1}{\partial \tau} \tag{7.2}$$

   - Boundary conditions:

$$\theta_1(0, \tau) = 1, \quad \theta_1\left(\xi_{\mathrm{LS},1}, \tau\right) = 0, \quad \frac{\mathrm{d}\xi_{\mathrm{LS},1}}{\mathrm{d}\tau} = -Ste_1\left(\frac{\partial \theta_1}{\partial \xi}\right)_{\xi = \xi_{\mathrm{LS},1}} \tag{7.3}$$

2. Stage B: sensible heating
   - Both layers undergo heating, but no further melting occurs.
   - Governing equations:

$$\frac{\partial^2 \theta_1}{\partial \xi^2} = \frac{\partial \theta_1}{\partial \tau}, \quad \frac{\partial^2 \theta_2}{\partial \xi^2} = \frac{1}{\alpha_2}\frac{\partial \theta_2}{\partial \tau} \tag{7.4}$$

3. Stage C: melting of layer 2
   - The second PCM begins melting due to heat conducted through the fully melted first layer.
   - Governing equations for layers 1 and 2: Eq. (7.4).

Case II: $T_{f,1} > T_{f,2}$

This case involves simultaneous melting of both PCMs, requiring coupled solutions for the temperature fields and phase-change front locations. Two stages are considered:

1. Stage A: concurrent melting
    - Both layers begin melting, and the process continues until one layer melts completely.
2. Stage B: completion of melting
    - The remaining layer undergoes complete melting while the other layer continues sensible heating.

- **Key Findings**
    - Analytical solutions provide explicit expressions for temperature distributions and phase-change front locations.
    - Results are validated against numerical simulations and show excellent agreement under specific conditions.
    - Critical parameters such as Stefan number, Biot number, and relative PCM thicknesses influence the phase-change dynamics.

- **Criticism and Comparison**

While the study offers significant contributions to understanding multi-PCM systems, it has limitations:

- Assumptions of one-dimensional heat transfer and constant properties may not capture the complexities of real-world applications.
- The neglect of convective effects in the liquid phase may underestimate heat transfer rates in certain configurations.
- Analytical solutions, while insightful, are limited to specific cases and require numerical methods for broader applicability.

Compared to existing works, this section extends the scope of phase-change modeling by addressing multi-PCM systems analytically. However, advancements in numerical methods provide more general solutions that account for non-linearities and multi-dimensional effects. Future work could focus on hybrid approaches combining analytical and numerical techniques to enhance predictive accuracy.

7.3 Analytical Modeling and Thermal Performance Analysis of a Flat Plate Latent Heat Storage Unit (LHSU)

In this section, we investigate the thermal performance of a flat plate latent heat storage unit (LHSU) composed of parallel slabs of PCM [3]. An analytical model was developed to describe the transient heat transfer fluid temperature distribution

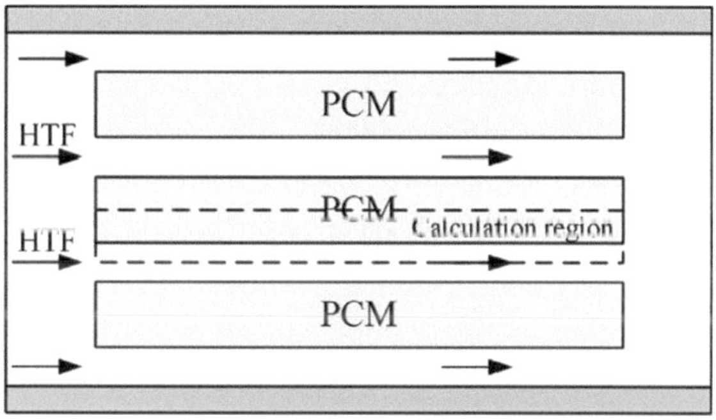

Fig. 7.2 Schematic of a flat plate LHSU [3]

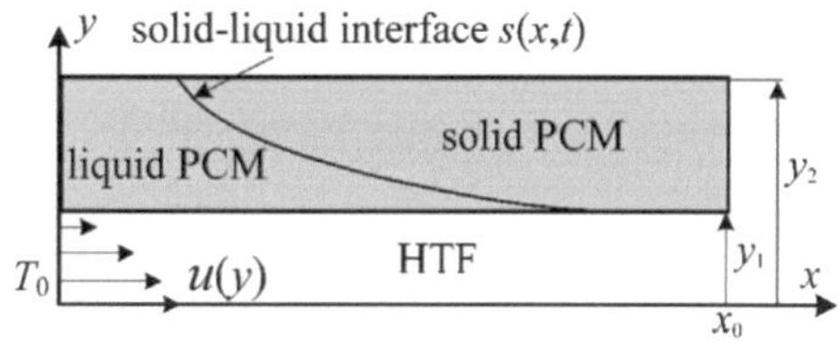

Fig. 7.3 Schematic of the analytical model in the melting process [3]

and the movement of the solid–liquid interface in the PCM. The model is validated against numerical simulations and previous analytical results, showing accuracy under specific conditions, such as low Stefan numbers. The work also introduces a dimensionless design criterion to optimize LHSU efficiency (Figs. 7.2 and 7.3).

7.3.1 Governing Equations and Assumptions

The system is modeled under several assumptions to simplify the analysis:

1. Thermophysical properties of HTF and PCM are constant.
2. Natural convection in the PCM is neglected due to constrained flow within thin slabs.
3. Convective heat transfer in the HTF is assumed quasi-steady-state.
4. Sensible heat of PCM during the phase change is ignored, valid for low Stefan numbers ($Ste < 0.15$).

- **Energy Conservation Equations**

For HTF:

$$u(y)\frac{\partial T_f(x,y,t)}{\partial x} = \alpha_f \frac{\partial^2 T_f(x,y,t)}{\partial y^2}, \quad 0 < x < x_0, \quad 0 < y < y_1 \tag{7.5}$$

Boundary conditions for HTF:

$$T_f(0, y, t) = T_0, \quad T_f(x, y_1, t) = T_m \tag{7.6}$$

For PCM during phase change:

$$\frac{\partial T_p(x, y, t)}{\partial t} = \alpha_p \frac{\partial^2 T_p(x, y, t)}{\partial y^2}, \quad 0 < x < x_0, \quad y_1 < y < y_2 \tag{7.7}$$

At the solid–liquid interface:

$$k_p \left. \frac{\partial T_p}{\partial y} \right|_{y=s(x,t)} = \rho_p L \frac{\partial s(x, t)}{\partial t} \tag{7.8}$$

Dimensionless variables: Dimensionless parameters were defined to streamline the analysis:

$$\eta = \frac{y}{y_1}, \quad \xi = \frac{x}{y_1}, \quad \theta = \frac{T - T_m}{T_0 - T_m}, \quad \tau = \frac{\alpha t}{y_1^2} \tag{7.9}$$

- **Analytical Solution**

The problem is divided into two stages:

1. *Sensible heating*: The PCM temperature increases uniformly to the phase-change temperature.
2. *Phase change*: The phase change occurs with a moving solid–liquid interface.

The analytical solution employs a quasi-steady-state assumption for HTF and simplifies the PCM behavior by neglecting sensible heat. The derived solutions for the HTF temperature and solid–liquid interface position are validated against numerical and previous analytical models, showing good agreement within acceptable error margins.

- **Key Findings**

  - *Thermal conductivity of PCM* (k_p): Increasing k_p enhances the effective latent heat storage ratio (E_r) up to a point, beyond which further increases yield diminishing returns.
  - *HTF channel height and PCM thickness*: Both parameters inversely impact E_r. Thicker PCM slabs and taller HTF channels increase thermal resistance, reducing efficiency.
  - *Stefan number (Ste)*: The model performs well for $Ste < 0.15$, beyond which errors due to the ignored sensible heat become significant.
  - *PCM mass distribution*: Increasing PCM length improves E_r by expanding the effective heat transfer area, while increasing thickness has a negative effect.

- **Design Recommendations**

A dimensionless criterion was proposed to optimize LHSU design:

$$\ln \theta^* + Bi\left(\eta_{y2} - 1\right)\left(\theta^* - 1\right) < -\frac{Nu \cdot \Gamma}{\Lambda}\xi \tag{7.10}$$

where Biot number (Bi) and Nusselt number (Nu) govern heat transfer efficiency.

- **Criticism and Comparison to Similar Studies**

While the analytical model provides a computationally efficient approach to evaluating LHSU performance, it has limitations:

1. *Neglect of sensible heat*: The assumption simplifies calculations but introduces errors at higher Stefan numbers or during rapid temperature changes.
2. *Natural convection omission*: This assumption underestimates heat transfer in larger PCM slabs or when convection is significant.
3. *Validation range*: The model is primarily validated for laminar flow regimes and may not generalize well for turbulent flows.

Compared to numerical methods, this model is computationally less demanding but sacrifices accuracy in capturing complex physical phenomena like convection or non-uniform phase-change behavior. Other analytical approaches, such as those incorporating iterative solutions for the moving interface, may provide more accuracy but at higher computational cost.

This section complements prior research by defining clear guidelines for LHSU design and offering insights into parameter impacts, making it a valuable tool for initial optimization in PCM-based thermal storage systems.

7.4 Analytical and Numerical Modeling of PCM and Hybrid PCM-Heat Sinks for High-Power Wireless EV Charging

This section investigates a passive thermal management system for high-power wireless electric vehicle (EV) charging, specifically targeting systems transferring 1 MW of power to heavy-duty vehicles within 35 min [5]. The main challenge lies in mitigating excessive heat dissipation, which can compromise the reliability and efficiency of power electronics. PCMs are explored as a key solution due to their capacity for absorbing and dissipating heat during phase transitions. The study develops both analytical and numerical models to analyze and optimize PCM-based heat sink designs.

7.4.1 Analytical Model Development

The analytical model is formulated for a semi-infinite PCM domain. Two boundary conditions are considered:

1. *Constant temperature*: One boundary is maintained at a fixed temperature above the melting point of the PCM.
2. *Constant heat flux*: Heat flux is imposed at one boundary, while the other maintains a constant temperature.

- **Governing Equations**

The heat transfer in the liquid and solid PCM regions is governed by:

$$\begin{aligned} \frac{\partial^2 T_L}{\partial x^2} &= \frac{1}{\alpha}\frac{\partial T_L}{\partial t} \quad 0 < x < x_{int} \\ \frac{\partial^2 T_S}{\partial x^2} &= \frac{1}{\alpha}\frac{\partial T_S}{\partial t}, \ x_{int} < x < L \end{aligned} \tag{7.11}$$

At the liquid–solid interface:

$$k\frac{\partial T_L}{\partial x} - k\frac{\partial T_S}{\partial x} = \rho L\frac{dx_{int}}{dt} \tag{7.12}$$

Boundary conditions for constant temperature:

$$\partial T_L(0,t) = T_0, \quad T_S(\infty,t) = T_i \tag{7.13}$$

Boundary conditions for constant heat flux:

$$-k\frac{\partial T_L(0,t)}{\partial x} = q'', \quad T_S(L,t) = T_i \tag{7.14}$$

These boundary conditions are illustrated in Figs. 7.4 and 7.5.

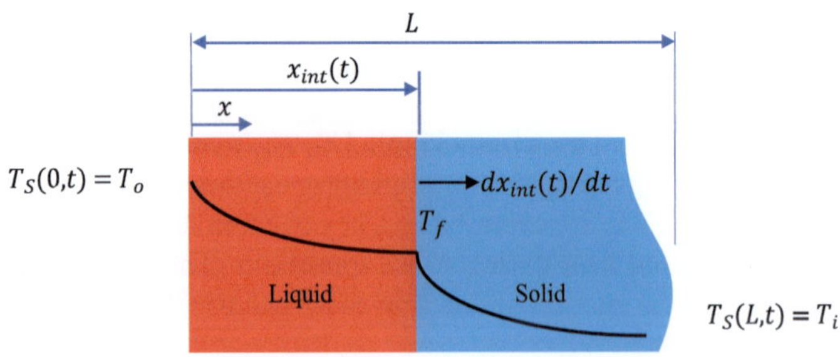

Fig. 7.4 Schematic of the semi-infinite PCM region undergoing melting, with both left and right surfaces maintained at constant temperatures [5]

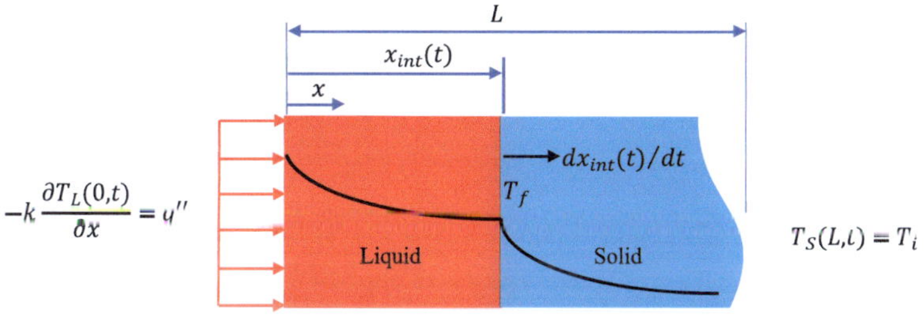

Fig. 7.5 Schematic of the semi-infinite PCM region undergoing melting, subjected to a constant heat flux on one surface and a constant temperature on the other [5]

7.4.2 Numerical Modeling

The numerical model employs computational fluid dynamics (CFD) to simulate heat transfer and phase transition. The enthalpy-porosity technique is used to model the mushy zone, representing partially solidified PCM as a porous medium. Assumptions include:

- Laminar and incompressible flow.
- Negligible thermal resistance at electronic interfaces.
- Constant thermophysical properties for the PCM and heat sink materials.

- **Results**

  1. *Heat sink configurations*:
     - Four designs with varying fin densities are compared. A configuration with higher fin density (HS-D) achieved lower surface temperatures due to reduced thermal resistance (see Fig. 7.6).
  2. *Thermal conductivity*:
     - Enhancing PCM thermal conductivity significantly reduces surface temperature, emphasizing the benefit of integrating expanded graphite into PCM.
  3. *Hybrid PCMs*:
     - Hybrid PCMs (e.g., combinations of n-Eicosane, RT70, and n-Tetracosane) improve temperature control and utilize latent heat effectively. A three-type hybrid PCM system achieves a 7.3% reduction in surface temperature compared to a single PCM.
  4. *Free versus forced convection*:
     - Forced convection improves heat dissipation, extending PCM utilization and reducing surface temperature, though at the cost of increased system complexity and energy consumption.

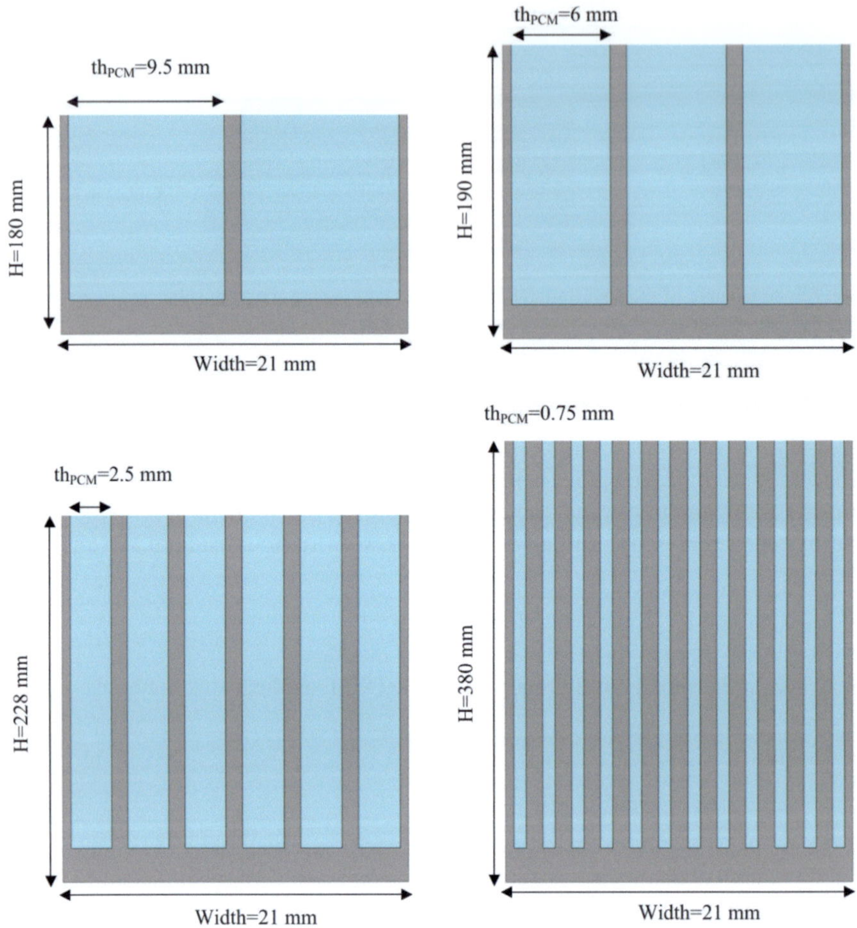

Fig. 7.6 The heat sink configurations for the thermal management system (the heights of heat sinks are not to scale) [5]

- **Criticism Compared to Similar Works**

While the study offers valuable insights into PCM-based thermal management, some limitations exist:

- The analytical model oversimplifies by ignoring free convection, which can be significant in high-power systems.
- The study relies on specific PCM combinations without exploring broader material possibilities.
- Unlike some studies, the impact of radiation heat transfer is not considered, potentially affecting accuracy in real-world applications.

- The focus on one-dimensional analytical modeling limits applicability to more complex geometries.
- The study lacks experimental validation, which is crucial for verifying numerical and analytical predictions.

Compared to similar works, this research uniquely integrates hybrid PCMs and high-fidelity numerical models. However, future studies should incorporate experimental data and explore three-dimensional models for more comprehensive solutions.

7.5 Enhancements in Thermal Energy Storage Systems Using Fins

Thermal energy storage systems are pivotal for harnessing renewable energy, particularly solar power, to counter its intermittent nature. Among the TES methods, latent heat storage stands out due to its high energy density and compactness, utilizing PCMs for energy storage during solid–liquid transitions. However, energy storage rates decline over time due to thermal resistance from the newly formed phase. Extended surfaces, such as fins, enhance heat transfer into PCMs. This section delves into a semi-analytical model for transverse and longitudinal fins in a cylindrical PCM setup, exploring their thermal behavior and design implications [8].

7.5.1 Theoretical Framework

- **Geometry and Configurations**

The study examines two fin types in an annular PCM bed surrounding a heat source:

1. *Transverse fins*: Radial fins with width w and a periodic pitch as shown in Fig. 7.7a.
2. *Longitudinal fins*: Axial fins extending radially with angular width $2\phi_f$ as shown in Fig. 7.7b.

7.5.2 Governing Equations

The heat transfer process comprises two distinct pathways:

- Direct heat transfer from the cylinder to PCM (q_1).
- Heat transfer through fins into PCM (q_2).

In this scenario, transverse fins are designed to extend circumferentially and radially from $r = R_i$ to $r = R_o$. The fin pitch and fin width in the axial direction are

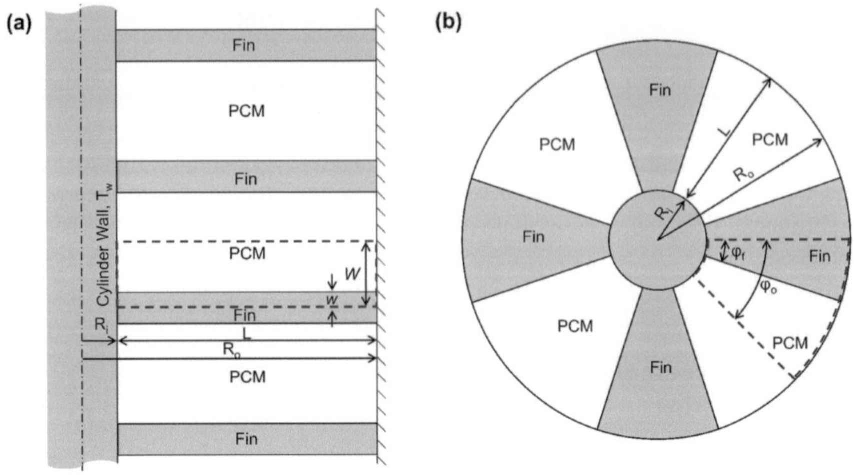

Fig. 7.7 **a** Cross-sectional schematic showing transverse fins extending into an annular PCM bed surrounding a cylindrical heat source. **b** Top-view schematic illustrating longitudinal fins extending into the annular PCM bed around the cylindrical heat source. Dashed lines indicate unit thermal cells based on symmetry for both configurations [8]

specified as $2W$ and $2w$, respectively. This configuration allows for a simplified heat transfer analysis by modeling an axisymmetric unit cell, as illustrated in Fig. 7.8. For transverse fins, the heat absorbed q_1 is derived from:

$$q_1(t) = -2\left(1 - \frac{w}{W}\right) \int_0^t \left.\frac{\partial \theta_\mathrm{p}}{\partial r}\right|_{r=R_\mathrm{i}} \mathrm{d}t \tag{7.15}$$

where θ_p represents the non-dimensional PCM temperature. The fin contribution q_2 is determined by solving the transient conduction equation:

$$\frac{c_\mathrm{f}}{k_\mathrm{f}} \frac{\partial \theta_\mathrm{f}}{\partial t} = \frac{1}{r} \frac{\partial}{\partial r}\left(r \frac{\partial \theta_\mathrm{f}}{\partial r}\right) - S_\mathrm{r} \tag{7.16}$$

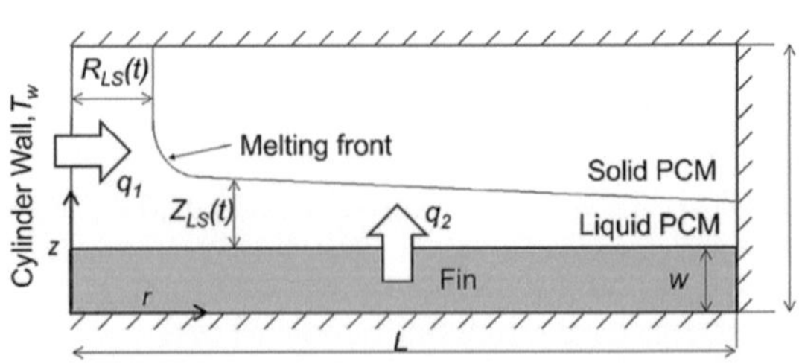

Fig. 7.8 Schematic representation of two distinct melting fronts within the unit thermal cell for transverse fin analysis [8]

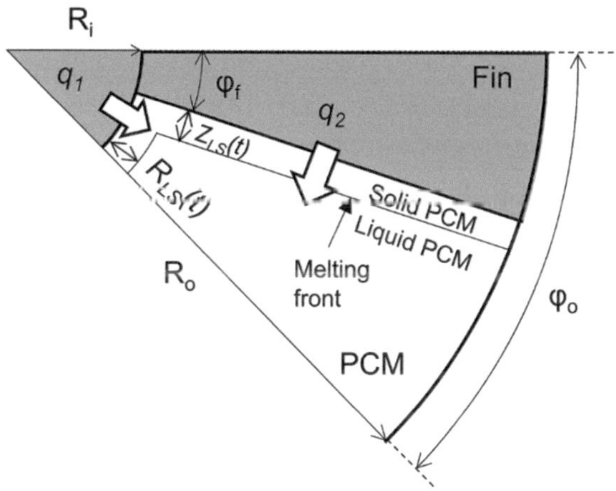

Fig. 7.9 Schematic of two melting fronts within the unit cell for longitudinal fin thermal analysis [8]

where S_r accounts for heat loss from the fin surface to PCM, incorporating phase-change dynamics.

For longitudinal fins, similar governing equations are adapted for axial symmetry. Based on symmetry in the problem, the unit cell in the circumferential direction that can be considered for analysis is shown in detail in Fig. 7.9.

7.5.3 Results and Design Insights

- **Key Findings**

  1. *Efficiency of thin fins*: Even minimal fin size significantly enhances heat transfer. However, increasing fin size beyond a threshold yield diminishing returns.
  2. *Thermal conductivity impact*: Higher thermal conductivity fins reduce the need for larger sizes, optimizing performance with minimal material.
  3. *Finite time constraints*: For practical applications with limited heat transfer durations, fins enhance performance. Infinite time renders fins counterproductive due to PCM displacement.

- **Practical Implications**

The insights aid in optimizing fin geometries for efficient energy storage, balancing heat transfer, and material costs. For transverse fins, the parameter *w*/*W* should be minimized, while longitudinal fins benefit from thin but adequately distributed structures.

- **Criticism and Comparative Analysis**

This section offers a comprehensive semi-analytical approach, validated against finite element simulations and experimental data. However, its assumptions, such as neglecting convective effects and idealized fin shapes, limit its applicability to real-world scenarios where non-linearities are pronounced. Unlike Cartesian fin studies, the cylindrical geometry introduces unique challenges but also nuances in performance trends. Notably, the simplicity of modeling two independent heat transfer fronts may overlook interactive complexities in tightly coupled systems.

Compared to similar studies:

- *Strengths*: Clear derivation of governing equations, robust validation, and practical design recommendations.
- *Limitations*: Less emphasis on dynamic environmental conditions or multi-material fin designs, which are increasingly relevant in advanced TES systems.

7.6 Analytical Exploration of PCM Solidification in Annular Systems

Phase-change materials have emerged as pivotal in energy storage solutions, especially in stabilizing intermittent renewable energy sources. This section focuses on the solidification dynamics of PCMs in annular configurations, utilizing a quasi-steady-state model [7]. Such geometries are widely adopted in heat storage systems due to their potential for enhanced thermal performance. A key novelty in this research lies in deriving the variable heat transfer coefficient at the solidification interface, which varies with the position of the moving front.

7.6.1 Theoretical Model

The system comprises an annular PCM layer confined between an inner cooling cylinder (radius R_1) and an outer boundary (radius R_2). The solidification front progresses radially outward from R_1 as the PCM undergoes phase change. The subject of this work was the solidification of a liquid placed in an annular space, as shown in Fig. 7.10.

- **Governing Equations**

The analysis builds upon the Stefan problem, a classical moving boundary framework, where:

1. The Fourier heat equation governs heat conduction in both solid and liquid phases.
2. Energy conservation applies at the moving boundary, the solidification front.

The heat conduction in the solidified region $R_1 < \hat{r} < r(t)$ follows:

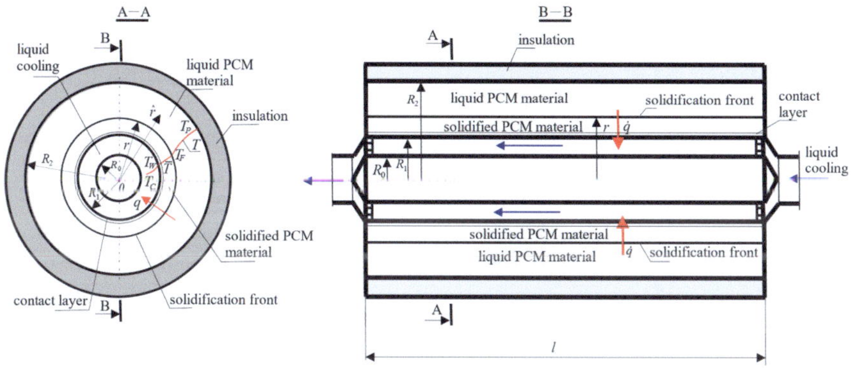

Fig. 7.10 Solidification in an annular space. PCM [7]

$$\frac{\mathrm{d}}{\mathrm{d}\hat{r}}\left(\hat{r}\frac{\mathrm{d}T}{\partial\hat{r}}\right) = 0 \tag{7.17}$$

with boundary conditions:

$$T = T_\mathrm{F} \quad @\hat{r} = r(t), \quad T = \overline{T} \quad @\hat{r} = R_1 \tag{7.18}$$

where T_F is the freezing temperature. The liquid region $r(t) < \hat{r} < R_2$ has a similar conduction framework:

$$T = T_\mathrm{F} \quad @\hat{r} = r(t) \tag{7.19}$$

- **Key Parameters**
  - *Heat flux in the solidified PCM*:

$$q_\mathrm{s} = -2\pi k_\mathrm{s} r \left.\frac{\partial T}{\partial \hat{r}}\right|_{\hat{r}=r(t)} \tag{7.20}$$

  - *Heat flux from overheated liquid PCM*:

$$q_\mathrm{L} = 2\pi r h(r)(T_\mathrm{P} - T_\mathrm{F}) \tag{7.21}$$

  where $h(r)$ is the local heat transfer coefficient, derived as:

$$h(r) = \frac{k_1}{r}\left(\frac{R_2^2}{R_2^2 - r^2\ln(R_2/r) - 1/2}\right) \tag{7.22}$$

- **Dimensionless Analysis**

To generalize findings, parameters are non-dimensionalized:

- *Radius*: $\tilde{r} = r/R_1$,
- *Time*: $\tau = Ste \cdot Fo$,
- *Overheating*: $\Theta_p = (T_p - T_F)/(T_{P0} - T_F)$.

The system of equations becomes:

$$\tilde{r}\frac{d\tilde{r}}{d\tau} = Bi_C Bi_{CON}/(Bi_C + Bi_{CON} + Bi_C Bi_{CON} \ln \tilde{r}) \tag{7.23}$$

with Bi_C and Bi_{CON} as Biot numbers for the cooling and contact layers.

7.6.2 Results

1. *Heat transfer dynamics*: The heat transfer coefficient increases near the end of solidification due to reduced PCM thickness, despite diminished overheating.
2. *Solidification time*: A thinner PCM layer reduces solidification time significantly. Also, for overheated liquids, solidification times increase with higher initial temperatures.
3. *Experimental validation*: Numerical simulations and experimental data for water as PCM show excellent agreement, validating the model's assumptions.

- **Case Study**

For RT64HC PCM, results show that solidification times can exceed 100 h under typical conditions, underscoring PCM's low conductivity challenge.

- **Design Implications**
  - Employing thinner annular segments accelerates heat transfer but may reduce total energy storage.
  - Increasing the number of annular layers improves system efficiency by enlarging heat transfer surfaces.

- **Criticism and Comparative Analysis**

This section advances PCM modeling with novel variable heat transfer coefficients and robust validation. However:

- *Strengths*: Provides practical design tools for annular PCMs in heat accumulators; offers detailed insights into the interplay between geometry and thermal performance.

- *Limitations*: Neglects convection effects, assuming conduction-dominated scenarios. Real-world systems may demand incorporating turbulent or forced convection influences.
- *Comparative analysis*: Unlike prior studies, this work integrates a dynamic heat transfer coefficient, enhancing model accuracy. Nevertheless, compared to active heat transfer approaches like fin-enhanced PCMs, passive geometries exhibit slower thermal response times.

7.7 AI-Driven Optimization of PCM Systems AI-Driven Optimization of PCM Systems

Incorporating artificial intelligence (AI) and machine learning (ML) into PCM systems has transformed the optimization of thermal management techniques. Researchers have optimized the phase-change prediction process, material selection, and system design with ease, which permits the energy to be stored and heat to be dissipated efficiently in electronics cooling and renewable energy systems. This section highlights recent developments about the AI-optimization of PCM systems with an emphasis on more practical computational techniques as well as their effects and operations in real life.

7.7.1 AI-Driven Phase-Change Prediction

To optimize PCM-based systems, accurate assessment of phase-change behavior is a necessity. The classical numerical techniques, like finite element analysis (FEA), are time-consuming, particularly in terms of multi-parameter-based optimization. Machine learning techniques, including artificial neural networks (NNs), have diversified by becoming faster alternatives so so-called surrogate models to perform evaluations. Bhatasana et al. [2], for example, used NNs to predict thermal performance parameters such as peak temperature and transient oscillations for PCM-embedded electronic packages (see Fig. 7.11). They used 250–2380 datasets to train their neural network models and were able to reduce the computational time from several weeks to hours without losing accuracy greater than 2.97% when compared to FEA. Similarly, Farahani et al. [4] utilized the group method of data handling (GMDH) algorithm to predict thermal resistance in microchannel heat sinks (MCHS) integrated with PCM and porous media. The GMDH model achieved an R^2 value of 0.94, demonstrating its capability to capture complex relationships between geometric parameters, fluid properties, and thermal performance. These approaches highlight how ML can replace or augment traditional physics-based models, enabling rapid exploration of phase-change dynamics under varying boundary conditions.

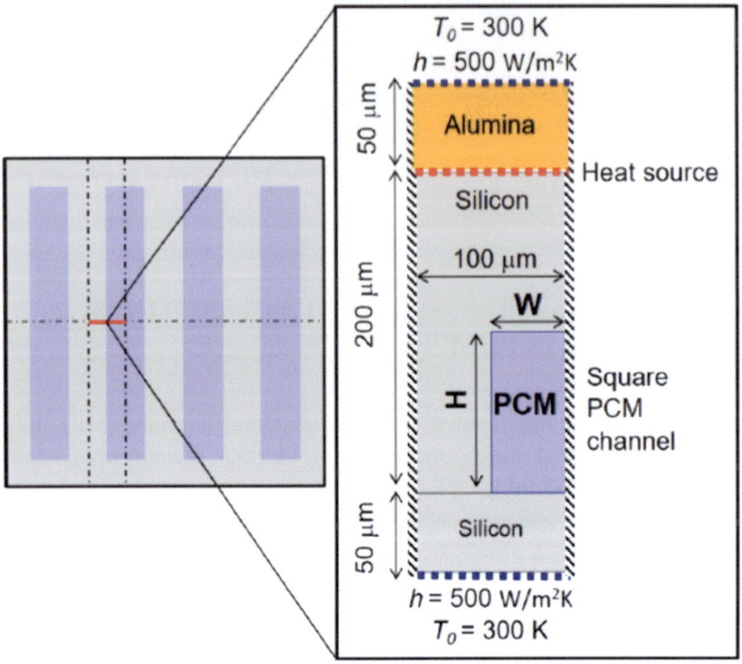

Fig. 7.11 Schematic of the PCM channels in the silicon chip [2]

7.7.2 Material Selection via Machine Learning

Choosing a material for PCM systems involves balancing various properties, including latent heat, thermal conductivity, and melting temperature. Machine learning tools can aid in the selection of materials by analyzing datasets of commercial and synthetic PCMs. In their study, Bhatasana et al. [2] evaluated seven metallic and organic PCMs and found that Solder 174, which has a melting point of 77 °C, is the most suitable for electronic cooling. This is due to its high thermal conductivity of 35.8 W/m K and significant latent heat. Their optimization process, using genetic algorithms, allowed for the effective determination of the optimal melting temperature for power density, highlighting the importance of adaptive material selection.

In the context of solar thermal systems, Alqaed et al. [1] utilized eco-friendly nanoparticles (Ag/CuO) synthesized from tea leaves to enhance the thermal conductivity of nanofluids while ensuring minimal impact on the ecosystem. Their machine learning optimization indicated that a pipe diameter of 8 mm is ideal for suspending PCM to optimize melting time and reduce panel temperature drop. Additionally, Farahani et al. [4] demonstrated that non-Newtonian fluids with a power-law index of 0.53 produce lower shear stress in microchannel heat sinks (MCHS), resulting in 2.8%

improved thermal performance compared to Newtonian fluids. These non-Newtonian fluids enhance the thermal efficiency of recirculating cooling systems.

These studies collectively underscore the role of machine learning in balancing competing system requirements and material properties.

7.7.3 System Design Optimization

Bhatasana et al. [2] utilized genetic algorithms and particle swarm optimization on PCM-embedded microchannel geometries, achieving a 19% reduction in peak temperature and 93% lower temperature fluctuations. Neural network-assisted strategies reduced computational costs by an additional 50% compared to exhaustive searches. Alqaed et al. [1] cooled solar panels by incorporating PCM and adjusting the pipe diameters from 4 to 8 mm. While the larger 8 mm diameters reduced panel temperatures, they also inhibited PCM melting and required machine learning to resolve conflicting objectives. Farahani et al. [4] examined wavy and converging microchannel designs and determined that wavy configurations performed 10.6% better thermally than smooth channels. Heat transfer was improved by an additional 15% upon integration of porous media with optimal porosity ($\varepsilon = 0.4$) and Darcy numbers ($Da = 10^{-6}$) determined through machine learning. These cases show that AI-driven design optimization enhances the operational effectiveness of PCM systems. In Fig. 7.12, thermal resistance is plotted against Reynolds number for various microchannel geometries, illustrating the dependence of performance on design. In Fig. 7.13, output flow temperature is plotted against time for differing pipe diameters, illustrating design trade-offs for system performance.

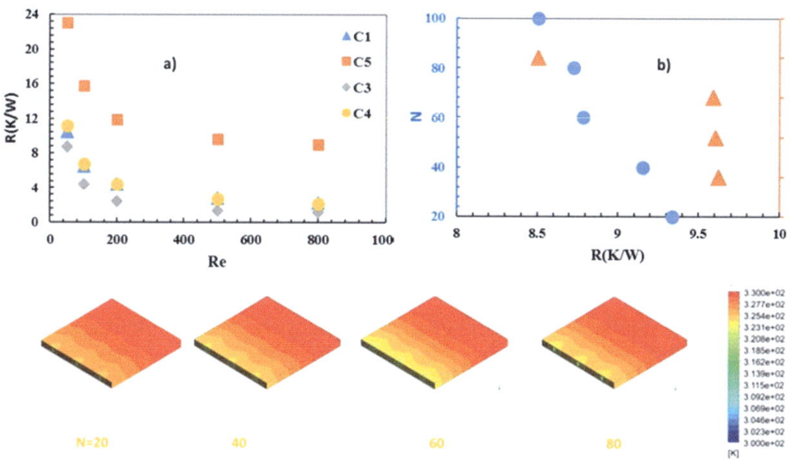

Fig. 7.12 Thermal resistance versus Reynolds number for microchannel geometries [4]

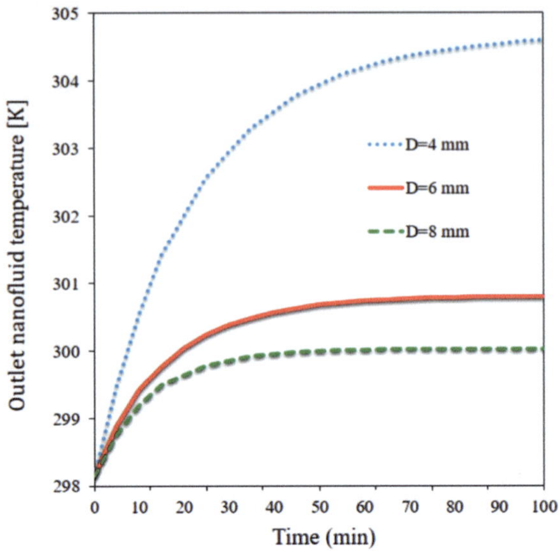

Fig. 7.13 Output flow temperature versus time for varying pipe diameters [1]

AI and ML have become essential for optimizing PCM systems, allowing for the rapid prediction of phase-change behavior, driven material choice, and efficient design space exploration. The integration of surrogate models with genetic algorithms and neural nets has allowed for enhanced thermal management of electronics, solar energy systems, and microfluidics. Further work should focus on extending ML datasets with new PCM composites and hybrid systems to enhance the reconciliation of simulations with real-world performance.

7.8 Summary

This chapter surveys analytical solutions for phase-change problems, emphasizing their role in understanding thermal behaviors and optimizing energy storage systems. It covers various configurations, including composite PCM walls, flat plate latent heat storage units (LHSUs), hybrid PCM heat sinks, and enhanced storage designs using fins. Analytical models are praised for their computational efficiency and ability to predict temperature distributions and phase-change front propagation, with validation against numerical simulations showing good agreement under controlled conditions.

Key findings highlight the influence of parameters like Stefan and Biot numbers, PCM thermal conductivity, and geometric factors on system performance. Design recommendations include optimizing fin dimensions, PCM thickness, and hybrid material integration to improve thermal efficiency. While insightful, these models are limited by simplified assumptions, such as neglecting convective effects and constant thermophysical properties. The chapter concludes with suggestions for combining analytical and numerical methods to improve accuracy and expand applicability to real-world scenarios.

References

1. Alqaed S, Mustafa J, Almehmadi FA, Alharthi MA, Sharifpur M, Cheraghian G (2022) Machine learning-based approach for modeling the nanofluid flow in a solar thermal panel in the presence of phase change materials. Processes 10:2291
2. Bhatasana M, Marconnet A (2021) Machine-learning assisted optimization strategies for phase change materials embedded within electronic packages. Appl Therm Eng 199:117384
3. Ding C, Niu Z, Li B, Hong D, Zhang Z, Yu M (2020) Analytical modeling and thermal performance analysis of a flat plate latent heat storage unit. Appl Therm Eng 179:115722
4. Farahani SD, Mamoei AJ, Alizadeh A (2023) Thermal performance of microchannel heat sink integrated with porous medium, slip coefficient and phase change material and machine learning approach. J Energy Storage 74:109357
5. Ghorbani M, Wang H, Roberts N (2024) Analytical and numerical modeling of phase change material and hybrid PCM-heat sinks for high-power wireless EV charging. Int Commun Heat Mass Transf 154:107460
6. Hasrati E, Krishnan G, Jain A (2024) Analytical modeling of phase change in a composite wall comprising two distinct phase change materials in series. Int J Heat Mass Transf 222:125055
7. Lipnicki Z, Małolepszy T (2020) Analytical study of the solidification of a phase change material in an annular space. Energies 13:5561
8. Mostafavi A, Parhizi M, Jain A (2020) Semi-analytical thermal modeling of transverse and longitudinal fins in a cylindrical phase change energy storage system. Int J Therm Sci 153:106352

Chapter 8
Latent Heat Thermal Storage and Its Applications

8.1 Introduction

The rapid growth of the global population, coupled with a significant rise in energy demand, has led to a notable environmental crisis. According to the International Energy Agency (IEA), primary energy production has surged by 49%, and CO_2 emissions have increased by 43% over the past two decades. Research indicates that buildings are responsible for nearly 41% of the world's energy consumption and 30% of annual greenhouse gas emissions. Projections suggest that energy demand in the building sector will increase by approximately 50% by 2050, with space cooling demand expected to triple between 2010 and 2050. Therefore, optimizing building envelopes to minimize cooling loads in hot climates is crucial. Energy-efficient cooling applications have the potential to save between 10 and 40% of energy.

Recently, phase-change materials, which utilize the principle of latent heat thermal storage, have garnered significant interest as a promising technology. PCMs have a high thermal energy storage capacity near their transition temperature and exhibit nearly isothermal behavior during the charging and discharging processes. Proper use of PCMs can reduce peak heating and cooling loads, maintain indoor temperatures within a comfortable range by minimizing temperature fluctuations, and decrease the size and energy consumption of related technical equipment. The primary advantage of using PCMs is their ability to enhance thermal storage potential with minimal changes to existing building designs. Numerous studies have explored the use of PCMs in buildings, demonstrating their potential to significantly improve building energy performance. However, challenges remain, particularly regarding the efficient use and practical application of PCMs.

A. Mosaffa, *Phase Changing Materials*, Materials Horizons: From Nature to Nanomaterials, https://doi.org/10.1007/978-981-95-3742-6_8

8.2 Harnessing PCM Technology for Thermal Efficiency in Buildings

The world is facing a serious problem with too much energy being used, leading to energy shortages. A big part of this energy is consumed by buildings, so we need to find solutions in this area. One effective way to collect energy from nature is by using solar energy, which helps keep buildings comfortable and saves energy. Latent Heat Thermal Storage is an important method that allows buildings to use solar energy as heat. The main benefit of using LHTS in solar systems for buildings is that it can help manage the energy that comes and goes at different times, matching the needs for heating and cooling.

To significantly reduce the amount of energy that buildings consume, it is crucial to design walls that use PCMs effectively. Choosing the right wall materials is key, and this depends on thoroughly analyzing the thermal performance of the PCM. Though the effectiveness of PCMs in walls has been studied a lot, more analysis is still needed.

In this part, we will look at how thermal resistance, thermal capacity, and phase changes in the material affect energy savings when a PCM wall is used in an air-conditioned room. However, studies have often not looked closely at how each of these factors works alone. Many researchers only look at how well PCMs perform during specific seasons, but it is important to analyze their performance throughout the entire year [6]. Additionally, there is a method for analyzing how well PCMs work all year round. Most studies focus on how PCMs can reduce energy use without considering whether they meet building energy efficiency standards for different times of the year. It is also important to evaluate how effective PCMs are in different locations, which can help researchers determine the best places and times to use them.

For example, Xie et al. [53] simulated to study how well PCM wallboards work in the outer wall of an air-conditioned room in Beijing. They considered four key factors: (1) how thermal performance varies by season, (2) when phase changes significantly affect energy efficiency, (3) whether the PCM meets energy efficiency standards, and (4) how the same PCM wallboard performs in different climates.

8.2.1 Building Model and Methodology

- **Structural Description**

The PCM wallboards were applied to the exterior south-facing wall of a simplified building model representing an air-conditioned room. The room, with dimensions 5 m × 5 m × 3 m, is situated on the second floor of a hypothetical multi-story building. Only the south wall is exposed to the external environment, constructed with three layers:

- Insulation layer: thickness 4 cm.
- Brick layer: thickness 30 cm.
- PCM wallboard: thickness 3 cm.

The construction of the PCM wall is shown in Fig. 8.1. Five PCM wallboard designs with distinct phase-change temperature ranges and latent heat capacities were examined. Table 8.1 summarizes the physical and thermal properties of the wall components.

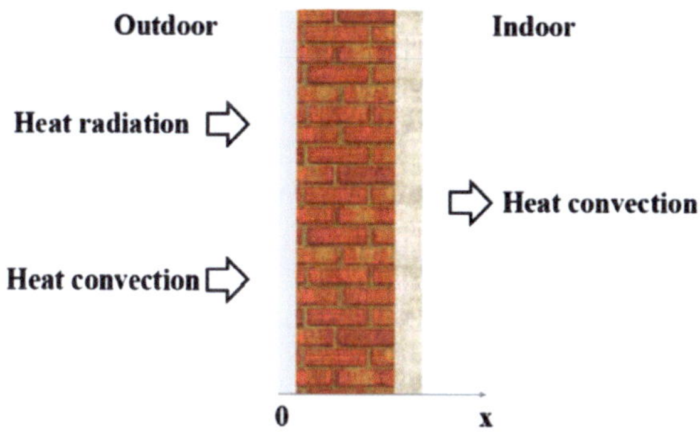

Fig. 8.1 Configuration and boundary conditions of the wall [53]

Table 8.1 Thermophysical properties of wall layers [53]

Component	Density (kg/m^3)	Specific heat (J/kg K)	Thermal conductivity (W/m K)	Latent heat (kJ/kg)	Phase-change range (°C)
Insulation layer	30	1500	0.04	–	–
Brick layer	1800	900	0.80	–	–
Gypsum wallboard	1000	1000	0.50	–	–
PCM Wallboard I	893	1340	0.20	34	18–23
PCM Wallboard II	893	1340	0.20	34	22–27
PCM Wallboard III	893	1340	0.20	34	25–28
PCM Wallboard IV	893	1340	0.20	34	24–26
PCM Wallboard V	893	1340	0.20	102	18–23

- **Heat Transfer Framework**

The analysis considers one-dimensional transient heat transfer through the wall, described by the following equation:

$$\rho\frac{\partial H}{\partial t}=\lambda\frac{\partial^2 T}{\partial x^2} \tag{8.1}$$

where

- ρ: density of the material,
- H: enthalpy, accounting for latent and sensible heat,
- λ: thermal conductivity,
- T: temperature.

For PCMs, the enthalpy (H) incorporates both sensible and latent heat contributions:

$$H=\int_{T_0}^{T} c_p(T)\cdot dT+\Delta H_{PCM}\cdot\chi(T) \tag{8.2}$$

Here, ΔH_{PCM} represents the latent heat of the phase change, and $\chi(T)$ is a fractional function reflecting the phase state.

Boundary conditions for the system include:

External wall surface:

$$\lambda\frac{\partial T}{\partial x}=h_{out}\left(T_{out}-T_{s,out}\right)+q_{solar} \tag{8.3}$$

Internal wall surface:

$$-\lambda\frac{\partial T}{\partial x}=h_{in}\left(T_{s,in}-T_{in}\right) \tag{8.4}$$

where q_{solar} is the absorbed solar radiation, h_{out} and h_{in} are the external and internal convection coefficients, respectively.

- **Numerical Implementation**

The governing equations were discretized using a finite-difference scheme and solved iteratively with MATLAB. Validation experiments on a similar PCM wallboard setup demonstrated excellent agreement, with a correlation coefficient of 0.955 between simulated and observed temperatures.

Spatial Discretization

$$\left.\frac{\partial^2 T}{\partial x^2}\right|_{i,j} = \frac{T_{i-1}^j - 2T_i^j + T_{i+1}^j}{\Delta x^2} \tag{8.5}$$

Time Discretization

$$\left.\frac{\partial H}{\partial t}\right|_{i,j} = \frac{H_i^j - H_i^{j-1}}{\Delta t} \tag{8.6}$$

Heat Transfer Equation

By combining Eqs. (8.5) and (8.6), the heat transfer equation is expressed as:

$$\rho_i \frac{H_i^j - H_i^{j-1}}{\Delta t} = \lambda \frac{T_{i-1}^j - 2T_i^j + T_{i+1}^j}{\Delta x^2} \tag{8.7}$$

Here, the indices i, $i - 1$, and $i + 1$ refer to the spatial coordinates, while j and $j - 1$ refer to the time coordinates.

8.2.2 Results and Seasonal Analysis

- **Monthly Thermal Performance**

The convective heat transfer (Q) through the PCM wallboard to the indoor air was a key metric for evaluating energy savings. Q was calculated as:

$$Q = \int_0^\tau h_{\text{in}}\left(T_{\text{s,in}} - T_{\text{in}}\right) \cdot \mathrm{d}t \tag{8.8}$$

The heat ratio (n) compared PCM wallboards to a reference gypsum wallboard:

$$n = \frac{Q_{\text{PCM}}}{Q_{\text{gypsum}}} \tag{8.9}$$

June (Summer Performance)

Figure 8.2 illustrates the thermal performance of various PCM wallboards during June. In summer, PCM wallboards with phase-change range above the indoor air temperature (> 25 °C) showed superior thermal performance. Notable findings include:

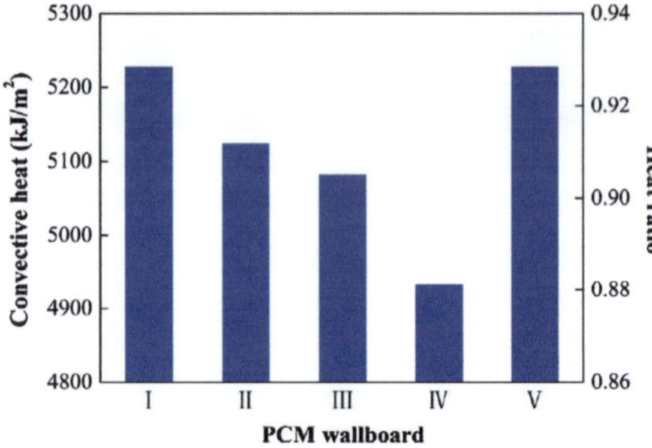

Fig. 8.2 Convective heat transfer and heat ratio for the five different wallboards in June [53]

- PCM Wallboard III reduced convective heat by 146 kJ/m^2 compared to Wallboard I.
- PCM Wallboard IV, with a narrower phase-change range, performed best, maintaining a heat ratio of $n = 0.92$.

December (Winter Performance)

Figure 8.3 displays the thermal performance of various PCM wallboards during December. In winter, PCM wallboards with higher latent heat capacity and phase-change ranges below the indoor temperature excelled in minimizing heat loss:

- PCM Wallboard V reduced heat transfer by 524 kJ/m^2 compared to Wallboard I, demonstrating enhanced insulation during cold months.

- **Yearly Heat Ratio Trends**

The yearly heat ratio analysis revealed seasonal fluctuations:

– *March and May*: Heat ratios ($n \approx 1$) indicated negligible phase-change effects.

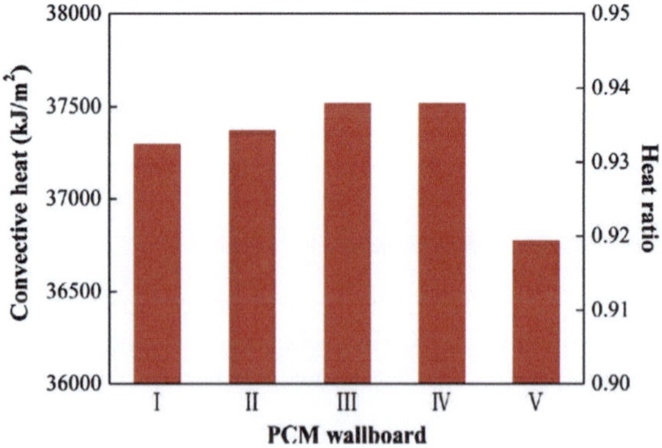

Fig. 8.3 Convective heat transfer and heat ratio for the five different wallboards in December [53]

- *June and September*: Maximum performance improvement, with heat ratios dropping by up to 0.08 compared to gypsum walls.
- *December*: PCM wallboards with high latent heat (e.g., Wallboard V) outperformed others.

- **Regional Performance and RTC Analysis**

The relative thermal conductivity (*RTC*) was introduced to quantify the influence of phase change on heat transfer:

$$\text{RTC} = \left(\frac{A}{n} + B\right)^{-1} \tag{8.10}$$

where *A* and *B* are constants derived from wall properties. Lower RTC values signified reduced heat transfer, aligning with improved insulation. Regional highlights include:

- *Beijing*: Significant RTC reductions in June and September.
- *Harbin*: Improved performance in July and August.
- *Guangzhou*: Best results observed in May.

- **Heat Transfer Coefficient and Energy Standards**

The effective heat transfer coefficient (K_p) was calculated to evaluate compliance with China's building energy standards ($K_{max} \leq 0.5$ W/m^2 K):

$$K_p = n \times K_g \tag{8.11}$$

where K_g is the coefficient for gypsum walls. PCM wallboards initially failed the standard but met it upon increasing the insulation layer to 4.5 cm.

8.2.3 *Criticism and Comparative Analysis*

The study on PCM wallboards for building applications provides valuable insights into optimizing energy efficiency through advanced material design. However, there are areas where the study could be critiqued, particularly when compared to similar works in the field.

8.2.3.1 Limited Scope of PCM Materials

While the study evaluates five PCM wallboards with varying phase-change ranges and latent heat capacities, it relies heavily on a limited set of configurations and materials. Comparable studies, such as those by Navarro et al. [36] and Pomianowski et al. [40], explore a broader range of PCM compositions, including microencapsulated PCMs and fatty acid-based materials. These studies highlight that varying PCM types can influence thermal behavior and durability in diverse climates.

8.2.3.2 Focus on Steady-State Energy Savings

The analysis assumes a consistent indoor air temperature (25 °C), which simplifies the simulation but limits the understanding of dynamic thermal performance under real-world conditions. In contrast, Kalnæs and Jelle [11] emphasized dynamic modeling, capturing fluctuations in indoor air temperatures and external heat loads, which revealed more nuanced insights into PCM effectiveness during extreme temperature variations.

8.2.3.3 Regional Bias in Simulation

The study is primarily tailored to Chinese climatic conditions, examining cities such as Beijing, Harbin, and Guangzhou. While this regional focus is valid for localized applications, it restricts the generalizability of the findings. Works like those by Kenisarin and Mahkamov [22] examine PCM integration across global climatic zones, providing more comprehensive recommendations for PCM use in various geographic contexts, including regions with high humidity or extreme desert climates.

8.2.3.4 Lack of Long-Term Performance Metrics

The study does not account for the long-term thermal degradation or material stability of the PCMs, which are critical in assessing their practical lifespan. Research by Dutil et al. [10] discusses how repeated phase transitions can degrade PCM performance, impacting both latent heat capacity and thermal conductivity over time. Incorporating such metrics would enhance the robustness of the conclusions.

8.2.3.5 Inadequate Discussion of Cost and Feasibility

While the study effectively analyzes energy savings and thermal performance, it does not evaluate the economic feasibility of PCM wallboards. Comparisons to Barzin et al. [4] show the importance of integrating cost–benefit analyses, including the payback period and manufacturing costs. Without such data, it is difficult to advocate for PCM adoption, particularly in regions where upfront material costs are a significant barrier.

8.2.3.6 Advancements Over Similar Studies

Despite these criticisms, the study introduces some novel contributions compared to existing works:

- *Year-round heat ratio analysis*: The use of heat ratio and relative thermal conductivity (RTC) to quantify PCM performance across months and regions is more refined than many existing studies that focus solely on single-season performance.
- *Numerical validation*: Experimental validation of the numerical model ensures credibility, which is not always a focus in studies like Mi et al. [33], where simulations often lack corroborating experimental data.

8.2.3.7 Recommendations for Improvement

- *Expand material scope*: Include a variety of PCM types, such as bio-based PCMs or nano-enhanced PCMs, to compare their thermal performance and long-term viability.
- *Dynamic and extreme conditions*: Simulate varying indoor air temperatures and incorporate extreme weather conditions to evaluate PCM adaptability.
- *Global applications*: Extend the analysis to include diverse climatic regions worldwide, offering broader applicability of findings.
- *Economic analysis*: Integrate cost modeling to highlight the financial benefits and feasibility of PCM implementation for end users.
- *Long-term studies*: Evaluate the aging and thermal degradation of PCMs over multiple years, assessing their suitability for long-term building integration.

8.2.4 Conclusion

While the study provides valuable insights into PCM wallboards' thermal performance, it falls short in addressing broader material diversity, economic implications, and long-term viability. By integrating approaches seen in global studies, it could enhance its contribution to energy-efficient building design. However, its innovative heat ratio and RTC methodologies set a benchmark for future research in PCM evaluation.

8.3 Innovations in Sustainable Building Heating Systems

In the pursuit of reducing energy consumption and greenhouse gas emissions, the building sector has increasingly embraced technologies that enhance energy efficiency and integrate renewable energy sources. A promising solution for cold climates, particularly in Canada, lies in the use of PCMs for thermal energy storage, integrated into radiant heating systems. This section delves into a groundbreaking study exploring such a system, designed for residential buildings in Toronto, aiming to achieve nearly zero energy building (NZEB) standards that have been conducted by Hosseini et al. [20].

8.3.1 *The Proposed System Design*

- **System Configuration**

The system comprises several key components as it is shown in Fig. 8.4:

- *Solar collector loop*: Arrays of flat-plate solar collectors capture solar energy and heat a thermal energy storage tank. This tank supplies heated water to the building's active layers for radiant heating. The loop operates during sunny periods, storing surplus energy for later use.
- *Active layer*: Embedded pipes in the walls and floors circulate the heated water, transferring thermal energy directly to the building's interior.
- *Phase-change material*: A layer of PCM encases the pipes within the walls, acting as a thermal buffer. The PCM absorbs excess heat when temperatures exceed its melting point (charging phase) and releases it as the ambient temperature drops (discharging phase).

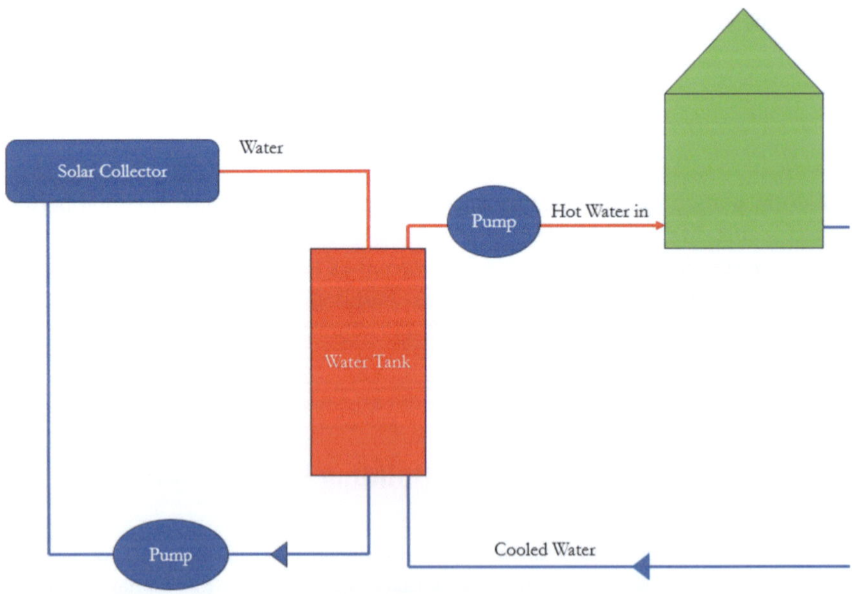

Fig. 8.4 Schematic of the studied system [20]

– *Thermal storage tank*: Designed to retain heat captured by the solar collectors, this tank ensures continuity in heat supply during non-sunny periods, particularly at night or during overcast conditions.

- **Building Case Study**

The study models a single-zone residential building in Toronto. Constructed using standard Canadian building materials, the structure includes insulated walls, a reinforced roof, and a thermally optimized floor. The simulated environment considers Toronto's seasonal weather patterns, emphasizing energy performance during winter.

- **Modeling and Simulation**

The system was modeled in TRNSYS software, an industry-standard thermal and energy system simulation tool. Each component of the system was individually validated for accuracy:

– *Solar collectors*: Tested for thermal efficiency against catalog specifications.
– *PCM layers*: Simulated to track phase transitions and thermal dynamics.
– *Storage tanks*: Modeled to predict heat loss and energy transfer efficiency. The overall system was fine-tuned through parametric studies to identify optimal configurations for pipe diameter, spacing, and PCM thickness.

8.3.2 Key Results and Findings

- **Energy Performance**
  – *Active layer effectiveness*: Embedding pipes within the building's walls and floors yielded a 65% reduction in heating energy consumption compared to traditional systems (see Fig. 8.5).
  – *PCM integration*: Adding a 12 cm PCM layer further reduced energy use, achieving an overall annual savings of 92%. This corresponds to a reduction from 14,400 MJ to just 1012 MJ per year.
  – *Temperature regulation*: The PCM layer moderated indoor temperature fluctuations, maintaining more consistent thermal comfort. During cold months, the system raised indoor temperatures by an average of 1.5–2 °C compared to buildings without the PCM-enhanced active layer.
- **Thermal Storage Performance**
  – The PCM layer's thermal storage capacity significantly reduced reliance on auxiliary heating systems.
  – During months with moderate solar availability (e.g., October and April), the PCM system eliminated the need for supplemental heating.
  – Optimal PCM performance was observed when its melting temperature aligned with indoor setpoints (20–24 °C).

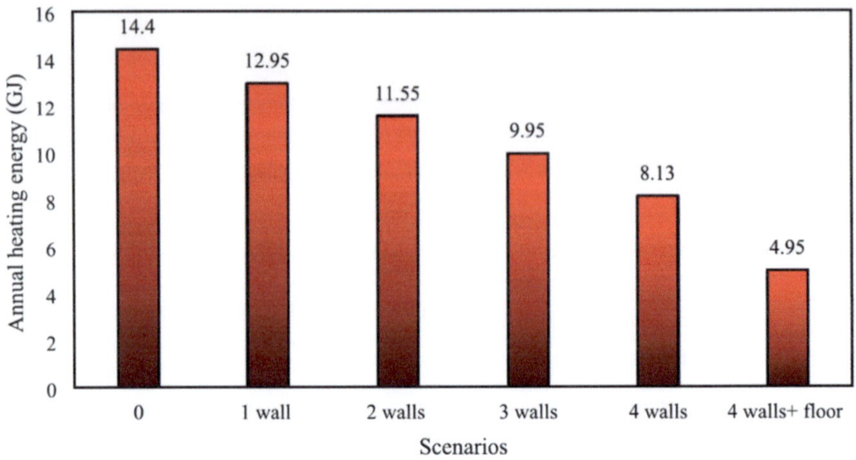

Fig. 8.5 Integrating active layers in building walls and floors impacts annual heating energy consumption [20]

- **Solar Energy Utilization**

  The solar collectors supplied sufficient energy during sunny months, with surplus heat stored in the PCM and thermal storage tank. For example:

  – In January, 43% of the solar collector's energy output was effectively used for heating.
  – During peak months like May, minimal heating was required, allowing the system to store most of the captured solar energy.

8.3.3 *Economic Evaluation*

- **Cost Analysis**

The initial costs include installation of solar collectors, a thermal storage tank, PCM layers, and active layer piping. While the upfront investment is substantial, the long-term savings justify the expenditure:

– *Initial investment*: Approximately $21,000 CAD, broken down as:

  Solar collectors: $10,000
  PCM materials: $6000
  Thermal storage tank: $3000
  Active layer components: $2000.

– *Operating costs*: Maintenance costs are estimated at 1% of the system's total value annually, or $210/year.

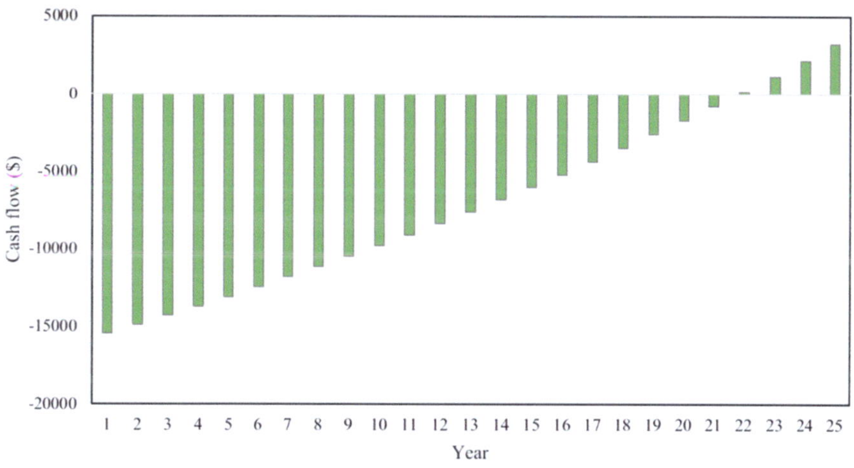

Fig. 8.6 Cash flow analysis of the system over a 25-year lifespan [20]

- **Payback Period**

Figure 8.6 depicts the cash flow of the proposed system over a 25-year lifespan. Energy savings translate into annual cost reductions of $1050, based on Ontario's electricity rate of $0.14/kWh. Accounting for operational expenses, the system achieves a payback period of 22 years. While longer than similar systems in warmer climates, this extended period reflects the additional challenges posed by Toronto's colder weather.

8.3.4 Challenges and Mitigation

1. *Climatic dependency*: The PCM's performance depends on the climate. In extremely cold conditions, incomplete phase transitions may limit thermal storage capacity. Mitigation includes selecting PCMs tailored to local conditions and integrating adaptive materials that adjust to varying temperatures.
2. *Material degradation*: Over time, PCM can lose efficiency due to repeated thermal cycling. Improved encapsulation techniques and composite PCMs are promising solutions to enhance durability.
3. *Economic constraints*: Long payback periods may deter adoption. Government incentives or subsidies could significantly improve feasibility for homeowners.

- **Future Directions**

  1. *Adaptive PCMs*: Research on materials that dynamically adjust their melting points could improve efficiency in variable climates.

2. *Smart controls*: Integrating IoT-enabled thermostats and sensors could optimize energy use, further reducing costs.
3. *Expanded applications*: Scaling the system for multi-zone buildings or pairing it with other renewable technologies like heat pumps could increase its appeal.

8.3.5 Criticism and Comparative Analysis

The study on using PCMs for radiant heating in Canadian residential buildings is innovative, yet it warrants critical examination, especially when juxtaposed with similar research in the field. Below, we analyze its strengths and limitations in comparison to other studies focused on energy-efficient building systems.

8.3.5.1 Energy Savings Efficiency

1. *Current study (Canada, PCM integration in walls)*
    - Achieved an impressive 92% reduction in annual heating energy demand.
    - Focused on a single-zone residential building with harsh winter conditions in Toronto.
    - Results are context-specific, reflecting the climatic challenges of cold environments where extended heating periods and lower solar irradiance dominate.
2. *Simou et al. (Morocco, solar-powered radiant floors)* [45]
    - Reported a 96% reduction in annual heating energy during colder months.
    - Utilized solar-powered radiant floor systems but in a much warmer climate.
    - The performance benefitted from prolonged solar availability, which is less applicable to colder climates like Toronto.
3. *Gou et al. (Europe, PCM in concrete floors)* [18]
    - Achieved a 67% reduction in heating energy demand using microencapsulated PCM in radiant floors integrated with solar thermal systems.
    - The study highlights higher efficiency when PCMs are embedded in floors rather than walls, as heat is distributed more effectively through direct contact with occupants.

While the current study demonstrates substantial energy savings, its reliance on wall-based PCM integration could limit its applicability to climates or buildings where radiant floors might be more effective. Radiant floors often have more direct interaction with living spaces, potentially offering greater efficiency for temperature regulation compared to walls, which primarily impact overall thermal mass.

8.3.5.2 Economic Feasibility

1. *Current study*
   - Initial investment of approximately $21,000 CAD, with a payback period of 22 years.
   - This extended period reflects the significant upfront costs of solar collectors, PCM materials, and thermal storage tanks, compounded by the high heating demand in Toronto's cold climate.
2. *Simou et al.* [45]
   - Lower initial costs due to simpler system configurations and favorable climatic conditions.
   - A shorter payback period of 13 years highlights better economic feasibility in warmer regions with less reliance on auxiliary heating systems.
3. *Larwa et al. (hydronic radiant floor heating with PCM, Poland)* [26]
   - Initial costs comparable to the Canadian study but achieved a payback period of 15 years.
   - Benefits from PCM's synergy with hydronic systems, which provide efficient heat distribution and lower operational costs.

The economic feasibility of the current study lags behind similar works. The 22-year payback period may deter adoption unless supplemented by subsidies or technological advancements that reduce costs. Furthermore, floor-based systems, as seen in Larwa's work, often show better cost-performance ratios due to higher heating efficiency.

8.3.5.3 System Adaptability

1. *Current study*
   - Optimized for Toronto's cold climate, but the PCM's performance is sensitive to temperature ranges.
   - The study assumes PCM's melting point is ideally matched to heating requirements, which might not be practical for regions with varying temperature profiles.
2. *Wu et al. (China, adaptive NZEBs with PCM)* [52]
   - Explored the use of PCMs with variable melting points optimized for multiple climate zones in China.
   - Employed dynamic building information modeling to adjust system configurations for seasonal and regional variations.

3. *Mao and Zhang (cascaded PCM storage systems)* [29]
    - Designed a cascade of PCMs with different melting points to enhance performance across a broader temperature range.
    - Demonstrated improved efficiency by layering PCMs in sequence, reducing the risk of performance drops in extreme conditions.

The Canadian study does not address the adaptability of its PCM to fluctuating temperatures, particularly during transitional seasons. Cascaded or adaptive PCMs could significantly enhance the system's efficiency and broaden its applicability across diverse climatic zones.

8.3.5.4 Durability and Long-Term Viability

1. *Current study*
    - Acknowledges the challenges of PCM degradation and leakage over time but provides limited mitigation strategies.
    - Relies on laboratory-grade PCM properties, which may not translate directly to real-world conditions.
2. *Xu et al. (composite PCMs for enhanced durability)* [55]
    - Investigated encapsulated composite PCMs to prevent leakage and maintain thermal stability over repeated cycles.
    - Found that improved encapsulation significantly extended the PCM's lifespan, reducing maintenance costs.
3. *Baetens et al. (thermal stability of PCMs)* [3]
    - Emphasized the importance of testing PCMs under real-world conditions, particularly in climates with large diurnal temperature variations.

This study lacks practical implementation details to address PCM durability issues, such as encapsulation techniques or the impact of cycling on thermal properties. Adopting strategies from Xu et al. could mitigate long-term performance concerns.

8.3.5.5 Strengths of the Current Study

- *Climatic specificity*: Provides a tailored solution for cold climates, where NZEB strategies are often underrepresented.
- *Comprehensive modeling*: Utilizes advanced simulation tools (TRNSYS) and parametric studies to optimize system performance.
- *Integrated approach*: Combines multiple renewable energy technologies (solar collectors, PCM, thermal storage tanks) into a cohesive system.

8.3.6 Conclusion

The integration of PCMs and radiant heating systems represents a groundbreaking step toward sustainable residential construction. By leveraging solar energy and advanced thermal storage, this system drastically reduces energy use while enhancing indoor comfort. Though the technology faces challenges in cost and scalability, its potential to revolutionize heating in cold climates is undeniable. As research continues, materials and system design innovations are poised to make NZEBs a reality in even the most demanding environments.

8.4 Advanced Thermal Insulation: PCM-Integrated Foam Concrete Panels

The integration of phase-change materials into foam concrete has emerged as a groundbreaking solution for achieving energy-efficient and sustainable buildings which was presented by Dora et al. [9]. This study investigates the development, testing, and simulation of foam concrete panels enhanced with PCM, nano silica, and coconut fiber, emphasizing their role in thermal insulation and heat storage. The resulting composite material offers a dual advantage of structural integrity and energy efficiency, paving the way for innovative building solutions.

8.4.1 Materials and Methodology

- **PCM Development**

The PCM used in this study is a composite of capric acid (CA) and exfoliated vermiculite (V), stabilized through a vacuum impregnation process. Key properties include:

- Melting point: 31.42 °C
- Freezing point: 21.08 °C
- Latent heat capacity: Melting: 17.55 J/g; Freezing: 9.72 J/g.

The vacuum impregnation method ensured uniform PCM distribution within the vermiculite matrix. This process leverages the porous nature of vermiculite to maximize PCM absorption while preventing leakage during phase transitions.

- **Foam Concrete Composition**

The foam concrete comprises:

- Cement–sand matrix: A standard mix of Portland Pozzolana Cement (PPC) and manufactured sand (M-sand) for structural strength.

– Nano silica (5%): Enhances compressive strength by reacting with calcium hydroxide to form calcium silicate hydrate (C-S-H).
– Coconut fiber (2%): Improves tensile strength and reduces shrinkage.
– PCM (5%): Replaces fine aggregates to introduce thermal storage properties.

The optimized mix, termed PNC-5%, exhibits enhanced thermal and mechanical properties.

- **Testing and Simulation**

  The study involved experimental and computational approaches:

  1. *Mechanical testing*:
     – Compressive strength was measured following IS 4031–6:1988 standards.
  2. *Thermal testing*:
     – Thermal conductivity: Assessed using the guarded hot plate method as per ASTM C177.
     – Differential scanning calorimetry (DSC): Measured latent heat and phase transition temperatures.
     – Thermogravimetric analysis (TGA): Evaluated material stability across temperature ranges.
  3. *Numerical simulation*:
     – Heat transfer and thermal performance were modeled using COMSOL Multiphysics® to simulate building environments with and without PCM panels.

- **Governing Equations**

1. *Heat transfer in PCM panels*

The general heat conduction equation governs the thermal performance of PCM-integrated panels:

$$\sum_{n=x,y,z}\left[\frac{\partial}{\partial n}\left(k\frac{\partial T}{\partial n}\right)\right]=\rho c_{\mathrm{p}}\frac{\partial T}{\partial t} \tag{8.12}$$

where

– k: Thermal conductivity (W/m K)
– T: Temperature (K)
– ρ: Density (kg/m^3)
– c_{p}: Specific heat capacity (J/kg K)
– t: Time (s).

During phase change, latent heat effects are included:

$$q = \Delta H \cdot \mathrm{d}t \cdot \mathrm{d}\phi \tag{8.13}$$

where

q: Heat flux (W/m^2)
- ΔH: Latent heat (J/kg)
- ϕ: Phase fraction (0 for solid, 1 for liquid).

2. *Energy balance in building envelopes*

Heat transfer into the building model was calculated using:

$$Q_\mathrm{i} = Q_\mathrm{w} + Q_\mathrm{d} + Q_\mathrm{r} \tag{8.14}$$

where

- Q_i: Total heat flux into the building (W)
- Q_w: Heat transfer through walls (W)
- Q_d: Heat transfer through doors (W)
- Q_r: Heat transfer through the roof (W).

Heat transfer through the wall (including PCM panel) is governed by:

$$Q_\mathrm{w} = h_\mathrm{in}(T_\mathrm{wall} - T_\mathrm{indoor}) + h_\mathrm{out}(T_\mathrm{outdoor} - T_\mathrm{wall}) \tag{8.15}$$

where

- h_in, h_out: Convective heat transfer coefficients (W/m^2 K)
- T_wall: Wall surface temperature (K)
- T_indoor, T_outdoor: Indoor and outdoor temperatures (K).

3. *Thermal storage capacity*

The energy stored/released by the PCM is expressed as:

$$E = m_\mathrm{PCM} \cdot \Delta H \tag{8.16}$$

where

- E: Thermal energy stored/released (J)
- m_PCM: Mass of PCM (kg)
- ΔH: Latent heat of fusion (J/kg).

8.4.2 *Results and Findings*

- **Mechanical and Thermal Performance**

  1. *Compressive strength* (Fig. 8.7):
     - PNC-5% achieved a compressive strength of 7.96 MPa, surpassing the minimum requirement for wall panels.
     - Nano silica reduced brittleness by filling voids and enhancing C-S-H formation.
  2. *Thermal conductivity*:
     - PNC-5% exhibited a low thermal conductivity of 0.15 W/m K, ideal for insulation.
  3. *Latent heat storage*:
     - PNC-5% stored 17.55 J/g during melting, contributing to effective thermal regulation.

- **Simulation Insights**
  - PCM panels reduced peak indoor temperatures by 6 °C compared to uninsulated buildings.

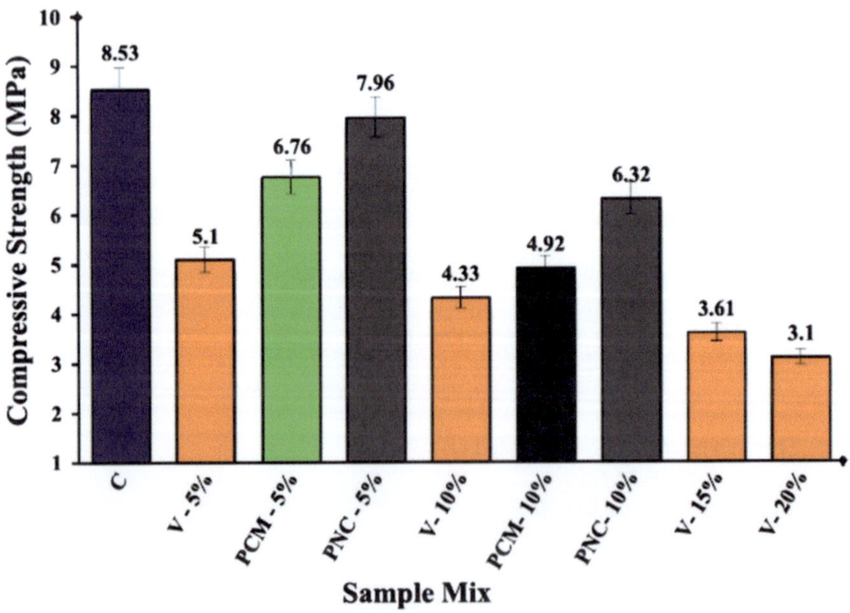

Fig. 8.7 Compressive strength of the samples [9]

  - Energy demand for heating/cooling decreased significantly, indicating potential HVAC savings.

- **Environmental Impact**

The use of natural and recycled materials like vermiculite and coconut fiber reduced the composite's carbon footprint. The system aligns with global goals for reducing greenhouse gas emissions.

8.4.3 Criticism and Comparative Analysis

The development of PCM-integrated foam concrete panels presents a promising advance in energy-efficient building materials. However, it is essential to critically examine its performance and contributions by comparing it to similar research in the field. Below, the study's outcomes are assessed against key comparable works.

8.4.3.1 Thermal Insulation and Energy Performance

1. *Current study*:
   - The PNC-5% composite demonstrated a thermal conductivity of 0.15 W/m K and latent heat storage of 17.55 J/g during melting.
   - The PCM's phase transition temperatures (21.08 °C freezing, 31.42 °C melting) align well with thermal comfort needs, stabilizing indoor temperatures and reducing HVAC energy use by approximately 6 °C during peak heat.
2. *Royon et al.* [43]:
   - Investigated PCM floor panels for lightweight building envelopes, achieving similar thermal conductivity but higher latent heat storage (~ 35 J/g).
   - Floor integration provided more direct interaction with indoor environments, leading to superior energy performance compared to wall-based systems.
3. *Gencel et al.* [13]:
   - Foam concrete with basalt powder and PCM achieved a thermal conductivity of 0.12 W/m K, outperforming the current study. However, the PCM's melting point (27.8 °C) was less optimal for broader climate adaptability.

While the PNC-5% mix shows effective insulation, its thermal conductivity is slightly higher than other PCM-based systems. Additionally, wall-based PCM integration may limit the direct interaction with indoor spaces, making floor or roof applications potentially more efficient for thermal regulation.

8.4.3.2 Compressive Strength and Mechanical Performance

1. *Current study*:
   - PNC-5% achieved a compressive strength of 7.96 MPa, sufficient for non-load-bearing wall panels. Nano silica addressed the brittleness caused by PCM and vermiculite.
   - The inclusion of coconut fibers reduced shrinkage and improved tensile strength.
2. *Kocyigit et al.* [23]:
   - Achieved a compressive strength of 11 MPa in foam concrete with a PCM and basalt powder composite, exceeding the strength observed in this study. This was attributed to basalt's superior structural reinforcement.
3. *Pasupathy et al.* [38]:
   - Explored PCM-integrated roofs and found compressive strength less critical in such applications, allowing for higher PCM proportions and better thermal storage.

The compressive strength of PNC-5% panels, though adequate for wall applications, remains lower than similar PCM composites utilizing stronger additives like basalt. For broader applications, such as load-bearing components, this limitation could restrict its use.

8.4.3.3 Economic Feasibility

1. *Current study*:
   - Utilizes cost-effective materials like vermiculite and coconut fibers but includes nano-silica, which is relatively expensive.
   - PCM integration adds to the overall cost, potentially resulting in longer payback periods for adoption.
2. *Gencel et al.* [13]:
   - Focused on recycled concrete powder as a base material, significantly reducing costs while maintaining performance.
   - The study highlighted cost savings from waste reutilization and scalability of PCM foam concrete.
3. *Alawadhi et al.* [1]:
   - Found that economic feasibility improves with PCM encapsulation techniques that minimize material usage while maximizing thermal storage.

This study's reliance on nano silica, despite its advantages, increases costs. Adopting alternative, lower-cost reinforcements or encapsulated PCM methods could improve affordability and market adoption.

8.4.3.4 Environmental Sustainability

1. *Current study*:
    - Demonstrates low carbon emissions due to natural materials like coconut fibers and vermiculite. However, the carbon footprint of nano silica (2.3 kg CO_2e/kg) slightly offsets these gains.
2. *Gencel et al.* [13]:
    - Foam concrete with recycled concrete powder and shape-stable PCM exhibited a smaller carbon footprint by using entirely recycled inputs.
3. *Zhang et al.* [58]:
    - Explored bio-based PCMs, which reduced environmental impacts but required further testing for durability under real-world conditions.

While this study integrates sustainable materials, the use of energy-intensive nano silica could be reconsidered. Employing recycled materials or bio-based PCMs may provide better alignment with environmental goals.

8.4.3.5 Strengths of the Current Study

1. *Innovative material composition*: The combination of PCM, nano silica, and coconut fibers offers a novel mix that balances thermal efficiency with structural integrity.
2. *Realistic testing and validation*: Extensive experimental tests and numerical simulations provide robust validation of the material's performance in real-world scenarios.
3. *Targeted climate suitability*: The PCM's phase transition temperatures are well-suited for tropical and subtropical climates, offering precise thermal regulation.

8.4.4 Conclusion

The integration of PCM into foam concrete panels demonstrates a multifaceted approach to building energy efficiency. Governing equations and simulation results underline the system's capability to enhance thermal regulation and structural performance. The PNC-5% composite, validated experimentally and computationally, represents a viable solution for sustainable construction, addressing both climate challenges and energy demands. Future research should focus on scalability and long-term stability, further pushing the boundaries of energy-efficient building materials.

8.5 Packed Bed Cold Storage Using Composite PCM Capsules for High-Temperature Solar Cooling

Cheng et al. [7] explored a packed bed cold storage unit (CSU) utilizing composite phase-change material capsules designed for high-temperature solar cooling applications. Solar cooling technology, while eco-friendly and widely available, suffers from the mismatch between solar energy supply and cooling demand. Integrating cold storage systems can mitigate this issue, with latent cold storage units (LCSUs) being particularly advantageous due to their high energy density and near-isothermal operation during phase transitions.

The study specifically examines a composite PCM—capric and lauric acid eutectic mixture with oleic acid additive (C-L/O PCM)—tailored for phase-change temperatures between 13 and 15 °C, ideal for solar cooling systems supplying chilled water at higher temperatures (16–19 °C). Experimental and mathematical analyses evaluate the system's feasibility, focusing on thermal performance under varying conditions.

8.5.1 Experimental Setup and Procedures

- **PCM Characteristics**

The composite PCM exhibits a latent heat of 93 kJ/kg within the phase-change temperature range, accounting for 63.3% of the total heat capacity. Differential scanning calorimetry (DSC) and T-history methods confirmed its phase transition temperatures—onset at 14.1 °C and peak at 12.6 °C—suitable for solar cooling requirements.

- **Design and Setup**

The CSU comprises 450 spherical capsules (38 mm diameter) filled with composite PCM, arranged within a cylindrical storage tank with cone-type ends to facilitate plug flow. The setup includes a homo-thermal circulator, a chilled-water storage tank, and extensive temperature monitoring with PT100 sensors. Experimental validation was conducted under controlled conditions to measure heat transfer and system efficiency. The design of the packed bed cold storage unit is depicted in Fig. 8.8.

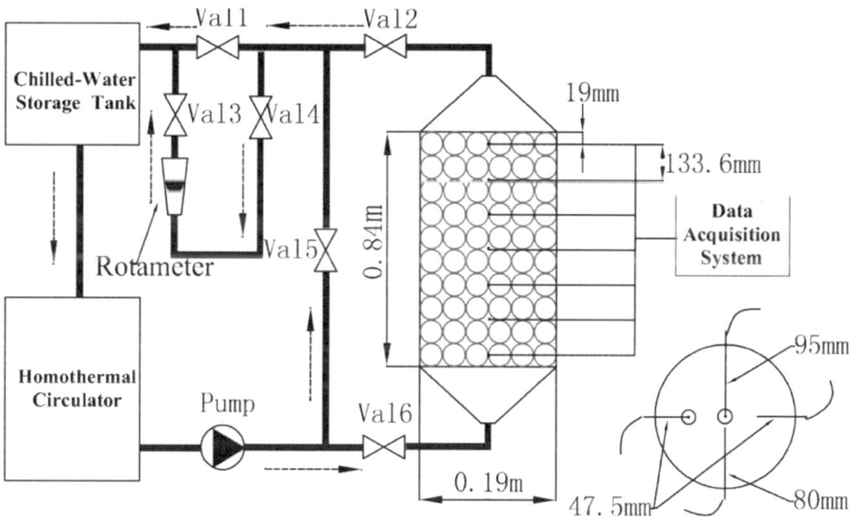

Fig. 8.8 Schematic of the experimental setup [7]

8.5.2 Mathematical Model

A numerical model based on the enthalpy method simulates the system's performance. Key assumptions include:

1. Plug flow and negligible radial heat dissipation.
2. Constant PCM properties outside the phase-change range, varying linearly within it.
3. Perfect thermal insulation of the system.

- **Governing Equations**

Energy conservation in the heat transfer fluid and PCM capsules is modeled by:

– *HTF energy equation*:

$$\varepsilon \rho_f c_f \left(\frac{\partial T_f}{\partial t} + u \frac{\partial T_f}{\partial x} \right) = k_f \frac{\partial^2 T_f}{\partial x^2} + h a_p \left(T_{P|r=R} - T_f \right) \tag{8.17}$$

where

- ε: void fraction of the bed
- T_f and T_p: temperature of HTF and PCM, respectively
- u: mean velocity in the storage unit
- a_p: superficial capsule area per unit bed volume.

– *PCM energy equation (enthalpy form)*:

$$\rho_{\mathrm{p}} \frac{\partial H_{\mathrm{p}}}{\partial t} = k_{\mathrm{p}} \left(\frac{\partial^2 T_{\mathrm{p}}}{\partial r^2} + \frac{2}{r} \frac{\partial T_{\mathrm{p}}}{\partial r} \right) \tag{8.18}$$

Here, $H_{\mathrm{p}}(T)$ is derived from DSC curves.

- **Performance Metrics**

Key metrics include charging rate ($\dot{q} = \dot{v} \rho_{\mathrm{f}} c_{\mathrm{f}} [T_{\mathrm{out}} - T_{\mathrm{in}}]$), accumulated cold stored, and exergetic efficiency. The latter evaluates thermodynamic quality:

$$\eta_{\mathrm{ex}} = \frac{\mathrm{Ex}_{\mathrm{stored}}}{\mathrm{Ex}_{\mathrm{sup}}} = \frac{\mathrm{Ex}_{\mathrm{f}} + \mathrm{Ex}_{\mathrm{p}}}{\mathrm{Ex}_{\mathrm{sup}}} \tag{8.19}$$

where

- $\mathrm{Ex}_{\mathrm{stored}}$: total exergy stored
- $\mathrm{Ex}_{\mathrm{sup}}$: exergy supplied by HTF.

8.5.3 Results and Discussion

- **Validation**

Experimental and numerical data align closely, confirming the model's validity. Small deviations arise from assumed plug flow behavior versus actual turbulent flow in the bed's upper regions.

- **Effect of HTF Temperature**

Lower inlet HTF temperatures (7–12 °C) yield higher charging rates but reduce exergetic efficiency, which improves by 10.4% as temperature increases. The system's maximal charging capacity decreases nonlinearly with temperature due to the PCM's enthalpy distribution. Figure 8.9 illustrates the effects of different inlet HTF temperatures on thermal performance.

- **Effect of Flow Rate**

Higher flow rates (50–250 L/h) improve charging rates up to a threshold but decrease exergetic efficiency. For practical applications, lower flow rates are recommended to prevent chiller overloading during initial charging. Figure 8.10 demonstrates how varying inlet HTF flow rates influence thermal performance.

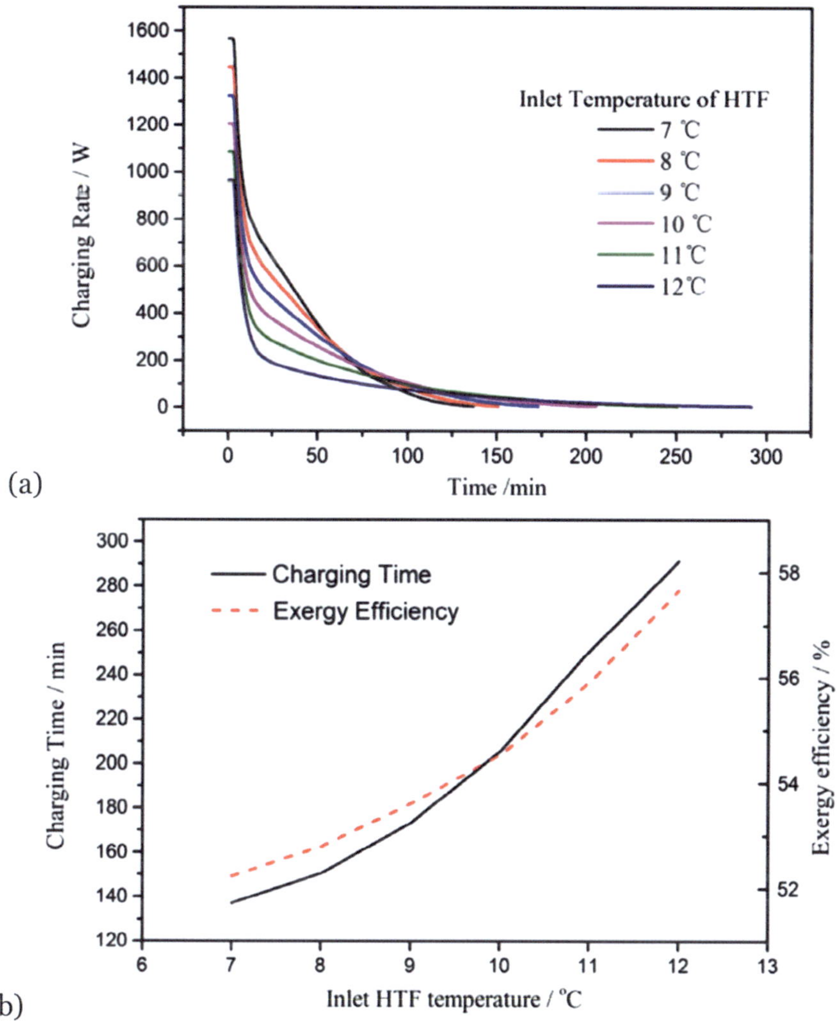

Fig. 8.9 Thermal performance affected by inlet HTF temperature: **a** charging rate, **b** charging time and exergy efficiency [7]

- **Dynamic Solar Conditions**

Simulations with fluctuating chilled water temperatures from a solar adsorption system reveal the CSU's ability to stabilize output, essential for maintaining consistent cooling. Exergetic efficiency favors low flow rates, while the stored cold remains sufficient for typical solar cooling periods.

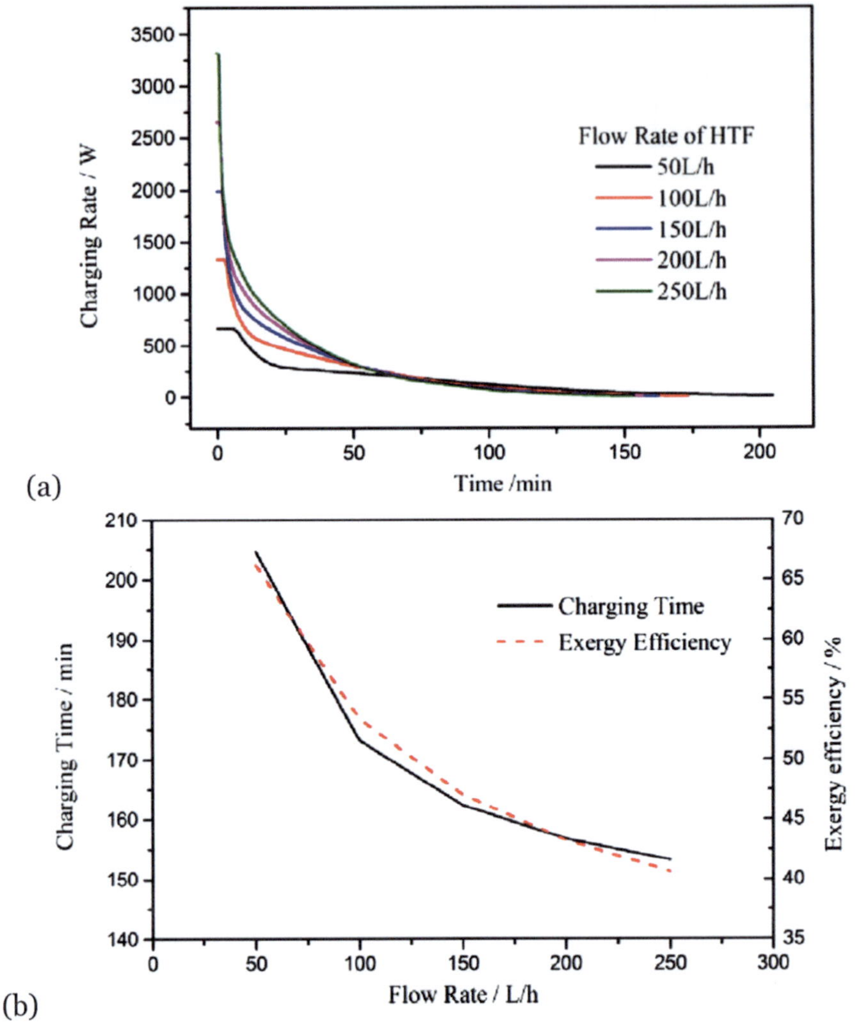

Fig. 8.10 Thermal performance affected by inlet HTF flow rate: **a** charging rate, **b** charging time and exergy efficiency [7]

8.5.4 Criticism in Comparison to Similar Works

The study presents a comprehensive and innovative approach to thermal storage using composite phase-change material capsules in a packed bed design. However, when compared to similar works in the field, a few criticisms and potential areas for improvement can be identified.

8.5.4.1 Focus on Composite PCM Versus Pure PCM Systems

While the composite PCM (C-L/O PCM) addresses the limited availability of pure organic PCMs with appropriate phase-change temperatures, the study narrowly focuses on this specific composite.

- **Criticism**

Other research has demonstrated the viability of eutectic PCMs or multiple PCMs in cascaded designs to improve charging rates and exergy efficiency. For example, Wu et al. [51] showed that cascaded phase-change materials with varying melting points could significantly enhance thermal stratification and system flexibility. The current work does not explore whether alternative PCM configurations or additives might further optimize performance.

8.5.4.2 Simplistic Modeling Assumptions

The mathematical model assumes plug flow behavior and negligible radial heat dissipation, which simplifies the analysis.

- **Criticism**

Similar studies, such as Anderson et al. [2], have employed more complex three-dimensional computational fluid dynamics (CFD) models to capture turbulent flow and non-uniform thermal distributions within the packed bed. These models often reveal nuanced thermal behaviors, such as localized overheating or underutilization of PCM. The current model, while computationally efficient, may oversimplify real-world dynamics, particularly in large-scale or non-ideal flow conditions.

8.5.4.3 Lack of Detailed Economic Analysis

The study primarily evaluates thermal and exergetic performance without delving into cost implications or life-cycle assessments.

- **Criticism**

The practical adoption of thermal storage systems often hinges on cost-effectiveness. Works such as Mehling and Cabeza [32] have emphasized the importance of comparing the cost-to-performance ratio of composite versus pure PCMs, insulation materials, and system designs. The absence of economic metrics, such as payback periods or cost per kWh stored, limits the practical relevance of the study for engineers and decision makers.

8.5.4.4 Limited Discussion of Dynamic Operation

While the study includes a dynamic simulation using fluctuating solar chilled water, it does not deeply explore how the system might perform under broader, real-world scenarios.

- **Criticism**

Systems like those described by Louis et al. [28] in multi-phase thermal energy storage systems evaluate extended operational scenarios, including diurnal and seasonal variations. A deeper exploration of long-term operational stability, especially under intermittent solar conditions, would add practical value to the findings.

8.5.4.5 Underemphasis on Environmental Impact

The environmental benefits of using composite PCMs are acknowledged but not quantified.

- **Criticism**

Recent works, such as Sharma et al. [44], have explored the carbon footprint of PCMs in thermal energy storage systems, showing how material selection impacts overall sustainability. While the study emphasizes the non-toxic and environmentally friendly aspects of composite PCMs, a comparative life-cycle analysis against alternative storage media (e.g., water, salts, or other PCMs) would strengthen its environmental argument.

8.5.4.6 Scale-Up and Integration Challenges

The experimental setup is well controlled but relatively small scale.

- **Criticism**

Works by Yang et al. [56] have demonstrated that scaling up packed bed systems introduces complexities, such as uneven heat distribution, pressure drops, and structural considerations. The current study does not address how its findings would translate to larger installations or systems operating under non-ideal conditions.

8.5.5 Recommendations for Improvement

1. Incorporate a broader comparison of composite and alternative PCM systems, including cascaded or hybrid designs.
2. Employ more detailed numerical simulations, such as CFD, to capture complex flow and heat transfer phenomena.

3. Include a cost–benefit analysis to enhance the practical relevance of the findings.
4. Explore broader dynamic operational scenarios, such as seasonal variations, to assess long-term stability.
5. Quantify environmental impacts through life cycle and carbon footprint analyses.
6. Address scale-up challenges and propose design adjustments for larger or commercial-scale systems.

By addressing these areas, the study could offer more holistic insights and further strengthen its contribution to the field of solar thermal energy storage.

8.5.6 Conclusion

This study validates the packed bed CSU's effectiveness for high-temperature solar cooling, highlighting key findings:

1. Optimal inlet HTF temperature improves system efficiency and capacity.
2. Lower flow rates enhance exergetic efficiency without significantly impacting cold storage.
3. Integration with solar cooling systems ensures stability and adaptability under dynamic conditions.

Future applications can scale the system with multiple units in parallel, optimizing energy use in solar-driven cooling solutions.

8.6 Experimental Evaluation of Low-Temperature PCMs for Refrigeration and Air Conditioning

The refrigeration and air conditioning sector contributes significantly to global electricity consumption (25–30%), with projections indicating rapid growth due to electrification trends. To enhance system efficiency and leverage renewable energy, latent thermal energy storage systems incorporating phase-change materials have been explored. PCMs offer superior energy density compared to sensible storage systems and easier management than thermochemical options.

The study presented by Guarda et al. [17] focuses on low-temperature PCMs (2–9 °C), critical for refrigeration and HVAC applications, particularly in the residential sector and the cold chain for fresh products. Existing studies have addressed PCMs for temperatures exceeding 20 °C but lack data on materials with phase change in the 2–10 °C range. This research fills that gap by systematically analyzing seven commercial PCMs, and evaluating their material compatibility, thermal properties, and phase-change behavior under varying conditions.

8.6.1 Material Selection

Seven PCMs with phase-change temperatures between 2 and 9 °C were selected. These included paraffin waxes, fatty acids, and vegetable oil-based materials. Their thermophysical properties—latent heat, density, specific heat capacity, and thermal conductivity—were compared against water (used as a reference PCM). Notably, paraffin-based PCMs exhibited low thermal conductivity, necessitating enhancements to improve their phase-change efficiency. Table 8.2 lists the PCMs, their nominal phase-change temperature (t_{PC}), PCM type, and main thermophysical properties.

The study highlighted the significant variability in PCM properties, such as dynamic viscosity, which influences heat transfer during the phase-change process. Higher viscosity materials like CR9 exhibit suppressed natural convection, limiting heat transfer.

8.6.2 Compatibility Tests

To ensure system reliability, the compatibility of PCMs with construction materials (aluminum, copper, stainless steel, and plastic) was tested. Samples were immersed in PCMs at 25 °C for two months, with weight and visual inspections conducted pre- and post-testing.

- **Key Findings**
  - Most materials showed negligible changes, indicating compatibility.
  - Exceptions included OM8, which showed plastic absorption (e.g., PLA increased from 2.90 to 4.27 g) and slight copper oxidation, requiring additional testing for long-term use with plastics.

8.6.3 Experimental Setup

Experiments were conducted using aluminum containers (40 × 40 × 60 mm) with internal 3D lattice structures of varying porosities (10, 20, 40 mm, and a reference without lattice). The lattice enhanced thermal conductivity, promoting faster phase changes.

PCMs were subjected to cooling and heating in thermostatic baths set to temperatures 3, 6, or 9 K above or below the PCM phase-change temperature (t_{PC}). The charging phase (solidification) stored cold energy, while the discharging phase (melting) released it. Figure 8.11 presents the schematic view of the experimental setup.

Table 8.2 Main thermophysical properties of selected PCMs [17]

	H_2O/Ice	RT2	savE-OM3	RT4	savE-OM5P	RT5	savE-OM8	CR9
Type	Inorganic	Organic, paraffin	Organic, fatty acid	Organic, paraffin	Organic, paraffin	Organic, paraffin	Organic, fatty acid	Organic, vegetable oil
Nominal t_{PC} (°C)	0	2	3	4	5	5	8	9
t melting (°C)	0	2	4	4	6	5–6	7	9
Latent heat (kJ/kg)	334	115	229	135	215	230	175	220
Density (liq) (kg/m^3)	999	770	835	770	763	760	1060	858
Density (sol) (kg/m^3)	919	880	912	880	727	880	1190	963
Specific heat capacity (liq) (kJ/kg K)	4.22	2	1.91	2	5.95	2	2.11	2.1
Specific heat capacity (sol) (kJ/kg K)	2.03	2	1.76	2	3.86	2	1.71	2.2
Thermal conductivity (liq) (W/m K)	0.597	0.2	0.146	0.2	0.37	0.2	0.118	0.15
Thermal conductivity (sol) (W/m K)	2.2	0.2	0.224	0.2	ND	0.2	0.235	0.24

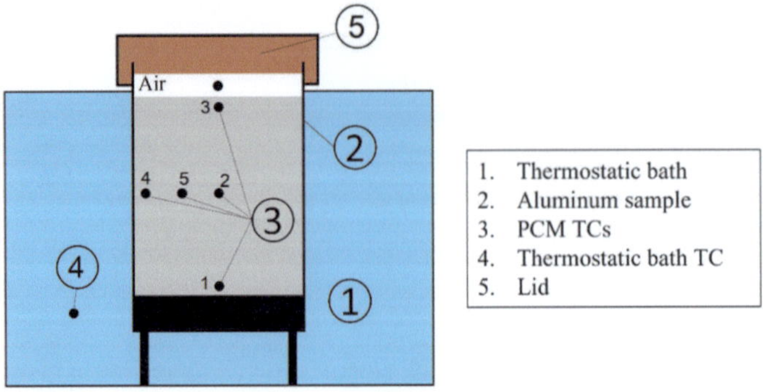

Fig. 8.11 2D sketch of the experimental setup [17]

8.6.4 Governing Equation: Carnot Efficiency

The theoretical efficiency (η) of PCM-based systems was evaluated using the Carnot cycle equation:

$$\eta_{\mathrm{ex}} = \frac{T_{\mathrm{e}}}{T_{\mathrm{c}} - T_{\mathrm{e}}} \tag{8.20}$$

where T_{e} and T_{c} are the evaporator and condenser temperatures, respectively. Selecting PCMs with higher t_{PC} improves energy efficiency by reducing the temperature lift required.

8.6.5 Results and Discussion

- **Phase-Change Behavior**

The thermal conductivity significantly influenced phase-change times:

- Water, with the highest thermal conductivity and latent heat, exhibited the fastest solidification (34 min at $\Delta T = 9$ K).
- Organic PCMs showed slower solidification, hindered by low thermal conductivity. Discharging times were generally faster due to natural convection.

- **Effects of 3D Lattice Structures**

Adding aluminum lattices reduced phase-change times:

- The 10 mm lattice reduced charging times by up to 8$\times$ compared to the reference sample and discharging times by 4$\times$.

- Smaller pore sizes (10 mm) enhanced solidification by counterbalancing the low thermal conductivity, while natural convection dominated melting processes.

- **Impact of Operating Conditions**

Lowering the temperature gradient (ΔT):

- Halving ΔT (from 6 to 3 K) increased charging time by 160%.
- A reduction to one-third ΔT (from 9 to 3 K) quadrupled the phase-change duration.

8.6.6 Criticism Compared to Similar Works

1. *Limited long-term testing*: While compatibility with construction materials was tested for two months, studies like Mehling et al. [31] emphasize longer-term evaluations to account for aging and performance degradation.
2. *Simplistic geometry analysis*: The study focuses on uniform lattice geometries. Research like Kalapala and Devanuri [21] suggests exploring non-uniform or adaptive geometries for optimized heat transfer.
3. *Absence of economic analysis*: No cost-performance assessment was provided, unlike works by Reddy et al. [41], which highlight the need for economic feasibility studies.
4. *Dynamic load analysis*: The study does not simulate variable operational conditions typical in real-world applications, such as fluctuating cooling loads or renewable energy integration, as done in Righetti et al. [42].

Addressing these limitations would enhance the practical relevance of the findings and align them more closely with industry needs.

8.6.7 Conclusion

The study provided critical insights:

1. Thermophysical properties like latent heat, thermal conductivity, and viscosity are crucial in PCM selection.
2. Metal lattice structures significantly enhance phase-change efficiency, with smaller pore sizes yielding faster transitions and uniform temperature fields.
3. Larger ΔT reduces operation times but increases energy costs, suggesting a trade-off for optimal system efficiency.

These findings underscore the importance of comprehensive PCM characterization for accurate thermal energy storage system design.

8.7 Characterization of Nano-Mixed PCM for Solar Cooker Applications

Papade and Kanase-Patil [37] explored the development and characterization of nano-mixed phase-change materials for use in solar cookers. Conventional solar cookers face challenges such as inability to cook during cloudy weather or at night. Energy storage materials, particularly PCMs, can address these limitations by storing solar energy as latent heat. However, traditional PCMs suffer from limitations, including low thermal conductivity and high melting points, which hinder their efficiency.

To enhance the thermal performance of PCMs, this study incorporates aluminum oxide (Al_2O_3) nanoparticles into a binary salt PCM consisting of sodium nitrate ($NaNO_3$) and potassium nitrate ($KNNO_3$) in a 60:40 weight ratio. The goal is to improve the thermal conductivity, heat storage capacity, and operational range of the PCM for effective thermal energy storage and cooking applications.

8.7.1 Materials and Characterization of Aluminum Oxide Nanoparticles

- **Material Selection**

Aluminum oxide (Al_2O_3) nanoparticles were chosen for their:

- High thermal conductivity (12 W/m K).
- Resistance to chemical attacks.
- High melting point and structural stability.
- Cost-effectiveness and availability.

- **Properties of Al_2O_3 Nanoparticles**

Table 8.3 highlights the key properties, including a particle size of 10–20 nm and a specific heat of 0.88 kJ/kg K. These nanoparticles were mixed with the binary PCM in weight fractions of 0.5, 1.0, and 1.5%.

- **Characterization Techniques**

1. *X-ray Diffraction (XRD)*: Used to analyze the crystalline structure and composition of nanoparticles and nano-mixed PCMs.
2. *Scanning Electron Microscopy (SEM)*: Examined the surface morphology, dispersion uniformity, and structural integrity of the nano-mixed PCMs.

Table 8.3 Properties of aluminum oxide (Al_2O_3) nanoparticles [37]

Property	Value
Purity	99.9%
Color	White
Nanoparticles' approximate crystalline size	10–20 nm
Surface area	120–140 m^2/g
Thermal conductivity	12 w/m k
Specific heat	0.880 kJ/Kg, K
Morphology	Spherical
Crystallographic structure	Rhombohedral
Atomic weight	101.96 g/mol
True density	3.97 g/cm^3

8.7.2 Experimental Results and Discussion

- **Thermal Behavior of Nano-Mixed PCMs**

The addition of Al_2O_3 nanoparticles significantly enhanced the heat storage capacity of the PCM. A 1.5% nanoparticle concentration increased the latent heat storage to 219 kJ, outperforming the unmodified PCM.

- **XRD Analysis**

XRD patterns confirmed the crystalline nature of the PCM and the effective dispersion of Al_2O_3 nanoparticles. The diffraction angles and peak intensities indicated minimal changes in the crystal structure, demonstrating the compatibility and stability of the nano-mixed PCM.

- **SEM Analysis**

SEM images revealed a smooth surface morphology for both pure and nano-mixed PCMs. Nano-mixed PCMs showed uniform dispersion of nanoparticles, with no significant agglomeration, ensuring consistent thermal performance.

8.7.3 Governing Equations

The performance of the PCM in the solar cooker system was governed by the principles of heat transfer and thermal storage, expressed as:

$$Q = mc_p\Delta T + mL \tag{8.21}$$

where

- Q: Total heat energy stored (J).

- m: Mass of the PCM (kg).
- c_p: Specific heat capacity (J/kg K).
- ΔT: Temperature change (K).
- L: Latent heat of fusion (J/kg).

The thermal conductivity improvement was modeled by incorporating nanoparticle enhancements using:

$$k_{eff} = k_{PCM}(1 + \phi\beta) \quad (8.22)$$

where

- k_{eff}: Effective thermal conductivity (W/m K).
- ϕ: Nanoparticle volume fraction.
- β: Enhancement factor dependent on nanoparticle properties.

8.7.4 Criticism Compared to Similar Works

8.7.4.1 Limited Range of Nanoparticles

While Al_2O_3 nanoparticles were effective, other nanoparticles like CuO or TiO_2 have shown higher thermal conductivity improvements in similar studies (e.g., [49]). A broader comparative analysis would enhance applicability.

8.7.4.2 Single-PCM System

The study focuses on binary salt PCMs. Works like Konuklu et al. [24] have explored hybrid PCMs with broader temperature ranges, offering more flexibility for varied climatic conditions.

8.7.4.3 Absence of Long-Term Stability Tests

The long-term thermal cycling performance of the nano-mixed PCM was not investigated. Stability under repetitive phase-change cycles is critical for practical deployment, as emphasized by Miqdam and Al-Ajeely [34].

8.7.4.4 Economic Feasibility

While the study highlights the material's effectiveness, it lacks a cost–benefit analysis to evaluate scalability. Studies by Li et al. [27] emphasize balancing performance improvements with economic constraints for large-scale applications.

8.7.4.5 Lack of Dynamic Operational Testing

Dynamic tests, including varying solar intensities and real cooking scenarios, were not performed. Integrating these could align results with real-world applications. Addressing these limitations could further validate the nano-mixed PCM's effectiveness and broaden its applicability across diverse solar thermal systems.

8.7.5 Conclusion

The incorporation of Al_2O_3 nanoparticles into the binary salt PCM enhanced its latent heat storage, thermal conductivity, and overall performance. This improvement enables the solar cooker to store sufficient energy for cooking during cloudy conditions and at nighttime. Key findings include:

1. A 1.5% nanoparticle concentration resulted in the highest improvement in thermal properties.
2. XRD and SEM analyses confirmed the compatibility and uniformity of the nano-mixed PCM.
3. The system efficiently supports both daytime and nighttime cooking.

8.8 Application of Active PCM Storage Systems for Building Energy Efficiency

Buildings are among the largest consumers of energy, and incorporating innovative energy storage systems is critical to reducing their environmental impact. Phase-change materials offer a promising solution, leveraging their ability to store and release latent heat during phase transitions. Gholamibozanjani and Farid [14] explored the application of an active PCM storage system for heating and cooling load reductions in buildings. Experimental results are presented, comparing a PCM-integrated system with a conventional setup under real environmental conditions.

8.8.1 Experimental Setup

- **PCM Heat Storage Units**

The PCM storage units consisted of 19 aluminum compact storage modules (CSMs) filled with RT25HC PCM. Each unit, weighing 9.5 kg, was insulated using PVC/NBR rubber foam. Key properties of the PCM included:

- *Melting range*: 22–26 °C
- *Latent heat of fusion*: 230 kJ/kg.

- **Solar Air Heater and Test Huts**
  - A solar air heater captured and transferred heat to the PCM units or directly to the building interior.
  - Two identical experimental huts, designed with standard lightweight materials and elevated to mimic real-world conditions, were used for testing. Hut 1 served as the reference, while Hut 2 incorporated PCM units.

- **Control System and Data Acquisition**

The system used CompactRIO and LabVIEW for data collection and real-time control, ensuring indoor temperatures remained within comfort ranges. Valve and fan operations directed heat transfer based on pre-programmed algorithms.

8.8.2 *Governing Equations*

- **Heat Transfer in PCM**

The transient heat conduction equation governs the PCM's behavior:

$$\frac{\partial}{\partial x}\left(k\frac{\partial T}{\partial x}\right) + \rho c_{\mathrm{p}}\frac{\partial T}{\partial t} = q \tag{8.23}$$

where

- k: Thermal conductivity
- T: Temperature
- ρ: Density
- c_{p}: Specific heat capacity
- q: Heat source term.

Latent heat effects are accounted for using the enthalpy method:

$$q = \Delta H\frac{\partial \varphi}{\partial t} \tag{8.24}$$

where φ represents the phase fraction of the PCM.

- **Energy Balance for Space Heating and Cooling**

The energy saved or consumed in the building is expressed as:

$$Q = Q_{solar} + Q_{PCM} - Q_{losses} \tag{8.25}$$

where

- Q_{solar}: Heat from the solar air heater
- Q_{PCM}: Heat storage/release by the PCM
  Q_{losses}: Thermal losses due to environmental factors.

8.8.3 *Results and Analysis*

- **Space Heating**
  1. *Solar and PCM integration*:
     - PCM systems stored excess solar heat during the day and released it at night, maintaining indoor temperatures within the comfort range 54% longer than the reference hut.
     - A daily energy savings of 54% was achieved, with an accumulative savings of 10.3% over 28 days of winter.
  2. *Combined electric and solar heating*:
     - PCM reduced peak electric heater use by storing daytime solar heat.
     - Energy savings reached up to 28%, particularly during high electricity cost hours.
- **Space Cooling**
  1. *Free night cooling*:
     - PCM captured cool nighttime air for use during the day, delaying the activation of air conditioning systems.
     - Energy savings ranged from 10 to 60%, depending on solar radiation and outdoor temperatures.
  2. *Seasonal variations*:
     - The system performed best during moderate cooling demands, with 30.4% energy savings in autumn and 10% during peak summer.

8.8.4 *Discussion*

- **Key Insights**
  - PCM's high latent heat capacity allowed for significant load shifting and energy savings.

- Proper charging and discharging cycles were critical for maximizing PCM performance.

- **Challenges**
  - High ambient temperatures in summer limited PCM's effectiveness due to incomplete solidification at night.
  - Optimizing PCM quantity, melting range, and solar collector capacity remains necessary for broader applicability.

8.8.5 Criticism Compared to Similar Works

- **Strengths**
  1. *Comprehensive testing*:
     - Unlike prior studies focused on artificial environments, this study tested PCM under real weather conditions, enhancing its practical relevance.
  2. *Active system design*:
     - The use of solar air heaters and fans for forced convection improved PCM charging efficiency, outperforming passive systems studied by Kuznik and Virgone [25].
- **Limitations**
  1. *Limited cooling performance*:
     - RT25HC PCM had suboptimal performance in high summer temperatures compared to studies by Barzin et al. [5], which used PCMs with adaptive melting points.
  2. *Economic feasibility*:
     - The integration of active components such as fans and controllers increases installation costs, potentially limiting adoption compared to simpler passive PCM systems.
  3. *Incomplete utilization*:
     - Compared to cascaded PCM systems used by Stathopoulos et al. [47], which operate across broader temperature ranges, this system is limited to specific climatic conditions.

8.8.6 Conclusion

This section demonstrates the effectiveness of active PCM systems in reducing heating and cooling loads in buildings. While the system shows promise, optimization of PCM properties and economic analysis are crucial for widespread adoption. By addressing these challenges, future systems could surpass the limitations observed here, further advancing sustainable building practices.

8.9 Numerical Modelling of High-Temperature Latent Heat Thermal Storage for Solar Applications

As global energy demand continues to surge, the need for sustainable and efficient energy management systems has become critical. In particular, addressing the energy supply–demand gap for solar applications poses unique challenges due to the inherent intermittency of solar energy. Latent heat thermal storage systems have emerged as a leading technology to mitigate these challenges, leveraging the high energy density and near-isothermal phase-change properties of phase-change materials. These systems are especially valuable for applications requiring precise thermal management, such as solar-driven cooling and industrial heat recovery.

This section focuses on a novel configuration that integrates an LHTS system with a solar-driven double-effect H_2O/LiBr absorption refrigeration system that was presented by Fan et al. [12]. By selecting hydroquinone as the PCM, the study targets the high-temperature range (150–200 °C), addressing gaps in current research where PCMs in this temperature range have been less explored. This integration aims to efficiently store surplus solar energy during peak periods and release it to meet cooling demands, minimizing reliance on auxiliary heating systems and fossil fuels. The system is modeled numerically using advanced thermodynamic and heat transfer principles to predict its performance under various operational scenarios.

8.9.1 Governing Equations

The mathematical model for the LHTS system incorporates the fundamental principles of transient heat transfer, phase-change dynamics, and thermodynamic efficiency. The governing equations are detailed below.

- **Heat Transfer in the PCM**

$$\frac{\partial h_{\text{PCM}}}{\partial t} = \frac{\partial}{\partial z}\left(\alpha_{\text{PCM}} \frac{\partial h_{\text{PCM}}}{\partial z}\right) + \frac{1}{r}\frac{\partial}{\partial r}\left(\alpha_{\text{PCM}} \cdot r \frac{\partial h_{\text{PCM}}}{\partial r}\right) \tag{8.26}$$

where

- h_{PCM}: Volumetric enthalpy
- α_{PCM}: Thermal diffusivity
- r and z: Radial and axial coordinates.

Equation (8.26) models heat diffusion and phase change in the PCM.

- **Heat Transfer in the Tube**

$$\rho_{\text{tube}} \cdot c_{\text{p,tube}} \frac{\partial T_{\text{tube}}}{\partial t} = \frac{\partial}{\partial z}\left(k_{\text{tube}} \frac{\partial T_{\text{tube}}}{\partial z}\right) + \frac{1}{r}\frac{\partial}{\partial r}\left(r \cdot k_{\text{tube}} \frac{\partial T_{\text{tube}}}{\partial r}\right) \quad (8.27)$$

Equation (8.27) describes the conductive heat transfer within the metallic tube walls of the heat exchanger.

- **Heat Transfer in the HTF**

$$\rho_{\text{HTF}} \cdot c_{\text{p,HTF}} \cdot V \frac{\partial T_{\text{HTF}}}{\partial t} = \dot{m}_{\text{HTF}} \cdot c_{\text{p,HTF}} \cdot dT_{\text{HTF}} - h_{\text{conv}} \cdot A_i (T_{\text{HTF}} - T_{\text{tube}}) \quad (8.28)$$

The HTF serves as the medium for energy transport, exchanging heat with the PCM through the tube walls.

- **Enthalpy–Temperature Relationship for Non-isothermal Phase Change**

$$h(T) = \begin{cases} \int_{T_0}^{T} \rho \cdot c_{\text{p,s}}(T)\text{d}T & \text{for } T \le T_{m1} \\ \int_{T_0}^{T_{m1}} \rho \cdot c_{\text{p,s}}(T)\text{d}T + \int_{T_{\text{m1}}}^{T} \rho \frac{\partial \Delta H}{\partial T}\text{d}T & \text{for } T_{m1} \le T \le T_{m2} \\ \int_{T_0}^{T_{m1}} \rho \cdot c_{\text{p,s}}(T)\text{d}T + \rho\Delta H + \int_{T_{\text{m2}}}^{T} \rho \cdot c_{\text{p,l}}(T)\text{d}T & \text{for } T > T_{m2} \end{cases} \quad (8.29)$$

Equation (8.29) captures the enthalpy dynamics of the PCM during solid–liquid phase transitions over a temperature range.

- **Boundary and Initial Conditions**
  - Initial conditions: Uniform system temperature T_{initial}.
  - Convective heat transfer between the HTF and the tube wall:

$$k \frac{\partial T^j_{\text{tube}}}{\partial r} = h_{\text{conv}}(T_{\text{tube}} - T_{\text{HTF}}) \quad \text{at } 0 \le z \le L \text{ and } r = r_i \quad (8.30)$$

  - Heat transfer between the PCM and tube is purely conductive:

$$k\frac{\partial T^j_{\text{tube}}}{\partial r} = k\frac{\partial T^j_{\text{PCM}}}{\partial r} = \overline{k}\frac{\partial}{\partial r}\left(T^j_{\text{PCM}} - T^j_{\text{tube}}\right) \quad \text{at } 0 \le z \le L \text{ and } r = r_o \tag{8.31}$$

- There is no heat transfer through the outer walls.

These equations are solved numerically using finite difference methods, with implicit time discretization for stability and accuracy.

8.9.2 *System Design and Operation*

- **System Description**

The solar-driven absorption system integrated with a LHTS unit is a sophisticated energy solution designed for efficient cooling. The system architecture, as illustrated in Fig. 8.12, comprises several key components: solar collectors, an LHTS system, an auxiliary gas burner, and the absorption refrigeration system. These elements work synergistically to harness solar energy, store excess heat, and meet variable cooling demands. The primary energy source for the system is solar radiation, captured by parabolic trough collectors. These collectors concentrate sunlight to generate high temperatures, converting solar radiance into thermal energy. This energy is absorbed by a heat transfer fluid, which flows through the absorber tubes within the collectors. The heated HTF is then transported to the high-pressure generator of the absorption system, where it facilitates the cooling process by driving the double-effect H_2O/LiBr refrigeration cycle.

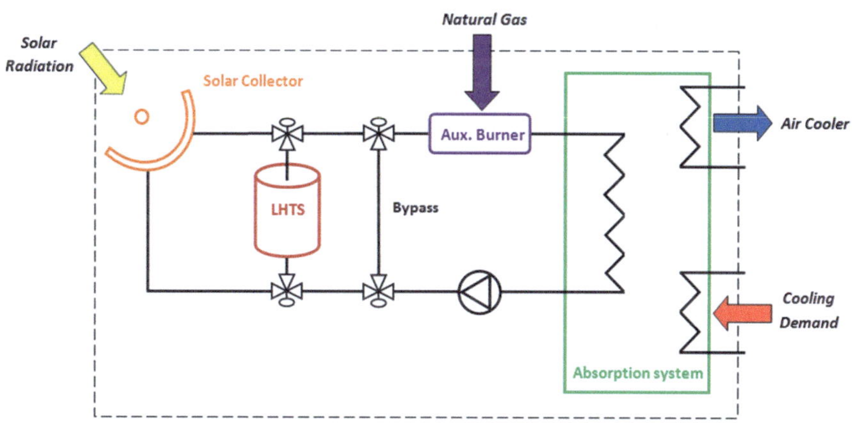

Fig. 8.12 The process flow diagram of the solar-driven absorption system, with the system boundary indicated by a dashed line [12]

To maximize efficiency, the system incorporates an LHTS unit to store surplus solar energy. When the solar energy supply exceeds the immediate cooling demand, a portion of the HTF is diverted to the LHTS system, where the energy is stored as latent heat in PCMs. This stored energy can be retrieved later to meet cooling requirements during periods of low solar radiation. The auxiliary gas burner serves as a backup, ensuring a reliable energy supply by heating the HTF when solar energy is insufficient.

The system is designed as a closed loop, with the HTF returning from the high-pressure generator to the solar collectors to complete the cycle. A bypass mechanism is included to regulate flow and maintain optimal operation under varying conditions. The operation of the system is influenced by multiple factors. The primary inputs include:

- *Cooling demand*: Estimated hourly requirements based on building specifications and environmental conditions.
- *Solar radiance*: The availability of direct sunlight, which varies with time and weather.
- *Ambient temperature*: Affects the efficiency of the solar collectors, as well as the absorber and condenser performance in the air-cooled absorption system.

The outputs of the system are:

- *Supplement energy demand*: The auxiliary gas burner provides additional energy when necessary.
- *LHTS temperature profile*: Reflects the stored energy and its readiness for retrieval.

- **Flow Scenarios and Operational Modes**

The system adapts to five distinct flow scenarios, categorized as switches 1–5, based on the interaction between solar energy availability and system requirements. These scenarios, depicted in Fig. 8.13, are as follows:

1. *Switch 1*: This mode activates when no solar energy is available, and the HTF temperature at the high-pressure generator outlet is lower than the LHTS average temperature. Heat is extracted from the LHTS to meet the cooling demand.
2. *Switch 2*: Also occurring in the absence of solar energy, this scenario is activated when the HTF temperature at the high-pressure generator outlet is higher than the LHTS average temperature, allowing for direct use of the HTF.
3. *Switch 3*: This mode represents a surplus of solar energy. Excess heat from the HTF is directed to the LHTS system, where it is stored for later use.
4. *Switch 4*: Activated when the available solar energy is insufficient to fully meet the high-pressure generator's energy demand, but the HTF temperature remains higher than the LHTS average temperature. Stored energy in the LHTS is supplemented with solar heat to fulfill the cooling requirement.
5. *Switch 5*: Occurs when solar energy is insufficient and the HTF temperature at the high-pressure generator outlet is lower than the LHTS average temperature.

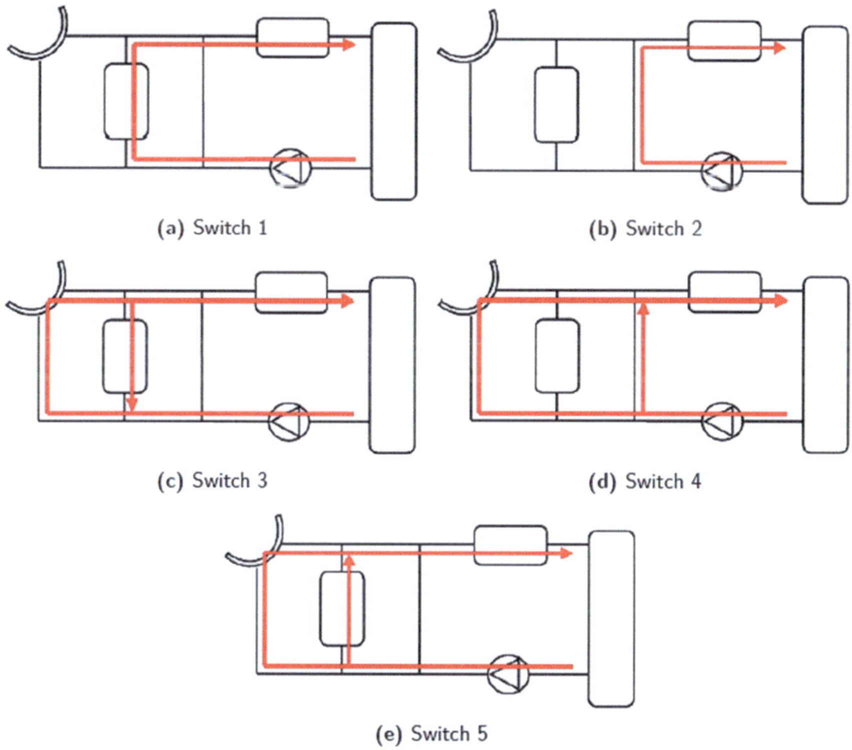

Fig. 8.13 Five possible flow distributions of the system [12]

In this case, stored energy from the LHTS and auxiliary heating from the gas burner are combined to maintain the required temperature.

- **LHTS System Description**

The LHTS system is based on a shell-and-tube heat exchanger as shown in Fig. 8.14. The HTF circulates through a bundle of tubes, while the PCM is stored in the shell. Hydroquinone was selected for its favorable thermal properties, including a latent heat of 230 kJ/kg and minimal supercooling during phase change. The system operates cyclically:

- *Charging phase*: Excess solar energy heats the HTF, causing the PCM to melt and store energy.
- *Discharging phase*: During reduced solar energy availability, the HTF absorbs heat from the solidifying PCM to maintain the cooling demand.

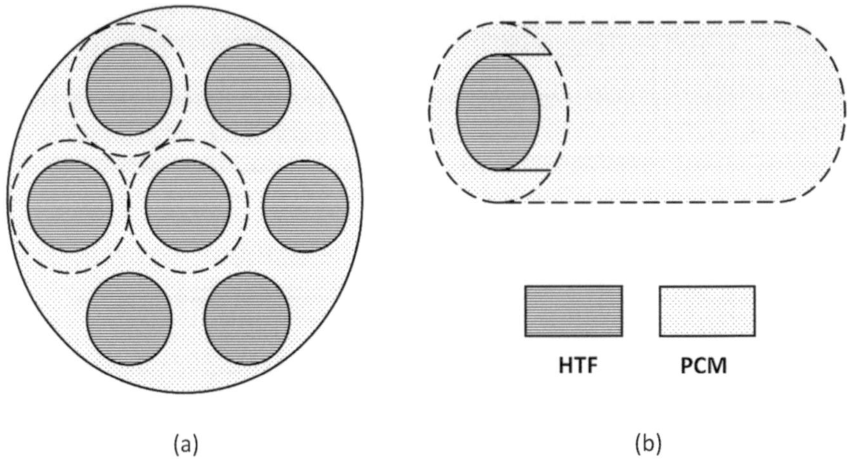

Fig. 8.14 **a** The LHTS unit and **b** the two concentric tubes system [12]

8.9.3 Performance Evaluation

- **Constant Cooling Load**

Under a constant 100 kW cooling load, the system efficiently stored solar energy during daylight hours and extended cooling operations by two hours post-sunset without auxiliary heating.

- **Variable Cooling Load**

With variable cooling demands, the system dynamically adjusted energy storage and release, minimizing auxiliary energy requirements. This flexibility highlights the potential of LHTS systems to adapt to real-world operational conditions.

- **Impact of Operating Temperatures**

Operating the HTF at lower temperatures improved solar collector efficiency but required more auxiliary energy during peak demand. Conversely, higher HTF temperatures enhanced system performance by reducing reliance on auxiliary heating.

- **Enhanced Storage Capacity**

By increasing the number of LHTS units from 24 to 44, the system provided continuous cooling without auxiliary energy inputs, demonstrating scalability and enhanced operational sustainability.

8.9.4 Criticism and Comparison to Similar Works

1. *Modeling assumptions*:
    - The study neglects long-term PCM degradation, which could affect thermal performance.
    - Natural convection was modeled using empirical approximations, which might not fully capture complex flow behaviors.
2. *Material selection*: While hydroquinone exhibits reliable phase-change properties, other materials such as sugar alcohols (e.g., galactitol) offer higher latent heat capacities and could enhance energy storage density.
3. *Comparison to existing studies*:
    - Gil et al. [15, 16] explored a broader range of PCMs but lacked detailed integration with solar absorption systems.
    - Medrano et al. [30] emphasized experimental validation but did not address high-temperature applications.

8.9.5 Conclusion and Future Directions

This study provides a comprehensive framework for designing and optimizing high-temperature LHTS systems for solar-driven cooling applications. The integration of hydroquinone-based PCMs with a double-effect absorption system demonstrates the potential for sustainable and efficient energy storage solutions. Future research should focus on:

- Developing advanced PCMs with higher thermal conductivity and latent heat capacities.
- Incorporating experimental validations for complex phenomena like natural convection.
- Scaling up the system for industrial applications and conducting long-term operational assessments.

By addressing these areas, LHTS systems can further solidify their role in the renewable energy landscape, contributing to a more sustainable and energy-efficient future.

8.10 Advanced Thermal Energy Storage: Cascaded PCM Configurations

With the escalating global energy demand, particularly in sectors like building, transportation, and industrial processes, managing renewable energy sources has become a paramount concern. While abundant, renewable energy sources such as solar and wind are inherently intermittent and unpredictable. This intermittency often creates mismatches between energy supply and demand. Latent heat thermal energy storage systems have gained prominence as a potential solution to bridge this gap, offering the ability to store energy during surplus periods and release it during peak demand. LHTES systems leverage phase-change materials, which store and release energy through phase transitions—typically between solid and liquid states. These systems are particularly valued for their high energy storage density and nearly isothermal operation. Despite these advantages, traditional single PCM systems exhibit limitations, including low thermal conductivity and inefficiencies when operating across broader temperature ranges.

The advent of cascaded PCM configurations represents a significant advancement in thermal energy storage. By utilizing multiple PCMs with different phase-change temperatures, cascaded systems can optimize energy storage and retrieval, enhance thermal performance, and extend operational applicability. Christopher et al. [8] explored the fundamental principles, thermodynamic evaluations, and optimization techniques for cascaded PCM systems, providing a detailed review of their potential and limitations.

8.10.1 Cascaded PCM Overview

The primary challenge of reducing the charging/discharging rate due to diminished temperature driving potential at the bottom of the LHTES tank can be mitigated by strategically placing multiple PCMs, a setup known as the cascaded configuration. PCMs are selected to ensure that the temperature driving potential is maintained, enabling uniform heat flux during charging and discharging, thereby enhancing the overall system performance. Compared to traditional latent heat energy storage systems, only a limited number of studies have explored the cascaded configuration [39]. It is evident that selecting PCMs with optimal phase-change properties and thermal transport characteristics is crucial for improving the thermal performance of the LHTES system. Additionally, arranging the PCMs in the correct sequence is vital to prevent temperature driving potential failure (the temperature difference between the melting temperature of PCMs and the heat transfer fluid), which can lead to reduced system performance [46].

Figure 8.15 illustrates a typical cascaded LHTES system, showing the charging and discharging processes for both hot and cold energy storage applications. In hot thermal energy storage, during the charging process, PCMs are arranged in decreasing

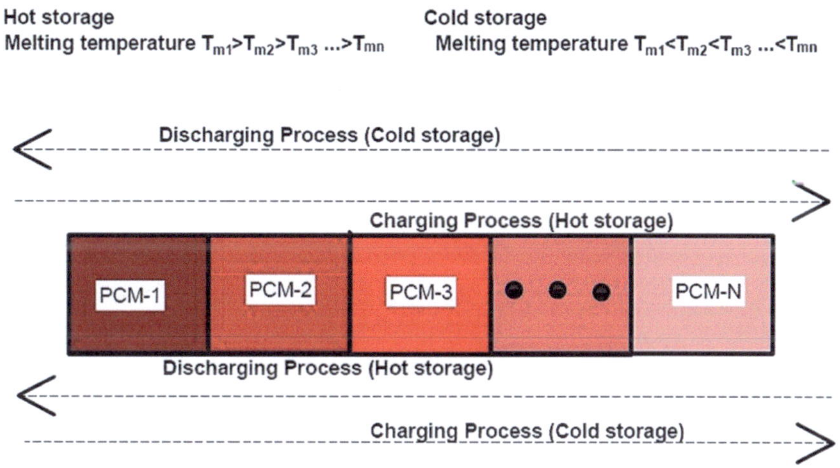

Fig. 8.15 Multiple layers of PCMs [8]

order of their phase-change temperature (melting) along the flow direction of the HTF, while the HTF flows in the reverse direction during discharging. For cool thermal energy storage, the process is reversed, as depicted in Fig. 8.15.

8.10.2 Governing Concepts and Equations

- **Energy and Exergy Analyses**

Energy analysis aims to evaluate the behavior, performance, and efficiency of systems for energy conversion. However, it is evident that energy analysis alone cannot identify and assess the irreversibility of the system. To address this limitation, the concept of exergy analysis has been introduced. Exergy analysis provides insights into the limitations of TES systems by directly indicating thermal exergy loss. This section aims to elucidate how energy and exergy concepts have been applied to cascaded/multiple PCMs and the practicality of such analyses. The overall energy balance during the charging process is given by:

$$\sum_{j=1}^{n} m_{\mathrm{HTF}} c_{\mathrm{p}} \left(T_{\mathrm{c},j+1} - T_{\mathrm{c},j} \right) - Q_{\mathrm{loss}} = \sum_{j=1}^{n} \left[m_{\mathrm{PCM},j} \left(\lambda T_{\mathrm{PCM},j} - \lambda T_{\mathrm{PCM,initial},j} \right) \right] \quad (8.32)$$

Heat loss between the TES and the ambient environment is considered through conductive loss via the insulation material. Convective heat transfers from the outer surface of the TES system are neglected, assuming the TES system is well insulated and the temperature difference between the TES tank surface and the ambient temperature is negligible.

$$Q_{\text{loss}} = Q_{\text{cod,ins}} = \frac{k_{\text{insulation}} \times A_{\text{TES section}} \times \left(\frac{\left(T_{\text{avg,HTF}} + T_{\text{avg,PCM}} \right)}{2} - T_{\text{e}} \right)}{l} \quad (8.33)$$

Energy efficiency during the charging process is defined by:

$$\eta_{\text{c}} = \frac{\sum_{j=1}^{n} \left[m_{\text{PCM},j} \left(\lambda T_{\text{PCM},j} - \lambda T_{\text{PCM,initial},j} \right) \right]}{\sum_{j=1}^{n} m_{\text{HTF}} c_{p} \left(T_{j+1} - T_{j} \right)} \quad (8.34)$$

where

- j: Stage number of the PCM
- T_{e}: Ambient temperature
- $T_{\text{c},j+1}$ and $T_{\text{c},j}$: Outlet and inlet temperatures of the HTF during the charging process
- T_{PCM}: Temperature of the PCM
- $k_{\text{insulation}}$: Thermal conductivity of the insulation
- A_{TES}: Cross-sectional area of the TES tank
- m_{PCM}: Mass of the PCM
- λ: Latent heat of the PCM
- m_{HTF}: Mass of the HTF
- l: Thickness of the insulation material.

The overall energy balance during the discharging process is expressed in Eq. (8.35). Similarly, energy efficiency during discharging is denoted by:

$$Q_{\text{PCM}} = \sum_{j=1}^{n} m_{\text{HTF}} c_{\text{p}} \left(T_{\text{d},j-1} - T_{\text{d},j} \right) + \sum_{\text{HTF}=1}^{n} \sum_{\text{PCM}=1}^{n} \frac{k_{\text{insulation}} \times A_{\text{TES section}} \times \left(\frac{\left(T_{\text{avg,HTF}} + T_{\text{avg,PCM}} \right)}{2} - T_{\text{e}} \right)}{l} \quad (8.35)$$

$$\eta_{\text{d}} = \frac{\sum_{j=1}^{n} m_{\text{HTF}} c_{\text{p}} \left(T_{\text{d},j-1} - T_{\text{d},j} \right)}{\sum_{j=1}^{n} \left[m_{\text{PCM},j} \left(\lambda T_{\text{PCM,initial},j} - \lambda T_{\text{PCM},j} \right) \right]} \quad (8.36)$$

where $T_{\text{d},j-1}$ and $T_{\text{d},j}$ are the outlet and inlet temperatures of the HTF during the discharging process.

Equation (8.37) represents the overall energy efficiency of the cascaded TES:

$$\eta = \frac{\sum_{j=1}^{n} \left(T_{\text{d},j-1} - T_{\text{d},j} \right)}{\sum_{j=1}^{n} \left(T_{\text{c},j+1} - T_{\text{c},j} \right)} \quad (8.37)$$

The overall exergy balance of the charging and discharging processes of the storage system is indicated by:

$$\mathrm{Ex_{in}} - \mathrm{Ex_{loss}} - I = \mathrm{Ex_{acc}} \tag{8.38}$$

where

- $\mathrm{Ex_{in}}$: Input,
- $\mathrm{Ex_{loss}}$: Loss,
- I: Consumption and
- $\mathrm{Ex_{acc}}$: Accumulation exergies.

The overall exergy balance and exergy efficiency during the charging process are respectively given by:

$$\begin{aligned}&\sum_{j=1}^{n} m_{\mathrm{HTF}}c_{\mathrm{p}}\left[\left(T_{\mathrm{c},j+1}-T_{\mathrm{c},j}\right)-T_{\mathrm{e}}\ln\frac{T_{\mathrm{c},j+1}}{T_{\mathrm{c},j}}\right]\\&\quad-\sum_{j=1}^{n} Q_{\mathrm{in},j}\left[\frac{1}{\frac{m_{\mathrm{PCM},j}}{m_{\mathrm{PCM}}}T_{\mathrm{PCM},j}}-\frac{1}{\frac{T_{\mathrm{c},j+1}+T_{\mathrm{c},j}}{2}}\right]T_{\mathrm{e}}-\mathrm{Ex_{loss}}\\&\quad=\sum_{j=1}^{n} Q_{\mathrm{PCM},j}\left[1-\frac{T_{\mathrm{e}}}{T_{\mathrm{PCM},j}}\right]\end{aligned} \tag{8.39}$$

$$\psi_{\mathrm{c}}=\frac{\sum_{j=1}^{n} Q_{\mathrm{PCM},j}\left[1-\frac{T_{\mathrm{e}}}{T_{\mathrm{PCM},j}}\right]}{\sum_{j=1}^{n}\left[m_{\mathrm{HTF}}c_{\mathrm{p}}\left(T_{\mathrm{c},j+1}-T_{\mathrm{c},j}\right)-T_{\mathrm{e}}\ln\frac{T_{\mathrm{c},j+1}}{T_{\mathrm{c},j}}\right]} \tag{8.40}$$

The same approach applied to the charging process is used for the discharging process. The overall exergy balance and exergy efficiency during discharging are given by Eqs. (8.41) and (8.42), respectively.

$$\begin{aligned}&\sum_{j=1}^{n} m_{\mathrm{HTF}}c_{\mathrm{p}}\left[\left(T_{\mathrm{d},j-1}-T_{\mathrm{d},j}\right)-T_{\mathrm{e}}\ln\frac{T_{\mathrm{d},j-1}}{T_{\mathrm{c},j}}\right]\\&\quad-\sum_{j=1}^{n} Q_{\mathrm{in},j}\left[\frac{1}{\frac{m_{\mathrm{PCM},j}}{m_{\mathrm{PCM}}}T_{\mathrm{PCM},j}}-\frac{1}{\frac{T_{\mathrm{d},j}+T_{\mathrm{d},j-1}}{2}}\right]T_{\mathrm{e}}-\mathrm{Ex_{loss}}\\&\quad=\sum_{j=1}^{n} Q_{\mathrm{PCM},j}\left[1-\frac{T_{\mathrm{e}}}{T_{\mathrm{PCM},j}}\right]\end{aligned} \tag{8.41}$$

$$\psi_{\mathrm{d}}=\frac{\sum_{j=1}^{n} m_{\mathrm{HTF}}c_{\mathrm{p}}\left[\left(T_{\mathrm{d},j-1}-T_{\mathrm{d},j}\right)-T_{\mathrm{e}}\ln\frac{T_{\mathrm{d},j-1}}{T_{\mathrm{d},j}}\right]}{\sum_{j=1}^{n} Q_{\mathrm{in},j}\left[1-\frac{T_{\mathrm{e}}}{T_{\mathrm{PCM},j}}\right]} \tag{8.42}$$

The overall exergy efficiency of the cascaded LHTES is presented as follows:

$$\psi = \frac{\sum_{j=1}^{n}\left[\left(T_{\mathrm{d},j-1} - T_{\mathrm{d},j}\right) - T_{\mathrm{e}} \ln \frac{T_{\mathrm{d},j-1}}{T_{\mathrm{d},j}}\right]}{\sum_{j=1}^{n}\left[\left(T_{\mathrm{c},j+1} - T_{\mathrm{c},j}\right) - T_{\mathrm{e}} \ln \frac{T_{\mathrm{c},j+1}}{T_{\mathrm{c},j}}\right]} \quad (8.43)$$

- **Entransy Dissipation Theory**

The theory of entransy and its dissipation has garnered significant attention for optimizing and evaluating the performance of various heat transfer modes and sources, including heat exchangers, steam hybrid thermal networks, finless evaporative air cooler systems, and HVAC systems.

First introduced by Guo et al. [19], the concepts of entransy and its dissipation measure the heat transfer capacity of a material. The term "entransy" combines "energy" (en-) and "transferability" (-transy). Entransy (G) is defined as the product of the material's internal energy (U) and temperature (T), divided by 2:

$$G = \frac{UT}{2} \quad (8.44)$$

To evaluate the thermal performance of the LHTES unit based on entransy and its dissipation, Guo et al. [19] conducted an entransy balance during the charging process, expressed as follows:

$$G_{\mathrm{E}} = G_{\mathrm{PCM}} + G_{\mathrm{loss},\phi} \quad (8.45)$$

where

- G_{E}: Overall entransy heat transfer rate
- G_{loss}: Heat dissipation loss rate
- ϕ: Entransy dissipation rate
- G_{PCM}: Entransy storage rate.

Figure 8.16 illustrates the entransy balance for the PCM medium during the charging process.

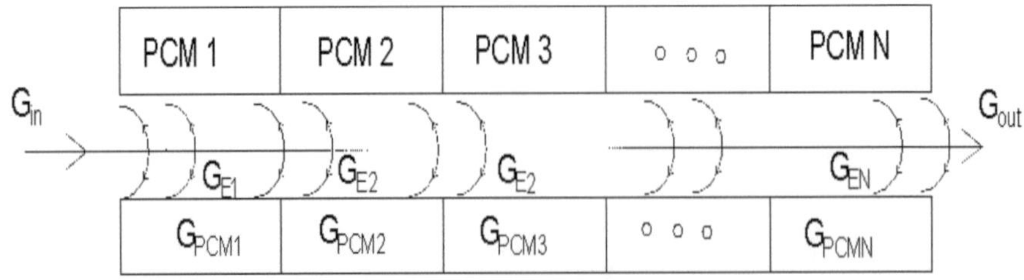

Fig. 8.16 Schematic diagram for the cascaded LHTES [8]

The entransy heat transfer rate from HTF to PCMs can be estimated using the equation provided by Guo et al. [19]:

$$G_{\mathrm{E}} = G_{\mathrm{out}} - G_{\mathrm{in}} = \frac{1}{2}\dot{m}cT_{\mathrm{out}}^2 - \frac{1}{2}\dot{m}cT_{\mathrm{in}}^2 \tag{8.46}$$

Guo et al. [19] further integrated Eq. (8.47) to evaluate the cumulative entransy transferred between HTF and PCMs (G_{CE}):

$$G_{\mathrm{CE}} = \int_0^t G_{\mathrm{E}}\mathrm{d}t \tag{8.47}$$

The stored entransy rate in the PCMs (G_{CPCM}) is evaluated as follows:

$$G_{\mathrm{CPCM}} = \sum_{j=1}^{n} G_{\mathrm{PCM},j} = \sum_{j=1}^{n}\left(\frac{1}{2}Q_{\mathrm{PCM},j}T_{\mathrm{PCM},j}\right) \tag{8.48}$$

The efficiency of entransy (η_{e}) is defined as the ratio of the cumulative entransy stored in PCMs to the cumulative entransy transferred from the HTF to PCMs:

$$\eta_{\mathrm{e}} = \frac{\int_0^t G_{\mathrm{CPCM}}}{G_{\mathrm{total}}} \tag{8.49}$$

These derived entransy equations describe the heat transfer capability during the process as a measure of irreversibility. Entransy is an alternative thermodynamic evaluation criterion for system performance analysis and optimization, distinct from entropy analysis. The use of entransy to optimize cascaded thermal storage systems is detailed in the study by Xu and Zhao [54]. This concept focuses on heat transfer rather than heat-work conversion.

8.10.3 Cascaded Configurations and Thermal Performance

- **Design Principles**

In cascaded LHTES systems, PCMs are strategically layered to maintain a high thermal gradient, enhancing heat transfer efficiency. Layers are arranged with progressively increasing phase-change temperatures to ensure effective energy transfer throughout the system. This configuration reduces thermal resistance and enables uniform charging and discharging rates.

- **Key Insights from Experimental Studies**

Experimental investigations reveal several performance enhancements:

- Cascaded systems with 3–5 PCM layers typically outperform single PCM systems by 20–40% in terms of energy efficiency.
- Properly selected PCM phase-change temperatures reduce thermal resistance, leading to faster energy storage and retrieval.
- Innovations such as fins, metal foams, and encapsulation in cylindrical or spherical shapes further augment heat transfer.

- **Optimization Techniques**

Numerical simulations using tools like FLUENT and COMSOL have been instrumental in optimizing cascaded PCM arrangements. These studies emphasize:

- *PCM layer count*: While increasing the number of PCM layers improves thermal performance, diminishing returns occur beyond five layers.
- *Thermal conductivity enhancements*: Incorporating additives like nanoparticles or employing advanced encapsulation geometries significantly enhances conductivity.

8.10.4 Governing Equations for Cascaded Optimization

Mathematical models are important for designing systems that store thermal energy using cascaded latent heat storage. For example, the best melting temperatures of phase-change materials are often arranged in a way that follows an exponential or linear pattern along the flow of the heat transfer fluid. This setup helps to get the most energy out while keeping the system stable.

Watanabe and Kanzawa [50] showed that when there is a big difference in temperature between the PCMs in each layer, the rates at which energy is stored (charging) and released (discharging) improve. However, they also found that if this temperature difference becomes too large, the time it takes for the PCMs to melt and solidify increases. This is because there are optimal melting temperatures that should not be exceeded. The best temperature differences between the PCMs can change depending on the temperature of the incoming HTF and the starting temperature of the PCM. They provided formulas to estimate the best melting temperatures for high, middle, and low PCMs during charging and discharging processes, based on the number of heat transfer units (NTU):

For the charging process:

$$T_{\mathrm{m(top)}} - T_{\mathrm{m(low)}} = \frac{\mathrm{NTU}}{1 + (\mathrm{NTU}/2)} \left(T_{\mathrm{HTF,in}} - T_{\mathrm{m(middle)}} \right) \tag{8.50}$$

$$\tau = t \times \frac{4h \times \left(T_{\mathrm{HTF,in}} - T_{\mathrm{m(middle)}}\right)}{L^{*}\rho D} \tag{8.51}$$

For the discharging process:

$$T_{\mathrm{m(top)}} - T_{\mathrm{m(low)}} = \frac{\mathrm{NTU}}{1 + (\mathrm{NTU}/2)}\left(T_{\mathrm{m(middle)}} - T_{\mathrm{HTF,in}}\right) \tag{8.52}$$

$$\tau = t \times \frac{4h \times \left(T_{\mathrm{m(middle)}} - T_{\mathrm{HTF,in}}\right)}{L^{*}\rho D} \tag{8.53}$$

where

- $L^{*} = L + c_{\mathrm{p}}\left(T_{\mathrm{m(middle)}} - T_{\mathrm{HTF,in}}\right)$: modified latent heat, including sensible heat
- D: diameter of the capsules
- ρ: density of the PCM in solid and liquid phases
- $T_{\mathrm{m(top,middle,low)}}$: melting temperatures of the PCMs placed at the top, middle, and bottom of the cascaded TES unit.

8.10.5 Criticism and Comparison

8.10.5.1 Challenges

Despite their advantages, cascaded PCM systems face several challenges:

1. *Material Limitations*: The selection of suitable PCMs with high latent heat and thermal conductivity remains a bottleneck.
2. *Complexity*: Designing cascaded systems with multiple layers and varying thermal properties is both technically challenging and costly.
3. *Operational Stability*: Ensuring the long-term stability of PCMs, particularly under repeated phase changes, requires further research.

8.10.5.2 Comparison to Similar Studies

Previous reviews, such as Narasimhan [35], provided broad insights into cascaded systems but lacked rigorous mathematical and experimental validation. Zayed et al. [57] focused on specific applications, like solar energy, but didn't explore diverse configurations. This chapter advances the field by integrating theoretical, numerical, and practical perspectives, offering a holistic approach to cascaded LHTES systems.

8.10.6 Conclusion and Future Directions

Cascaded PCM configurations have demonstrated significant potential in enhancing the efficiency and applicability of thermal energy storage systems. Future research should focus on:

- Developing novel PCMs with superior thermal properties.
- Exploring cost-effective manufacturing methods for complex arrangements.
- Conducting long-term field studies to validate theoretical predictions.

By addressing these challenges, cascaded LHTES systems could play a pivotal role in achieving sustainable energy storage solutions.

8.11 Scalability of Composite PCM Capsules

Composites containing phase-change materials, especially those with integrated liquid metals, stand out as effective solutions for thermal energy storage and management. The work of Wang et al. [48] showcases a scalable micro-encapsulated PCM–LM composite for CPU cooling, which indicates the potential for wider uses in systems like solar cooling. Still, scalability depends on overcoming fabrication, material, and integration hurdles. This discussion examines these challenges specifically regarding mass production and integration into solar cooling systems, informing the analysis from the highlighted work.

8.11.1 Fabrication Challenges in Mass Production

Encapsulation is the single most important feature in determining the scalability of PCM-LM composites. Encapsulation interfaces using techniques like polymerization or coacervation about PCM suffer due to low size uniformity, low encapsulation rates, and even toxicity from residual monomers. Wang et al. cured these problems in this case by fabricating monodisperse calcium alginate-shelled PCM microparticles in the co-flowing microfluidic chip. Achieving a droplet generation rate of one capsule every ~ 43.4 s, this method can be considered scalable. Even so, these processes developed at smaller scales are directly transferable and tackle throughput problems. Reproducibility also poses problems because microfluidic systems are notoriously poor for parallelization. Controlled linear flow becomes a problem when dealing with high volume.

Furthermore, the integration of LM into the composite introduces additional complexity. The filling process described by Wang et al. [48] (layering PCM capsules with LM and foam copper) requires precise control to prevent buoyancy-driven segregation. Scaling this process would demand automated systems to ensure uniform distribution and avoid voids, which could degrade thermal performance.

8.11.2 Material and Cost Considerations

Material selection critically impacts scalability. Gallium-based LMs, such as eutectic gallium-indium (EGaIn), offer high thermal conductivity (~ 27.7 W/m K) and low melting points, making them ideal for thermal management [48]. However, gallium is expensive and subject to supply chain volatility, posing economic challenges for large-scale deployment.

The calcium alginate shell, while cost-effective and non-toxic, may lack mechanical robustness under prolonged thermal cycling. Wang et al. [48] reported a PCM mass fraction of 84.4% within capsules, but long-term durability in solar cooling systems, where temperature fluctuations are frequent, remains untested. Additionally, the use of foam copper to suppress buoyancy adds material costs and complicates recyclability.

8.11.3 Integration with Solar Cooling Systems

Solar cooling systems require PCM composites to absorb and release heat cyclically, often under variable solar irradiance. The PCM-LM composite demonstrated a 414.3% prolongation in thermal buffering compared to traditional pin–fin heat sinks in passive cooling. However, solar applications demand higher latent heat capacities and compatibility with higher temperature ranges (e.g., 80–150 °C for solar thermal systems). The RT25 paraffin used in the study [48] has a phase-change temperature of ~ 25 °C, which is unsuitable for most solar cooling applications. Adapting the composite to higher-temperature PCMs, such as salt hydrates or fatty acids, would require re-engineering the encapsulation process and ensuring LM compatibility.

Moreover, the composite's thermal conductivity (27.7 W/m K) must be maintained in larger configurations. In solar cooling systems, thermal contact resistance between PCM modules and heat exchangers could negate the LM's benefits. Wang et al. [48] emphasized the importance of matching heat absorption capacity with thermal loads, suggesting that the PCM's latent heat should be at least half the total heat generated (Fig. 8.17). For solar systems, this necessitates custom-sizing composites to match daily insolation patterns, complicating standardization.

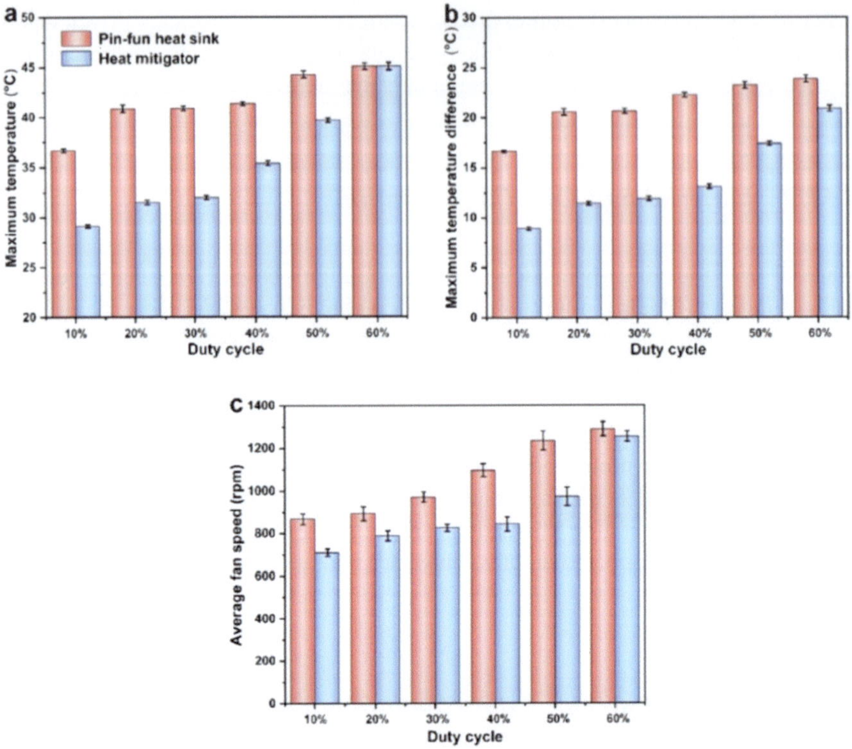

Fig. 8.17 Comparative thermal performance of the PCM-LM composite heat mitigator versus a traditional pin–fin heat sink under varying duty cycles. **a** Maximum temperature peaks, **b** peak-to-trough temperature differences, and **c** average fan speeds highlight the composite's superior thermal buffering and energy-saving potential [48]

8.11.4 Long-Term Stability and Environmental Impact

Long-term stability impacts scalability critically. The study noted that LM is corrosive to metals such as aluminum, which calls for corrosion-resistant shell materials. Calcium alginate stably binds with LM [48] but high temperatures and humid conditions (typical in solar systems) would have to be endured without the shell having to safely undergo thermal and hydrolytic aging. Correlated issues surrounding the environmental impact of corrosion also exist.

Another issue stems from the ability to recycle PCM-LM composites. Omega believes that directly removing LM from spent capsules for reuse would involve unproductive work processes, which might cancel out the energy-efficient value the composites provide. High-efficiency PCM–LM composite integration within solar cooling systems requires overcoming challenges regarding cost-effective fabrication and system integration. Although microfluidics presents a scalable solution, industrial implementation needs advances in segmental production and replacement of

LM. Combustion PCM-LHM focus, corrosion control, and life-cycle sustainability, routed eco-friendly analysis are the forefront of speculation for follow-up research.

8.12 Summary

This chapter delves into the applications of latent heat thermal storage systems, focusing on their integration with phase-change materials to improve energy efficiency in buildings. LHTS systems capitalize on the high energy storage density and isothermal behavior of PCMs during phase transitions, making them ideal for addressing the increasing global energy demand and environmental concerns.

The chapter highlights the role of PCMs in building envelopes, particularly in reducing cooling and heating loads. Studies demonstrate their ability to stabilize indoor temperatures, lower HVAC energy consumption, and meet seasonal energy efficiency standards. Comprehensive analyses reveal the impact of PCM properties, such as latent heat, thermal conductivity, and phase-change temperature, on thermal performance in various climates.

Innovative designs, including PCM-enhanced wallboards and foam concrete panels, are discussed, emphasizing their potential to improve thermal insulation, reduce peak energy demand, and support renewable energy integration. Advanced modeling techniques, such as finite difference methods and heat transfer simulations, underline the importance of precise PCM selection and system design for optimized performance.

Despite their advantages, challenges such as material stability, long-term performance, and economic feasibility remain. The chapter concludes by recommending expanded material research, life-cycle analysis, and dynamic operational studies to enhance PCM applications and broaden their impact in achieving sustainable energy solutions.

References

1. Alawadhi EM (2008) Thermal analysis of a building brick containing phase change material. Energy Build 40:351–357
2. Anderson R, Bates L, Johnson E, Morris JF (2015) Packed bed thermal energy storage: a simplified experimentally validated model. J Energy Storage 4:14–23
3. Baetens R, Jelle BP, Gustavsen A (2010) Phase change materials for building applications: a state-of-the-art review. Energy Build 42:1361–1368
4. Barzin R, Chen JJJ, Young BR, Farid MM (2015) Peak load shifting with energy storage and price-based control system. Energy 92:505–514
5. Barzin R, Chen JJJ, Young BR, Farid MM (2015) Application of PCM underfloor heating in combination with PCM wallboards for space heating using price based control system. Appl Energy 148:39–48
6. Cheng R, Wang X, Zhang Y (2014) Energy-efficient building envelopes with phase-change materials: new understanding and related research. Heat Transf Eng 35:970–984

7. Cheng X, Zhai X, Wang R (2016) Thermal performance analysis of a packed bed cold storage unit using composite PCM capsules for high temperature solar cooling application. Appl Therm Eng 100:247–255
8. Christopher S, Parham K, Mosaffa AH, Farid MM, Ma Z, Thakur AK, Xu H, Saidur R (2021) A critical review on phase change material energy storage systems with cascaded configurations. J Clean Prod 283:124653
9. Dora S, Kuznik F, Mini KM (2025) A novel PCM-based foam concrete for heat transfer in buildings—experimental developments and simulation modelling. J Energy Storage 105:114625
10. Dutil Y, Rousse DR, Salah NB, Lassue S, Zalewski L, Dutil Y, Rousse DR, Ben N, Salah NB, Lassue S, Zalewski L (2011) A review on phase-change materials: mathematical modeling and simulations. Renew Sustain Energy Rev 15:112–130
11. Edsjø S, Petter B, Kalnæs SE, Jelle BP (2015) Phase change materials and products for building applications: a state-of-the-art review and future research opportunities. Energy Build 94:150–176
12. Fan Z, Infante Ferreira CA, Mosaffa AH (2014) Numerical modelling of high temperature latent heat thermal storage for solar application combining with double-effect H_2O/LiBr absorption refrigeration system. Sol Energy 110:398–409
13. Gencel O, Ustaoglu A, Benli A, Hekimoğlu G, Sarı A, Erdogmus E, Sutcu M, Kaplan G, Yavuz Bayraktar O (2022) Investigation of physico-mechanical, thermal properties and solar thermoregulation performance of shape-stable attapulgite based composite phase change material in foam concrete. Sol Energy 236:51–62
14. Gholamibozanjani G, Farid M (2020) Application of an active PCM storage system into a building for heating/cooling load reduction. Energy 210:118572
15. Gil A, Oró E, Castell A, Cabeza LF (2013) Experimental analysis of the effectiveness of a high temperature thermal storage tank for solar cooling applications. Appl Therm Eng 54:521–527
16. Gil A, Oró E, Peiró G, Álvarez S, Cabeza LF (2013) Material selection and testing for thermal energy storage in solar cooling. Renew Energy 57:366–371
17. Guarda D, Righetti G, Longo GA, Zilio C, Mancin S (2023) Experimental assessment of low temperature phase change materials (PCM) for refrigerating and air conditioning applications. Int J Refrig 154:33–42
18. Guo J, Dong J, Wang H, Jiang Y, Tao J (2021) On-site measurement of the thermal performance of a novel ventilated thermal storage heating floor in a nearly zero energy building. Build Environ 201:107993
19. Guo Z-YY, Zhu H-YY, Liang X-GG (2007) Entransy—a physical quantity describing heat transfer ability. Int J Heat Mass Transf 50:2545–2556
20. Hosseini M, Tasnim SH, Mahmud S (2024) Application of PCM for radiant heating of residential buildings in Canada towards achieving nearly zero energy buildings. J Energy Storage 102:114030
21. Kalapala L, Devanuri JK (2018) Influence of operational and design parameters on the performance of a PCM based heat exchanger for thermal energy storage—a review. J Energy Storage 20:497–519
22. Kenisarin M, Mahkamov K (2016) Salt hydrates as latent heat storage materials: thermophysical properties and costs. Sol Energy Mater Sol Cells 145:255–286
23. Kocyigit F, Bayram M, Hekimoglu G, Cay VV, Gencel O, Ustaoglu A, Sarı A, Erdogmus E, Ozbakkaloglu T (2023) Thermal energy saving and physico-mechanical properties of foam concrete incorporating form-stabilized basalt powder/capric acid based composite phase change material. J Clean Prod 414:137617
24. Konuklu Y, Ersoy O (2017) Fabrication and characterization of form-stable phase change material/xonotlite microcomposites. Sol Energy Mater Sol Cells 168:130–135
25. Kuznik F, Virgone J (2009) Experimental assessment of a phase change material for wall building use. Appl Energy 86:2038–2046
26. Larwa B, Cesari S, Bottarelli M (2021) Study on thermal performance of a PCM enhanced hydronic radiant floor heating system. Energy 225:120245

27. Li M, Guo Q, Nutt S (2017) Carbon nanotube/paraffin/montmorillonite composite phase change material for thermal energy storage. Sol Energy 146:1–7
28. Louis AT, Lavine AS, Lakeh RB, Wirz RE (2015) Exergetic optimization and performance evaluation of multi-phase thermal energy storage systems. Sol Energy 122:396–408
29. Mao Q, Zhang Y (2022) Study on the thermal storage performance of a new cascade structure phase change thermal storage tank. J Energy Storage 56:106155
30. Medrano M, Yilmaz MO, Nogués M, Martorell I, Roca J, Cabeza LF (2009) Experimental evaluation of commercial heat exchangers for use as PCM thermal storage systems. Appl Energy 86:2047–2055
31. Mehling H, Brütting M, Haussmann T (2022) PCM products and their fields of application—an overview of the state in 2020/2021. J Energy Storage 51:104354
32. Mehling H, Cabeza LF (2008) Integration of active storages into systems. In: Heat and cold storage with PCM. Springer, Berlin. pp 181–189
33. Mi X, Liu R, Cui H, Memon SA, Xing F, Lo Y (2016) Energy and economic analysis of building integrated with PCM in different cities of China. Appl Energy 175:324–336
34. Miqdam C, Al-Ajeely M (2015) Thermal conductivity enhancement by using nano-material in phase change material for latent heat thermal energy storage systems. SAUSSUREA 5:48–55
35. Narasimhan NL (2019) Assessment of latent heat thermal storage systems operating with multiple phase change materials. J Energy Storage 23:442–455
36. Navarro L, de Gracia A, Castell A, Cabeza LF (2016) Experimental study of an active slab with PCM coupled to a solar air collector for heating purposes. Energy Build 128:12–21
37. Papade CV, Kanase-Patil AB (2023) Characterization of nano-mixed PCM for solar cooker application. Mater Today Proc. https://doi.org/10.1016/j.matpr.2023.03.415
38. Pasupathy A, Athanasius L, Velraj R, Seeniraj RV (2008) Experimental investigation and numerical simulation analysis on the thermal performance of a building roof incorporating phase change material (PCM) for thermal management. Appl Therm Eng 28:556–565
39. Peiró G, Gasia J, Miró L, Cabeza LF (2015) Experimental evaluation at pilot plant scale of multiple PCMs (cascaded) vs. single PCM configuration for thermal energy storage. Renew Energy 83:729–736
40. Pomianowski M, Heiselberg P, Zhang Y (2013) Review of thermal energy storage technologies based on PCM application in buildings. Energy Build 67:56–69
41. Reddy KS, Mudgal V, Mallick TK (2018) Review of latent heat thermal energy storage for improved material stability and effective load management. J Energy Storage 15:205–227
42. Righetti G, Zilio C, Guarda D, Feo D, Auerbach M, Butters M, Mancin S (2023) Experimental analysis of a commercial size bio-based latent thermal energy storage for air conditioning. J Energy Storage 72:108477
43. Royon L, Karim L, Bontemps A (2013) Thermal energy storage and release of a new component with PCM for integration in floors for thermal management of buildings. Energy Build 63:29–35
44. Sharma RK, Ganesan P, Tyagi VV, Metselaar HSC, Sandaran SC (2015) Developments in organic solid–liquid phase change materials and their applications in thermal energy storage. Energy Convers Manag 95:193–228
45. Simou Z, Hamdaoui S, Oubenmoh S, El Afou Y, Babaharra O, Mahdaoui M, Ait Msaad A (2024) Thermal evaluation and optimization of a building heating system: radiant floor coupled with a solar system. J Build Pathol Rehabil 9:24
46. Solomon L, Oztekin A (2016) Exergy analysis of cascaded encapsulated phase change material—high-temperature thermal energy storage systems. J Energy Storage 8:12–26
47. Stathopoulos N, El Mankibi M, Issoglio R, Michel P, Haghighat F (2016) Air–PCM heat exchanger for peak load management: experimental and simulation. Sol Energy 132:453–466
48. Wang J-X, Qian J, Wang N, Zhang H, Cao X, Liu F, Hao G (2023) A scalable micro-encapsulated phase change material and liquid metal integrated composite for sustainable data center cooling. Renew Energy 213:75–85
49. Wang J, Xie H, Guo Z, Guan L, Li Y (2014) Improved thermal properties of paraffin wax by the addition of TiO_2 nanoparticles. Appl Therm Eng 73:1541–1547

50. Watanabe T, Kanzawa A (1995) Second law optimization of a latent heat storage system with PCMS having different melting points. Heat Recov Syst CHP 15:641–653
51. Wu M, Xu C, He Y, Wu M, Xu CHY (2016) Cyclic behaviors of the molten-salt packed-bed thermal storage system filled with cascaded phase change material capsules. Appl Therm Eng 93:1061–1073
52. Wu X, Li X, Qin Y, Xu W, Liu Y (2023) Intelligent multiobjective optimization design for NZEBs in China: four climatic regions. Appl Energy 339:120934
53. Xie J, Wang W, Liu J, Pan S (2018) Thermal performance analysis of PCM wallboards for building application based on numerical simulation. Sol Energy 162:533–540
54. Xu HJ, Zhao CY (2019) Analytical considerations on optimization of cascaded heat transfer process for thermal storage system with principles of thermodynamics. Renew Energy 132:826–845
55. Xu Y, He Y-L, Zhu G-H, Lv S, Shan B (2017) Al/Al_2O_3 form-stable phase change material for high temperature thermal energy storage. Energy Procedia 105:4328–4333
56. Yang L, Zhang X, Xu G (2014) Thermal performance of a solar storage packed bed using spherical capsules filled with PCM having different melting points. Energy Build 68:639–646
57. Zayed ME, Zhao J, Elsheikh AH, Hammad FA, Ma L, Du Y, Kabeel AE, Shalaby SM (2019) Applications of cascaded phase change materials in solar water collector storage tanks: a review. Sol Energy Mater Sol Cells 199:24–49
58. Zhang C, Chen Y, Wu L, Shi M (2011) Thermal response of brick wall filled with phase change materials (PCM) under fluctuating outdoor temperatures. Energy Build 43:3514–3520

Chapter 9
Phase-Change Materials for Free Cooling Application

9.1 Introduction

Phase-change materials have emerged as a promising technology in the quest for sustainable and energy-efficient cooling solutions. Free cooling, which leverages the natural temperature differences between day and night, can significantly reduce the energy consumption of HVAC systems. PCMs, with their ability to store and release thermal energy during phase transitions, play a crucial role in enhancing the efficiency of free cooling applications.

This chapter explores the principles of free cooling, the properties and selection of suitable PCMs, and their integration into building systems. We will delve into the thermodynamic principles that govern the phase-change process and how these materials can be optimized for maximum energy savings. Additionally, we will examine the various types of PCMs available, their thermal properties, and the criteria for selecting the most appropriate PCM for specific applications.

By examining case studies and real-world applications, we aim to comprehensively understand how PCMs can contribute to greener and more cost-effective cooling strategies. The chapter will also discuss the challenges and limitations associated with PCM-based free cooling systems and propose potential solutions to overcome these hurdles. Ultimately, this chapter aims to equip readers with the knowledge and tools to implement effective free cooling solutions using PCMs in various settings.

A. Mosaffa, *Phase Changing Materials*, Materials Horizons: From Nature to Nanomaterials, https://doi.org/10.1007/978-981-95-3742-6_9

9.2 Thermal Performance Optimization of Multiple PCM Thermal Storage Units for Free Cooling

Thermal energy storage systems are a promising alternative for reducing peak cooling loads in buildings. TES units with PCMs offer high energy density and consistent thermal regulation during phase transitions. Free cooling systems using LHTS have gained attention for their ability to store and release cooling energy with minimal energy consumption. However, the low thermal conductivity of PCMs presents challenges in achieving high melting and solidification rates.

This section investigates a multiple PCM thermal storage unit (MPCM-TSU) designed for free cooling applications [4]. The system comprises rectangular air channels separated by PCM slabs. Multiple PCMs with varying melting temperatures are strategically arranged to enhance heat transfer rates during melting and solidification, maintaining near-constant heat fluxes and optimizing performance.

9.2.1 System Description

The MPCM-TSU consists of air channels and PCM slabs arranged in parallel. Key system specifications are shown in Table 9.1, and the layout is illustrated in Fig. 9.1. Air (heat transfer fluid, HTF) flows through the channels, while the PCMs absorb or release heat during phase transitions. PCMs with melting points of 18–30 °C were selected to ensure thermal comfort. Table 9.2 details the thermophysical properties of the selected PCMs: $CaCl_2{\cdot}6H_2O$, paraffin C18, and RT25.

To enhance the heat transfer, the system incorporates multiple PCMs arranged in decreasing melting point order along the airflow direction. During melting, this configuration ensures a nearly constant temperature difference between the HTF and PCMs, improving energy efficiency.

Table 9.1 System specifications of the storage used in the calculations [4]

Parameters	Value
Thickness of PCM slabs, a	10 mm
Thickness of air channels, b	3.6 mm
Length of PCM slabs, l	1.4 m
Height of storage	0.5 m
Width of storage	1.12 m
Air flow rate, $\dot{V}$	1200 m^3/h ($u = 2.5$ m/s)

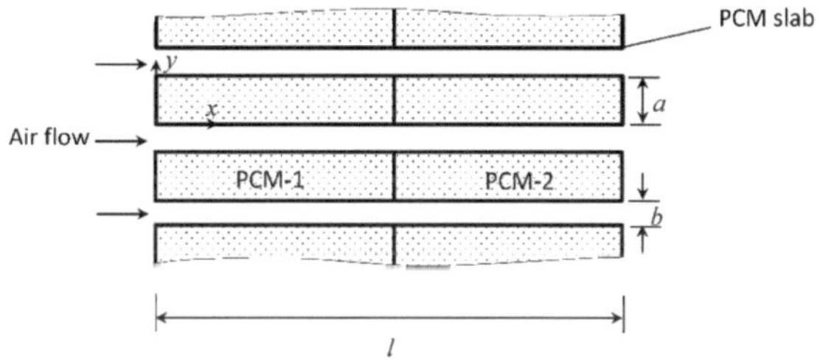

Fig. 9.1 Schematic diagram of the TES unit [4]

Table 9.2 Thermal properties of selected PCMs [4]

Property	$CaCl_2{\cdot}6H_2O$ (PCM-1)	Paraffin C_{18} (PCM-2)	RT25 (PCM-3)
Density, ρ (kg/m^3)	1530 (liquid) 1710 (solid)	774 (liquid) 814 (solid)	749 (liquid) 785 (solid)
Heat capacity at constant pressure, c_p (J/kg K)	1400 (liquid) 2200 (solid)	2160	1410 (liquid) 1800 (solid)
Thermal conductivity, k (W/m K)	0.54 (liquid) 1.09 (solid)	0.15 (liquid) 0.358 (solid)	0.18 (liquid) 0.19 (solid)
Latent heat of fusion, L (J/kg)	190,800	244,000	232,000
Melting temperature, T_m (°C)	29	27.5	26.6

9.2.2 *Mathematical Modeling of PCM Heat Transfer and System Performance*

This section details the mathematical model developed to analyze the heat transfer characteristics of the phase-change material and the heat transfer fluid within the studied system. The model considers the unsteady, two-dimensional nature of the solidification and melting processes within the PCM and the heat transfer dynamics of the HTF.

To simplify the model and focus on the dominant heat transfer mechanisms, several assumptions were made. Given the minimal temperature variation observed in the container wall, temperature gradients perpendicular to the flow direction are considered negligible. Furthermore, the influence of natural convection is assumed to be insignificant. This assumption is supported by observations from Laouadi and Lacroix [1], who demonstrated that natural convection becomes important only when PCM slabs are vertically oriented and subjected to lateral heating, a configuration different from this study. Another simplification is the assumption of temperature-independent thermophysical properties for the PCM, although distinct values are used for the solid and liquid phases.

9.2.2.1 Governing Equations

The energy equation governing heat transfer within the PCM, along with its associated initial and boundary conditions, can be expressed as follows:

$$\frac{\partial T}{\partial t} = \frac{k}{\rho c}\left(\frac{\partial^2 T}{\partial x^2} + \frac{\partial^2 T}{\partial y^2}\right), \quad t > 0 \tag{9.1}$$

- *Initial condition*:

$$T(x, y, 0) = T_{\mathrm{i}} \tag{9.2}$$

- *Boundary condition 1: T (wall boundaries)*

$$k\frac{\partial T(x, 0, t)}{\partial y} = -k\frac{\partial T(x, a, t)}{\partial y} = h(T_{\mathrm{w}} - T_{\mathrm{f}}) \tag{9.3}$$

- *Boundary condition 2: Heat flux at HTF interface*

$$\frac{\partial T(0, y, t)}{\partial x} = -\frac{T(l, y, t)}{\partial x} = 0 \tag{9.4}$$

The energy balance at the solid–liquid interface, defining the interface location (S) as a function of time, is given by:

$$\left[1 + \left(\frac{\partial S}{\partial x}\right)^2\right]\left[k_{\mathrm{s}}\frac{\partial T_{\mathrm{s}}}{\partial y} - k_{\mathrm{l}}\frac{\partial T_{\mathrm{l}}}{\partial y}\right] = \rho L\frac{\partial S}{\partial t} \quad \text{at} \quad y = S(x, t) \tag{9.5}$$

$$S(x, y, 0) = 0 \tag{9.6}$$

The heat transfer within the HTF is modeled using the energy conservation equations and boundary conditions for incompressible, laminar flow, neglecting viscous dissipation:

$$\frac{\partial T_{\mathrm{f}}}{\partial t} + u\frac{\partial T_{\mathrm{f}}}{\partial x} = \frac{k_{\mathrm{f}}}{\rho_{\mathrm{f}} c_{\mathrm{f}}}\left(\frac{\partial^2 T_{\mathrm{f}}}{\partial y^2}\right), \quad t > 0 \tag{9.7}$$

- *Initial boundary condition*:

$$T_{\mathrm{f}}(x, y, 0) = T_{\mathrm{i}} \tag{9.8}$$

- *Wall boundary condition*:

$$k_f \frac{\partial T_f(x, 0, t)}{\partial y} = -k_f \frac{\partial T_f(x, -b, t)}{\partial y} = h(T_w - T_f) \tag{9.9}$$

- *Inlet boundary condition (assuming fully developed flow)*:

$$T_f(0, y, t) = T_{inlet} \tag{9.10}$$

where u is the HTF velocity vector and k_f is the thermal conductivity of the HTF.

The local Nusselt number (Nu) is defined as:

$$\mathrm{Nu}(x) = \frac{q'' D_h}{k_f (T_{f,b} - T_w)} \tag{9.11}$$

where D_h is the hydraulic diameter.

The fluid bulk temperature ($T_{f,b}$) is defined as:

$$T_{f,b} = \frac{\int u(y) T_f(x, y) dy}{\int u(y) dy} \tag{9.12}$$

where the integrals are taken over the cross-sectional area of the HTF channel.

9.2.2.2 Phase-Change Modeling

In reality, a phase change occurs over a temperature range rather than at a discrete temperature. To account for this, the effective heat capacity method is employed. This method relates the material's effective heat capacity to the energy stored and released during phase transition, as well as the specific heat capacity. The effective heat capacity is incorporated into the energy equation (Eq. 9.1) as follows:

$$c = \begin{cases} c_s, & T < T_1 \\ \frac{L}{T_2 - T_1} + \frac{c_s + c_l}{2}, & T_1 \le T \le T_2 \\ c_l, & T > T_2 \end{cases} \tag{9.13}$$

While theoretical studies often depict solidification/melting as occurring at a constant temperature (isothermal), this model assumes a small melting range of 1 K to better represent real-world behavior.

9.2.2.3 System Performance Parameters

The power associated with frictional losses can be calculated as:

$$\dot{W}_{\text{fan}} = \left(\Delta P_{\text{total}} \dot{V}\right) / \eta_{\text{fan}} \tag{9.14}$$

where ΔP_{total} is the total pressure drop and $\dot{V}$ is the volumetric flow rate.

Pressure losses are calculated using established correlations for duct flows and losses due to expansions and contractions:

- *Frictional pressure loss*:

$$\Delta p_{\text{fric}} = \frac{64}{\text{Re}_{D_{\text{h}}}} \frac{l}{D_{\text{h}}} \frac{\rho u^2}{2} \tag{9.15}$$

- *Expansion cross section loss*:

$$\Delta p_{\text{ecs}} = 2\left(1 - \frac{A_{\text{s}}}{A_{\text{L}}}\right) \frac{\rho u^2}{2} \tag{9.16}$$

- *Contraction cross section loss*:

$$\Delta p_{\text{ccs}} = 1.1\left(1 - \frac{A_{\text{s}}}{A_{\text{L}}}\right) \frac{\rho u^2}{2} \tag{9.17}$$

The total fan power (P_{fan}), accounting for pressure drops across the air filter and distribution equipment, is the sum of the power consumed during the melting (day-time) and solidification (night-time) processes:

$$P_{\text{fan}} = \left(\dot{W}_{\text{fan}} \times t\right)_{\text{day time}} + \left(\dot{W}_{\text{fan}} \times t\right)_{\text{night time}} \tag{9.18}$$

The cooling load ($\dot{Q}_{\text{c}}$) of the system is calculated as:

$$\dot{Q}_{\text{c}} = \frac{Q_{\text{PCM}}}{t_0} = \frac{\int_0^{t_0} \dot{Q}_{\text{PCM}}(t)\text{d}t}{t_0} \tag{9.19}$$

where Q_{PCM} is the total heat absorbed by the PCM.

Finally, the system's coefficient of performance (COP) is defined as:

$$\text{COP} = \frac{Q_{\text{PCM}}}{P_{\text{fan}}} \tag{9.20}$$

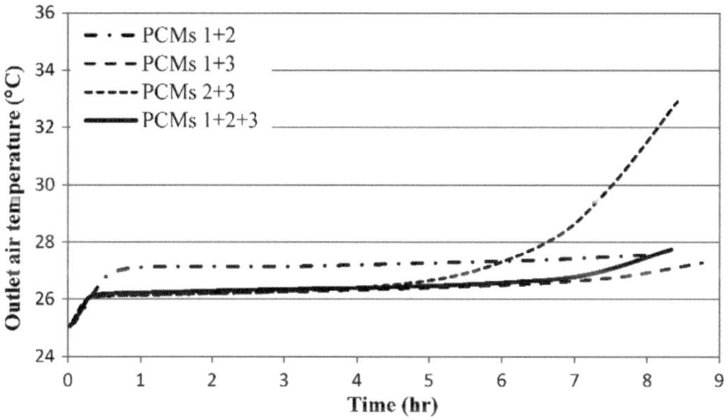

Fig. 9.2 Effect of different composition of PCMs in the slabs on outlet air temperature. PCMs 1, 2 and 3 [4]

9.2.3 *Results and Discussion*

9.2.3.1 PCM Selection

Three PCMs were evaluated for their suitability in free cooling applications. Based on melting temperature and thermal properties, a combination of $CaCl_2{\cdot}6H_2O$ and RT25 was found to deliver optimal cooling performance.

9.2.3.2 Thermal Performance

- *Effect of PCM composition*: Using a combination of $CaCl_2{\cdot}6H_2O$ and RT25 reduced air outlet temperatures below 27 °C for 8 h, maintaining thermal comfort (Fig. 9.2).
- *Effect of PCM geometry*: Increasing slab length and thickness reduced air outlet temperatures during melting but increased solidification time due to the lower thermal conductivity of liquid PCM (Fig. 9.3).
- *Heat transfer rates*: Solidification rates were slower than melting due to thermal resistance in liquid PCM. Higher airflow rates during solidification mitigated this effect (Fig. 9.4).

9.2.3.3 COP Analysis

The system COP was evaluated for varying slab dimensions. Shorter slabs (1.3 m length, 10 mm thickness) achieved the highest COP of 7.0, outperforming conventional cooling systems (Table 9.3).

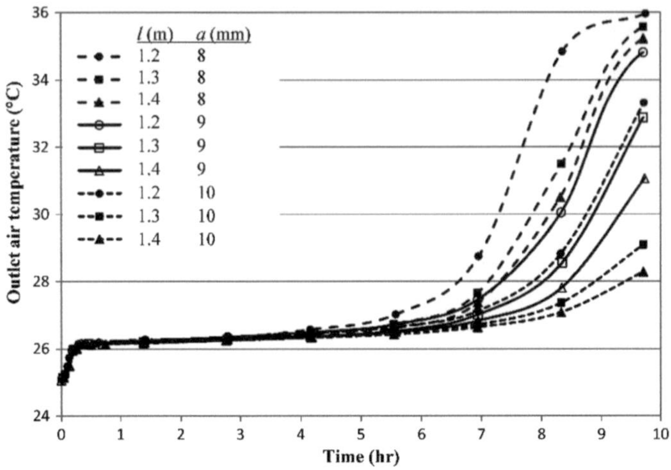

Fig. 9.3 Effect of the length and thickness of the PCM slabs on the outlet air temperature from the storage. For the geometry consult Fig. 9.2 [4]

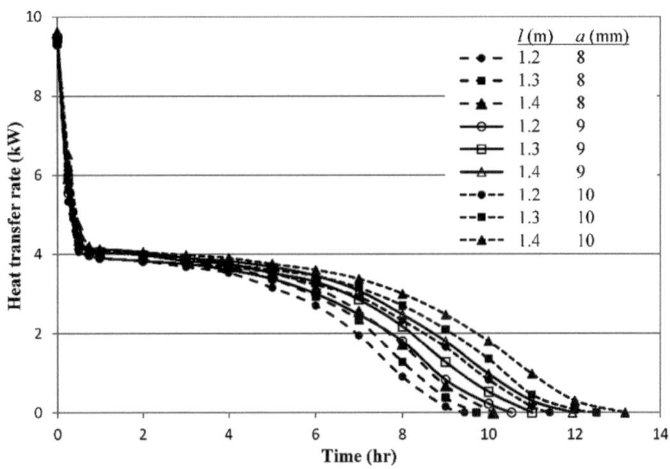

Fig. 9.4 Effect of the length and thickness of the PCM slabs on the heat transfer rate during the solidification process ($T_i = 36$ °C, $T_{inlet} = 25$ °C and $\dot{V} = 2250$ m^3/h) [4]

9.2.3.4 Variation of Wall, Air, and PCM Temperatures

Figure 9.5 illustrates the temperature distribution of the wall, air, and PCM along the centerline of a PCM slab at different time intervals during the melting process. Initially, the air temperature near the inlet is consistent with the PCM's initial temperature due to a significant temperature difference between the air and PCM. This causes rapid heat transfer, leading to a sharp temperature rise in the air. As the PCM starts

Table 9.3 COP for the cases in Figs. 9.3 and 9.4 (operating time during day is 8 h) [4]

Case	l (m)	a (mm)	PCM mass (kg)	Solidification time (h)	Absorbed heat by PCMs (MJ)	P_{fan} (kWh)	$\dot{Q}_c$ (kW)	COP
1	1.2	8	484.6	9.44	111.46	3.84	3.87	8.06
2	1.3	8	525.0	9.72	115.49	4.02	4.01	7.97
3	1.4	8	557.3	10.11	116.64	4.22	4.05	7.68
4	1.2	9	545.2	10.53	117.22	4.17	4.07	7.80
5	1.3	9	590.6	11.03	120.10	4.42	4.17	7.54
6	1.4	9	627.0	11.94	121.54	4.80	4.22	7.02
7	1.2	10	605.7	11.42	119.52	4.45	4.15	7.46
8	1.3	10	656.2	12.51	121.25	4.89	4.21	6.88
9	1.4	10	696.6	13.17	122.40	5.19	4.25	6.55

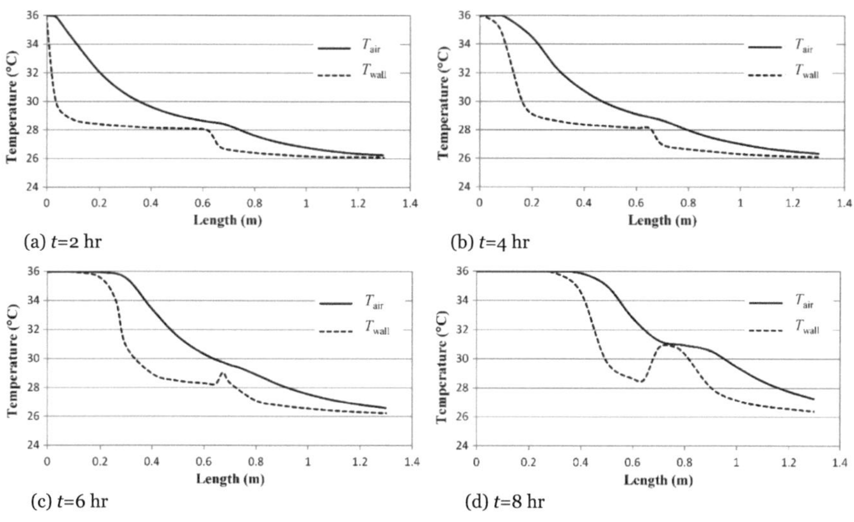

Fig. 9.5 Variation of wall, air and PCM temperature along the centerline of the slab through the thermal storage system (l = 1.3 m and a = 10 mm) [4]

melting, the wall temperature stabilizes close to the PCM's melting temperature. The air temperature rises gradually due to the latent heat absorption by the PCM, which moderates the heat transfer rate. As the liquid PCM layer thickens, its thermal resistance increases, reducing heat transfer efficiency. Consequently, the air temperature becomes higher over time, reflecting diminished cooling performance.

The melting process effectively maintains thermal comfort conditions for a significant duration, but the increasing thermal resistance of the liquid PCM limits long-term efficiency.

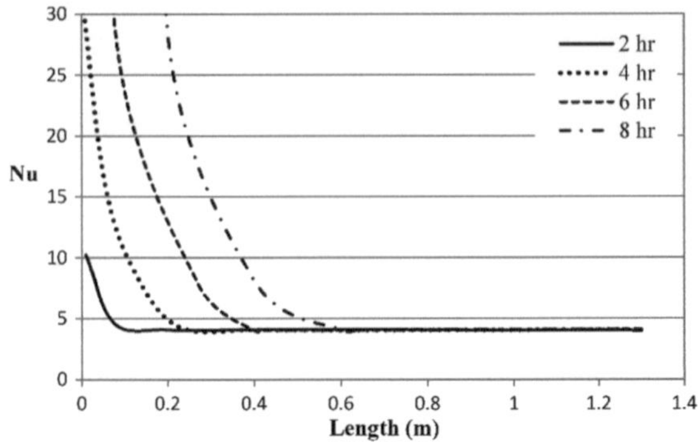

Fig. 9.6 Variation of local Nusselt number with time along the flow direction ($l =$ 1.3 m, $a = 10$ mm, $b =$ 3.6 mm and the number of plates = 80) [4]

9.2.3.5 Variation of Local Nusselt Number Along the Flow Direction

Figure 9.6 presents the variation of the local Nusselt number along the flow direction at different time intervals. The local Nusselt number decreases along the flow direction as the temperature difference between the air and the PCM diminishes due to heat transfer. Over time, the PCM melts further downstream, reducing the region where effective heat transfer occurs. After a certain distance, no heat transfer takes place, and the local Nusselt number approaches zero. Near the fully melted region, the air-wall temperature difference becomes small, but the heat flux remains considerable. This results in high local Nusselt numbers in this zone.

The reduction in the Nusselt number along the flow direction emphasizes the importance of PCM arrangement and airflow design to maintain efficient heat transfer throughout the thermal storage system.

9.2.3.6 Air Channel Thickness

Optimizing air channel thickness to 3.2 mm reduced energy consumption by 8.94%, with minimal impact on cooling capacity. Increasing thickness beyond this value reduced COP due to higher airflow resistance (Figs. 9.7 and 9.8).

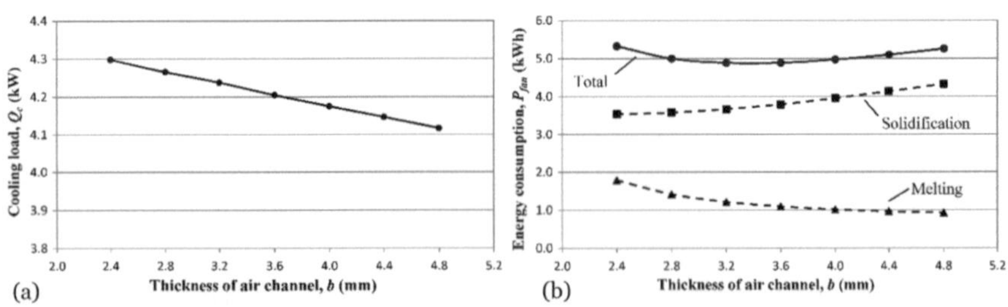

Fig. 9.7 Effect of thickness of air channels on: **a** cooling load, **b** energy consumption ($l = 1.3$ m, $a = 10$ mm and number of plates = 80) [4]

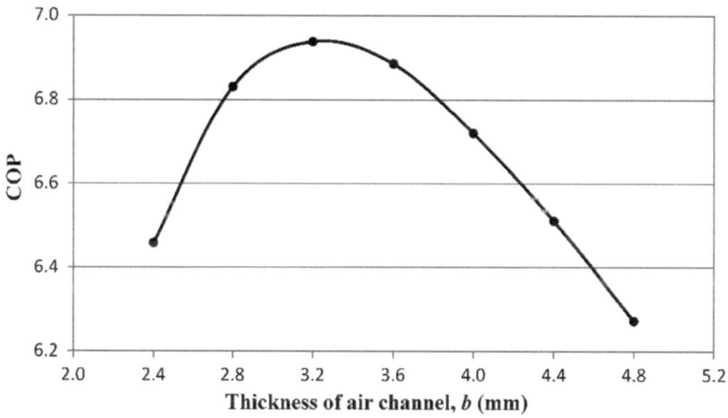

Fig. 9.8 Effect of thickness of air channels on COP (l = 1.3 m, a = 10 mm and number of plates = 80) [4]

9.2.4 Conclusion

A multiple PCM thermal storage unit was numerically analyzed for free cooling applications. Key findings include:

- Optimal PCM combinations and geometries significantly enhance cooling performance.
- The system achieves a high COP of 7.0, making it energy-efficient for climates like Tabriz, Iran.
- Air channel thickness and PCM arrangement are critical design parameters for achieving optimal performance.

The developed model provides a framework for optimizing TES systems in various climatic conditions, offering a sustainable alternative to conventional air conditioning systems.

9.3 Optimizing Free Cooling Systems Using Multiple PCM-Based Thermal Storage Units

The increasing demand for air conditioning systems, driven by rising comfort expectations and climate challenges, has motivated the adoption of TES systems. Among TES technologies, latent heat thermal storage offers advantages such as high energy density and minimal temperature fluctuation during phase transitions. Free cooling systems utilizing LHTS can store nighttime cool air and release it during the day, reducing energy consumption in buildings.

This section examines a free cooling system equipped with a multiple PCM LHTS unit. The system consists of layered PCM slabs with air channels for the heat transfer fluid [5]. The inclusion of multiple PCMs, arranged based on their melting temperatures, enhances thermal efficiency by maintaining a near-constant heat flux.

9.3.1 System Description

Figure 9.9 illustrates the operational concept of a free cooling system integrated with a LHTS unit. The system uses ambient air to store coolness during the night and releases it during the day to maintain thermal comfort in buildings.

- *Charging process (night)*: Cool ambient air flows through the LHTS unit, transferring heat to the PCMs, which solidify by releasing latent heat.
- *Discharging process (day)*: Warm indoor air passes through the LHTS unit, where the PCMs absorb heat and undergo melting, reducing the air temperature and cooling the indoor space.

This process eliminates the need for conventional cooling systems during the day, reducing energy consumption and providing an environmentally sustainable cooling solution. The figure highlights the role of PCMs in maintaining consistent thermal conditions while efficiently utilizing diurnal temperature variations.

The LHTS unit comprises flat PCM slabs arranged parallel to air channels, as shown in Fig. 9.10. PCMs with melting temperatures ranging from 15 to 30 °C are used to achieve thermal comfort during free cooling. Two PCMs, $KF{\cdot}4H_2O$ and Climsel C24, were selected for their suitability in maintaining comfort temperatures. During melting, PCMs are arranged in decreasing order of their melting points to sustain heat transfer. For solidification, the arrangement is reversed. The system's specifications and PCM properties are listed in Tables 9.4 and 9.5.

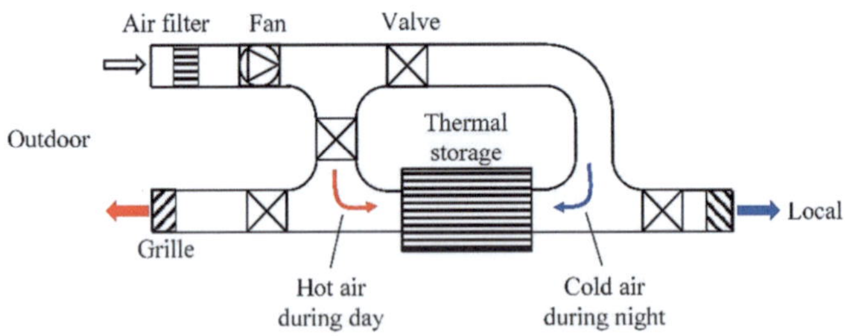

Fig. 9.9 Installation of thermal storage for free cooling [5]

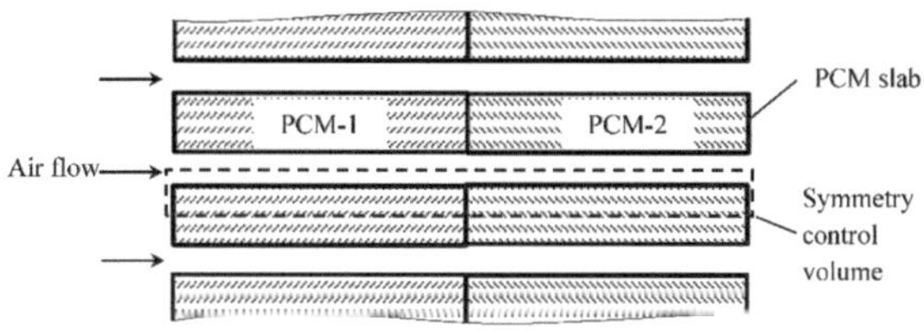

Fig. 9.10 Schematic of the LHTS unit [5]

Table 9.4 System specifications of the storage used in the calculations [5]

Parameters	Value
Number of slabs	37
Thickness of PCM slabs, δ	10 mm
Thickness of air channels, γ	3.2 mm
Length of PCM slabs, l	1.3 m
Width of storage	1.1 m
Height of storage	0.5 m
Air flow rate, $\dot{V}$	1200 m^3/h

Table 9.5 Thermal properties of selected PCMs [5]

Property	Climsel C24 (PCM-1)	$KF{\cdot}4H_2O$ (PCM-2)
Density (kg/m^3)	1380	1447 (liquid) 1455 (solid)
Heat capacity at constant pressure, c_p (J/kg K)	3600	2390 (liquid) 1840 (solid)
Thermal conductivity, k (W/m K)	0.5 (liquid) 0.7 (solid)	0.53
Latent heat of fusion, L (J/kg)	216,000	231,000
Melting temperature, T_m (°C)	24	18.5

9.3.2 *Mathematical Formulation*

9.3.2.1 Performance Evaluation of Latent Heat Thermal Storage Units

This section details the performance assessment of LHTS units, which typically comprise multiple flat PCM slabs, as depicted in Fig. 9.10. Within each slab, various PCMs are contained. The actual energy stored or released during the phase transition is intrinsically linked to the unit's effectiveness, directly impacting the usable stored energy. To evaluate the LHTS unit's performance, it is analyzed as a heat exchanger, and its effectiveness is a crucial metric.

9.3.2.2 Effectiveness of LHTS Units

The effectiveness (ε) of the LHTS unit, representing the ratio of actual heat transfer to the maximum possible heat transfer, is defined as:

$$\varepsilon = \frac{Q}{Q_{\text{max}}} = \frac{T_{\text{inlet}} - T_{\text{outlet}}}{T_{\text{inlet}} - T_{\text{m}}} \tag{9.21}$$

where Q represents the actual heat transferred during the phase-change process and Q_{max} signifies the maximum theoretical heat transfer achievable.

The average effectiveness (ε_{avg}) over the entire phase-change process (from the start to the end of melting or solidification) can be determined by integrating the instantaneous effectiveness over the duration of the phase change:

$$\overline{\varepsilon} = \varepsilon_{\text{avg}} = \frac{1}{t}\int_{0}^{t_{\text{pc}}} \varepsilon_{\text{PCM}-1} \cdot \varepsilon_{\text{PCM}-2} \cdot \text{d}t \tag{9.22}$$

where t_{pc} is the total time required for the phase-change process.

Within LHTS units, the energy exchange between the heat source and the heat sink is governed by the effectiveness during the melting and solidification phases. Consequently, the anticipated stored energy within an LHTS unit can be characterized by an energy storage effectiveness, which can be directly related to the volumetric latent heat of the PCM:

$$E = \overline{\varepsilon}_{\text{m}}\overline{\varepsilon}_{\text{s}}r \tag{9.23}$$

where r is the compactness factor, representing the ratio of the PCM volume to the total volume of the storage unit.

The compactness factor (r) is defined as:

$$r = \frac{V_{\text{PCM}}}{V_{\text{Total}}} = \frac{\delta}{\delta + \gamma} \tag{9.24}$$

where V_{PCM} is the total volume occupied by the PCM within the LHTS unit, V_{Total} is the total volume of the LHTS unit.

9.3.2.3 Modified Effectiveness Accounting for Energy Losses

To account for energy losses within the LHTS unit, such as those due to fan operation and heat losses to the surroundings, a modified effectiveness is introduced. This modified effectiveness considers the energy consumed by auxiliary components like fans and compares it to the maximum theoretical stored energy. The modified melting and solidification effectiveness is thus determined:

$$\overline{\varepsilon}^*_{\text{m or s}} = \overline{\varepsilon}_{\text{m or s}} - \frac{\dot{W}_{\text{fan}}}{Q_{\max}/t_{\text{m or s}}} \tag{9.25}$$

where $\dot{W}_{\text{fan}}$ represents the fan power required for air circulation within the LHTS unit.

Therefore, the modified energy storage effectiveness of the LHTS unit, which accounts for these energy losses, can be expressed as:

$$E^* = \overline{\varepsilon}^*_{\text{m}} \overline{\varepsilon}^*_{\text{s}} r \tag{9.26}$$

This modified effectiveness provides a more realistic representation of the LHTS unit's performance by incorporating the energy expenditure required for its operation. It offers a more accurate assessment of the usable stored energy.

9.3.3 Model Assumptions

- Thermophysical properties of PCMs are phase-dependent but temperature-independent.
- Natural convection within the liquid PCM is negligible.
- Airflow and PCM temperature variations are considered two-dimensional.

9.3.4 Results and Discussion

9.3.4.1 Effect of PCM Slab Geometry

Increasing PCM slab length and thickness reduces outlet air temperature during melting but prolongs solidification due to low thermal conductivity. A configuration with 1.3 m slab length and 9 mm thickness achieves optimal cooling for 8 h with a total solidification time of 12 h (Fig. 9.11).

9.3.4.2 Effect of PCM Slab Length and Thickness on Heat Exchanger Effectiveness

Figure 9.12 illustrates the influence of PCM slab length and thickness on the instantaneous heat exchanger effectiveness during the melting process.

- *Initial phase*: The effectiveness is initially high, often exceeding unity, due to the sensible energy stored in the PCM. The temperature difference between the air and the PCM melting temperature drives efficient heat exchange.

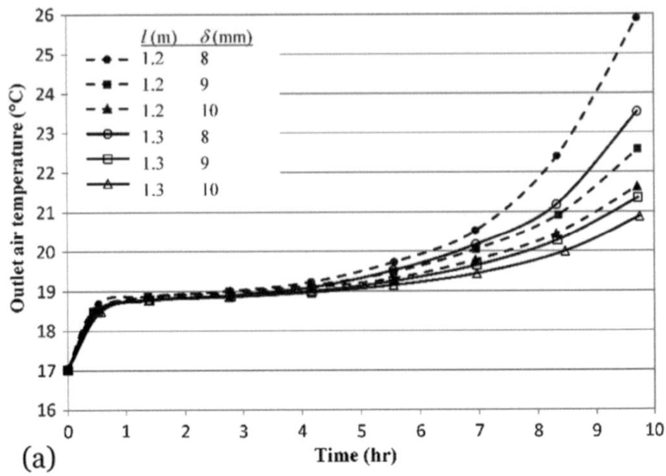

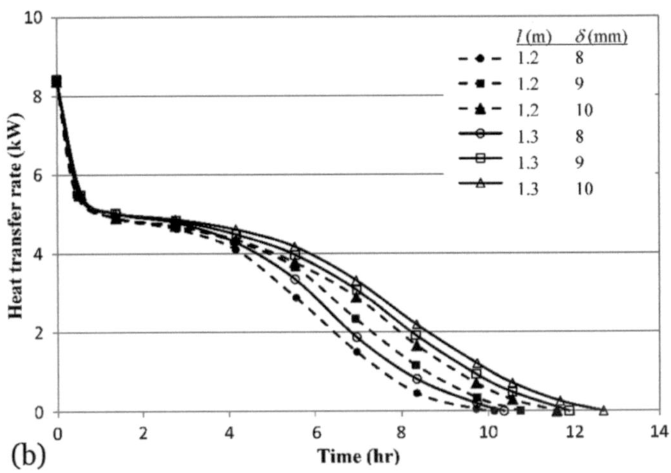

Fig. 9.11 Effect of length and thickness of the PCM slabs ($T_{inlet} = 29$ °C, $\gamma = 3.2$ mm): **a** on the outlet air temperature from the system ($\dot{V} = 1200$ m^3/h) and **b** on the solidification time of the total PCM during night-time ($\dot{V} = 1950$ m^3/h) [5]

- *Over time*: As melting progresses, the outlet air temperature gradually increases, reducing the effectiveness of the heat exchanger.
- *Impact of geometry*: Larger PCM slab lengths and thicknesses enhance effectiveness due to increased thermal mass, but they also introduce additional pressure drops, which can lower system efficiency.

Optimizing slab dimensions is crucial for balancing heat exchanger performance and minimizing energy losses due to pressure drop.

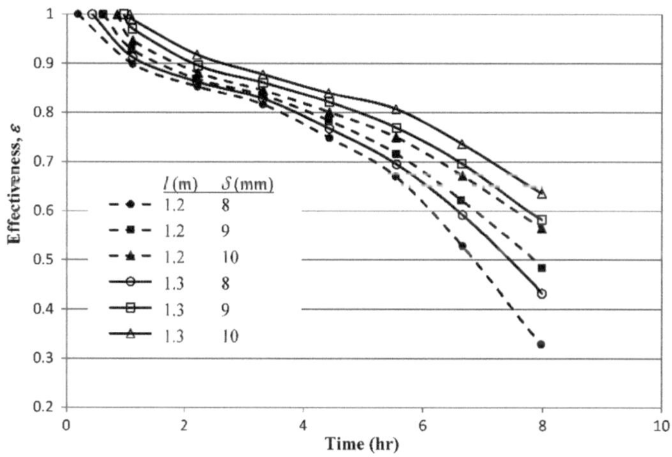

Fig. 9.12 Effect of length and thickness of the PCM slabs on the instantaneous heat exchanger effectiveness during the melting process ($T_{\text{inlet}} = 29$ °C, $\dot{V} = 1200$ m^3/h, $\gamma = 3.2$ mm) [5]

9.3.4.3 Effect of Air Flow Rate on Outlet Air Temperature

Figure 9.13 shows how varying air flow rates affect the outlet air temperature during the melting process.

- *Higher flow rates*: At higher air flow rates, the heat transfer fluid removes heat more quickly, resulting in higher outlet air temperatures due to reduced residence time in the PCM storage unit.

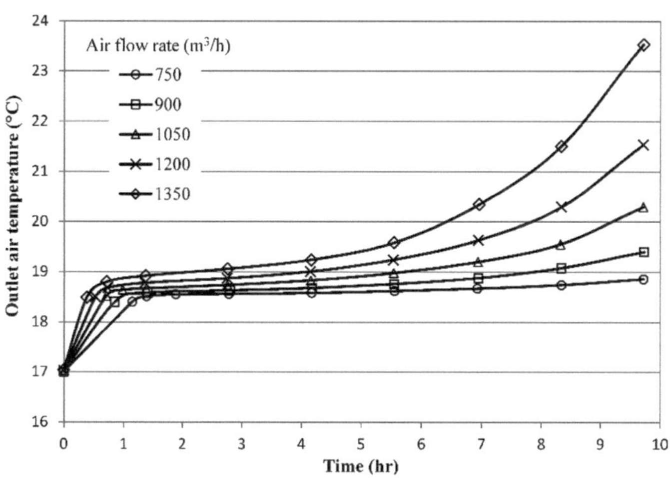

Fig. 9.13 Effect of air flow rate on the outlet air temperature ($T_{\text{inlet}} = 29$ °C, $l = 1.3$ m, $\delta = 9$ mm and $\gamma = 3.2$ mm) [5]

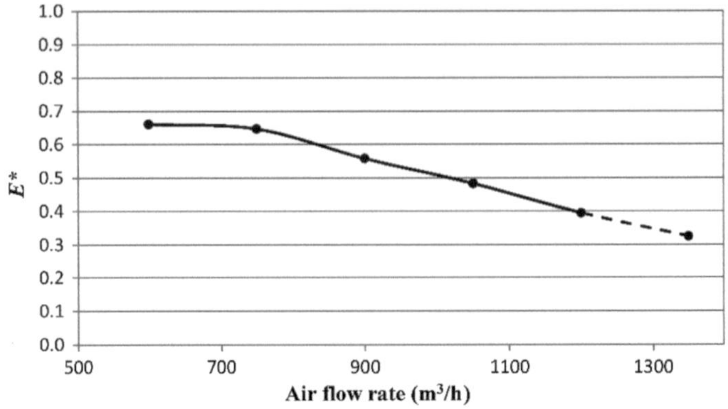

Fig. 9.14 Variation of the modified energy storage effectiveness with air flow rate ($T_{\text{inlet}} = 29$ °C, $l = 1.3$ m, $\delta = 9$ mm and $\gamma = 3.2$ mm). The dotted line indicates operating conditions outside the required range [5]

- *Lower flow rates*: Reduced flow rates allow for better heat transfer, maintaining the outlet air temperature below 20 °C for longer durations.
- *Optimal flow rate*: The system can achieve effective cooling for approximately 8 h with a maximum air flow rate of 1200 m^3/h, maintaining the desired thermal comfort.

Air flow rates must be carefully managed to optimize cooling duration while maintaining energy efficiency.

9.3.4.4 Effect of Airflow Rate

Higher airflow rates increase cooling load and energy consumption but improve COP due to enhanced heat transfer. The system achieves a maximum COP of 7.63 with an airflow rate of 1200 m^3/h during melting and 4.6 m/s during solidification (Figs. 9.14 and 9.15).

9.3.4.5 Air Channel Thickness Optimization

Reducing air channel thickness enhances energy storage effectiveness but increases energy consumption during solidification. An optimal thickness of 3.2 mm balances cooling performance and energy efficiency (Figs. 9.16 and 9.17).

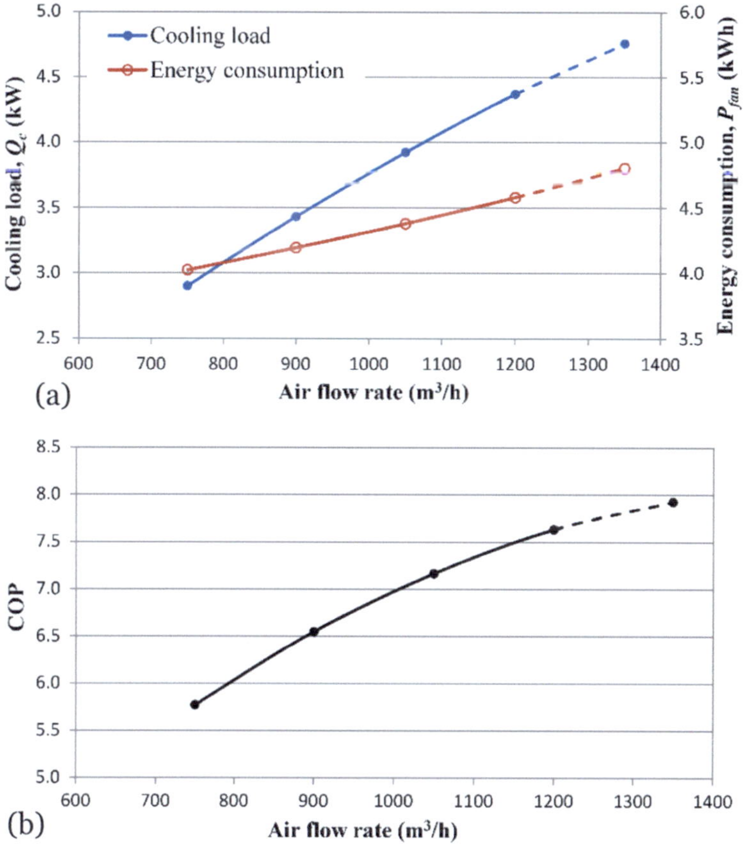

Fig. 9.15 Effect of air flow rate on the **a** cooling load and energy consumption and **b** COP with air flow rate ($T_{\text{inlet}} = 29$ °C, $l = 1.3$ m, $\delta = 9$ mm and $\gamma = 3.2$ mm). The dotted lines indicate operating conditions outside the required range [5]

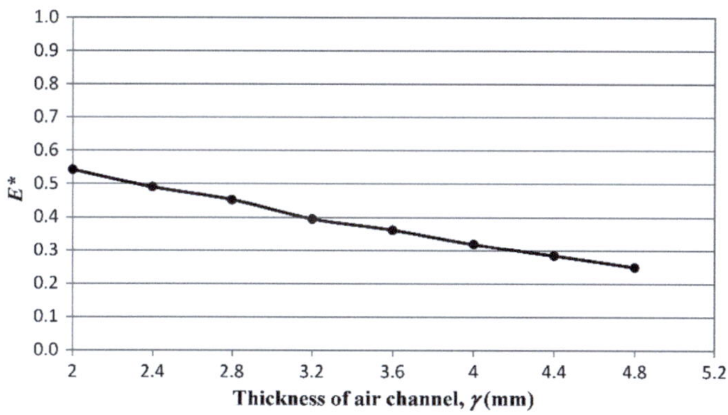

Fig. 9.16 Variation of the modified energy storage effectiveness with the thickness of air channels ($T_{\text{inlet}} = 29$ °C, $\dot{V} = 1200$ m^3/h, $l = 1.3$ m, $\delta = 9$ mm) [5]

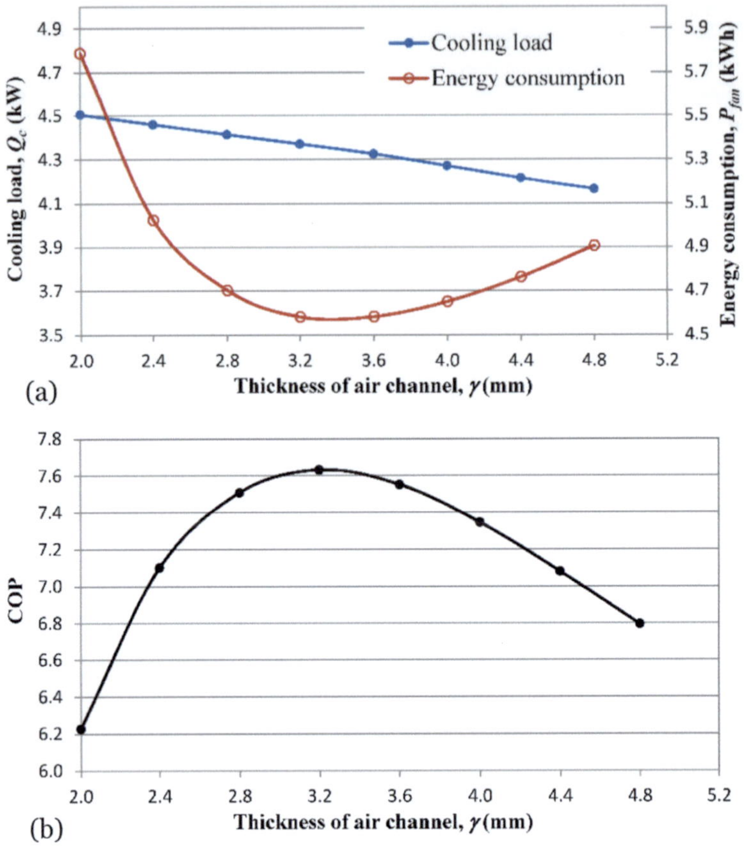

Fig. 9.17 Variation of the **a** cooling load and energy consumption and **b** COP with thickness of the air channel ($T_{\text{inlet}} = 29$ °C, $\dot{V} = 1200$ m^3/h, $l = 1.3$ m, $\delta = 9$ mm) [5]

9.3.4.6 Seasonal Performance

The proposed system maintains outlet air temperatures below 20 °C during peak summer days in Eindhoven. Daily COP variations reflect the system's capability to adapt to changing environmental conditions (Figs. 9.18 and 9.19).

9.3.5 Conclusions

This study demonstrates the potential of multiple PCM-based LHTS units in free cooling applications. Key findings include:

- Optimal configurations achieve a COP of 7.63 under peak summer conditions, with a cooling load of 4.38 kW.

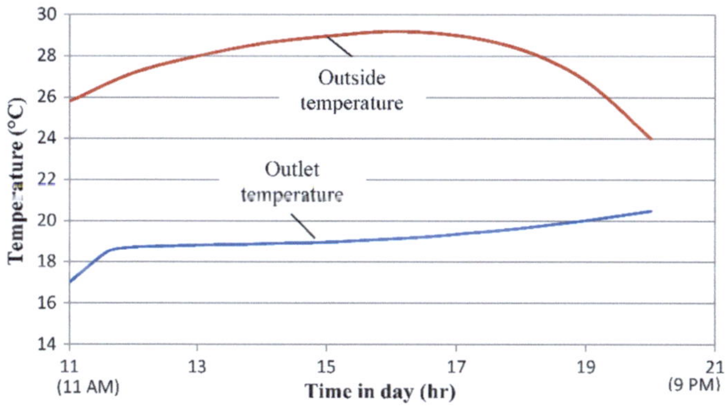

Fig. 9.18 Outlet air temperature for proposed free cooling LHTS unit subject to a variable outside air temperature during the peak summer day ($\dot{V} = 1200\ \mathrm{m}^3/\mathrm{h}$, $l = 1.3$ m, $\delta = 9$ mm and $\gamma = 3.2$ mm) [5]

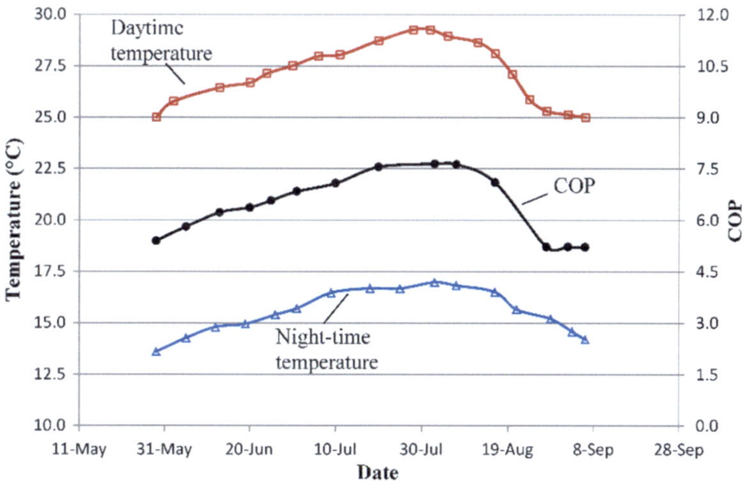

Fig. 9.19 Daily variation of COP and maximum temperatures during hot season in Eindhoven ($\dot{V} = 1200\ \mathrm{m}^3/\mathrm{h}$, $l = 1.3$ m, $\delta = 9$ mm and $\gamma = 3.2$ mm) [5]

- Best performance is achieved with PCM slab dimensions of 1.3 m length and 9 mm thickness, and an air channel thickness of 3.2 mm.
- While energy storage effectiveness provides valuable insights, COP-based optimization is more effective for energy-efficient design.

9.4 Energy and Exergy Evaluation of Multiple-PCM Thermal Storage for Free Cooling Applications

Thermal energy storage is essential for managing mismatched energy supply and demand, especially in cooling applications. Latent heat thermal storage systems, utilizing phase-change materials, are particularly effective for free cooling. In this approach, cool air is stored at night and released during the hottest hours of the day to maintain thermal comfort in buildings.

While energy analysis is widely used to evaluate TES performance, exergy analysis offers deeper insights into system efficiency by accounting for irreversibilities. This section presents a combined energy and exergy evaluation of a free cooling system using a multiple-PCM LHTS unit [7]. The study investigates the impact of air flow rate and inlet air temperature on performance metrics, highlighting the advantages of multiple-PCM configurations over single-PCM systems.

9.4.1 System Description

The LHTS unit consists of parallel PCM slabs separated by air channels for HTF flow. The slabs incorporate multiple PCMs with distinct melting temperatures arranged in decreasing order for the charging process to optimize heat transfer. The system schematic is shown in Fig. 9.9. Air passes through the channels, transferring heat to or from the PCMs, which undergoes phase change. Figure 9.20 shows the computational domain and primary parameters.

Key design used in the performance analysis parameters include:

- *PCM selection*: Calcium chloride hexahydrate ($CaCl_2{\cdot}6H_2O$) and RT25, chosen for their thermal compatibility.
- *Flow configuration*: The storage system specifications in Table 9.6 are used in the analyses.

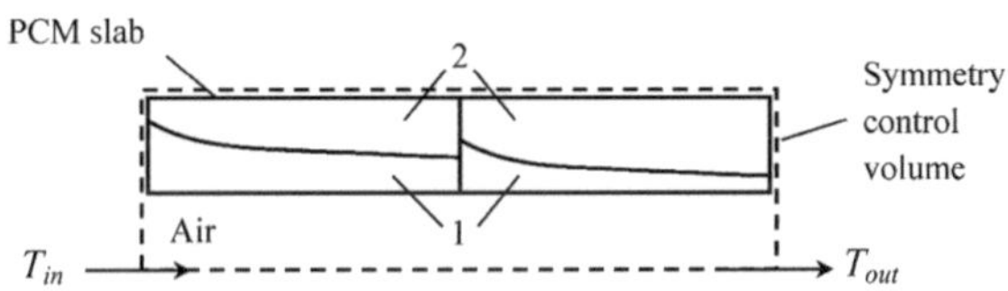

Fig. 9.20 Computational domain during charging: (1) liquid PCM with the mass of m_1 at $\overline{T}_1$ and (2) solid PCM with the mass of m_s at $\overline{T}_s$ [7]

Table 9.6 System specifications of the storage used in the calculations [7]

Parameters	Value
Thickness of PCM slabs	10 mm
Thickness of air channels	3.2 mm
Length of PCM slabs	1.3 m
Width of storage	0.5 m
Number of PCM slabs	80
Operating time during daytime	8 h
Efficiency of the fan	60%
Pressure drop for air filters and distribution equipment	180 Pa

9.4.2 Mathematical Formulation

9.4.2.1 Exergy Analysis

During the charging process, the air stream transfers exergy to the PCM, where a portion of this exergy is stored. The useful effect in this process is the change in exergy of the PCM. Conversely, during discharging, the stored exergy within the PCM is released and transferred to the air stream, becoming available for use. Based on this exchange, exergy efficiencies for charging and discharging can be defined as:

$$\eta_{\text{exe, charging}} = \frac{\text{Total exergy supplied by the HTF}}{\text{Net exergy change in the PCMs} + \text{Power input}} \tag{9.27}$$

$$\eta_{\text{exe, discharging}} = \frac{\text{Net exergy change in the PCMs}}{\text{Total exergy extracted by the HTF} + \text{Power input}} \tag{9.28}$$

- **Exergy Change in the PCM**

The net exergy change within each PCM element can be expressed as:

$$\begin{aligned} \text{Ex}_{\text{PCM}} = & \left\{ m_{\text{l}} c_{\text{p,s}} \left[(T_{\text{m}} - T_{\text{i}}) - \overline{T}_0 \ln \frac{T_{\text{m}}}{T_{\text{i}}} \right] \right\} + m_{\text{l}} L \left(1 - \frac{\overline{T}_0}{T_{\text{m}}} \right) \\ & + \left\{ m_{\text{l}} c_{\text{p,l}} \left[\left(\overline{T}_{\text{l}} - T_{\text{m}} \right) - \overline{T}_0 \ln \frac{\overline{T}_{\text{l}}}{T_{\text{m}}} \right] \right\} + \left\{ m_{\text{s}} c_{\text{p,s}} \left[\left(\overline{T}_{\text{s}} - T_{\text{i}} \right) - \overline{T}_0 \ln \frac{\overline{T}_{\text{s}}}{T_{\text{i}}} \right] \right\} \end{aligned} \tag{9.29}$$

where:

- m is the mass of the PCM.
- T_{i} is the initial temperature of the solid PCM.
- T_{s} is the temperature of the solid PCM at a given time $t > 0$.

- T_l is the temperature of the liquid PCM.
- T_m is the melting temperature of the PCM.
- c_{ps} and c_{pl} are the specific heat capacities of the solid and liquid PCM, respectively.
- T_0 is the dead state temperature.
- L is the latent heat of fusion.

This equation accounts for the exergy changes associated with sensible heating of the solid PCM, the phase-change process itself, and sensible heating of the liquid PCM.

- **Exergy Supplied/Extracted by the Air**

The total exergy supplied or extracted by the air can be expressed as:

$$\mathrm{Ex}_{\mathrm{air}} = \int_0^t \dot{m}\left[c_{\mathrm{p,air}}(T_{\mathrm{out}} - T_{\mathrm{in}}) - \overline{T}_0(s_{\mathrm{out}} - s_{\mathrm{in}})\right]\mathrm{d}t \tag{9.30}$$

where s_{in} and s_{out} are the specific entropies of the air at the inlet and outlet, respectively.

Also, the thermodynamic average temperature is defined as:

$$\overline{T}_0 = \frac{T_{\mathrm{out}} - T_{\mathrm{in}}}{\ln(T_{\mathrm{out}}/T_{\mathrm{in}})} \tag{9.31}$$

And the specific entropy change can be calculated using:

$$s_{\mathrm{out}} - s_{\mathrm{in}} = c_{\mathrm{p,air}}\ln\frac{T_{\mathrm{out}}}{T_{\mathrm{in}}} - R\ln\frac{p_{\mathrm{out}}}{p_{\mathrm{in}}} \tag{9.32}$$

- **Average Exergy Efficiency and Overall System Efficiency**

The average exergy efficiency during the phase-change process can be determined by integrating the instantaneous exergy efficiency over the duration of the process:

$$\overline{\eta}_{\mathrm{exe}} = \frac{1}{t}\int_0^t \eta_{\mathrm{exe}}\mathrm{d}t \tag{9.33}$$

The operation of an LHTS unit constitutes a cycle, consisting of an energy storage (charging) phase followed by an energy release (discharging) phase. Therefore, the overall exergy efficiency of a PCM free cooling unit, considering both charging and discharging processes, can be determined as:

$$\eta_{\mathrm{exe,overall}} = \frac{\text{Total exergy supplied by the HTF}}{\text{Total exergy input to the system}} = \frac{\mathrm{Ex}_{\mathrm{air,supplied}}}{W_{\mathrm{fan}}} \tag{9.34}$$

9.4.3 Results and Discussion

9.4.3.1 Thermal Performance

- *Air temperature profiles*: Air entering at 36 °C is cooled as it passes through the LHTS unit, with outlet temperatures remaining below 27 °C during operation (Fig. 9.21). The cooling rate decreases as PCMs approach full melting.
- *Effect of inlet air temperature on outlet air temperature*: Figure 9.22 shows the impact of varying inlet air temperatures on outlet air temperature during the charging process. Higher inlet temperatures result in increased heat transfer rates, causing the PCMs to melt faster and leading to higher outlet temperatures. The results highlight the need to carefully select inlet air temperatures to balance cooling performance and energy consumption.
- *Impact of air flow rate*: Higher air flow rates increase heat transfer but lead to higher outlet temperatures due to reduced residence time (Fig. 9.23).
- *Effect of inlet air temperature on heat transfer rate*: Figure 9.24 illustrates how the heat transfer rate in the LHTS unit changes with inlet air temperature during the day. As the inlet air temperature increases, the driving temperature difference between the air and PCMs grows, enhancing the heat transfer rate. However, higher inlet temperatures can accelerate PCM melting, reducing the heat transfer duration.
- *Effect of air flow rate on heat transfer rate*: Figure 9.25 examines the relationship between air flow rate and heat transfer rate in the LHTS unit. Higher air flow rates increase the heat transfer rate due to improved convection. However, rapid

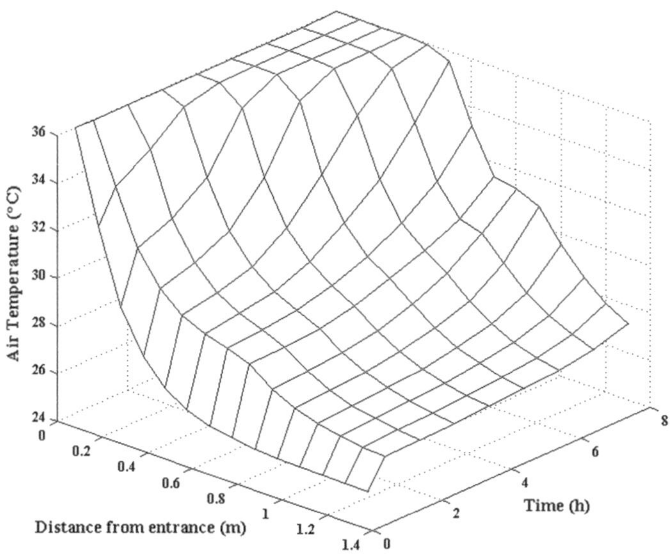

Fig. 9.21 Air temperature profiles during passage through the LHTS unit ($\dot{V} = 1200$ m^3/h) [6]

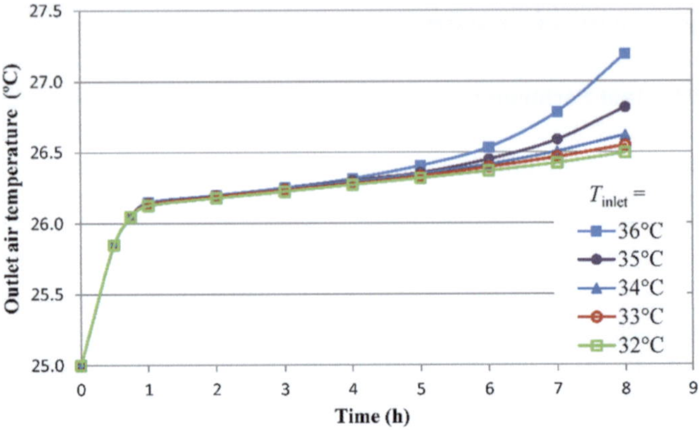

Fig. 9.22 Effect of inlet air temperature on the on the outlet air temperature during the day ($\dot{V}$ = 1200 m^3/h) [6]

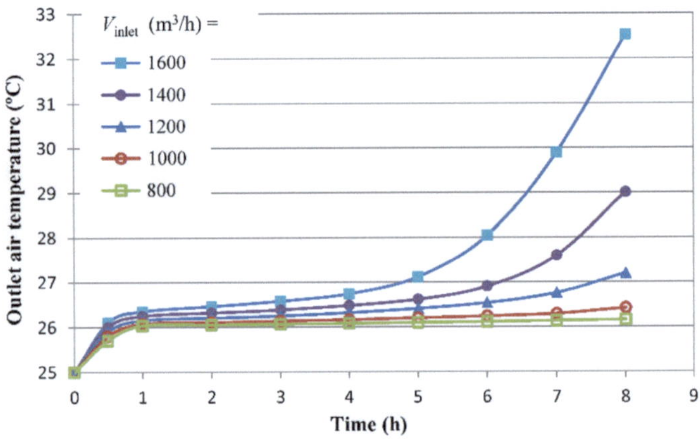

Fig. 9.23 Effect of air flow rate on the on the outlet air temperature during the day (T_{inlet} = 36 °C) [6]

melting at higher flow rates can reduce the system's cooling duration, indicating a trade-off between instantaneous and long-term performance.

9.4.3.2 Exergy Analysis

- *Exergy efficiency*: Exergy efficiency improves with lower inlet air temperatures and higher air flow rates, as these conditions reduce irreversibility (Figs. 9.26 and 9.27). However, the benefits of higher flow rates are offset by increased fan energy consumption.

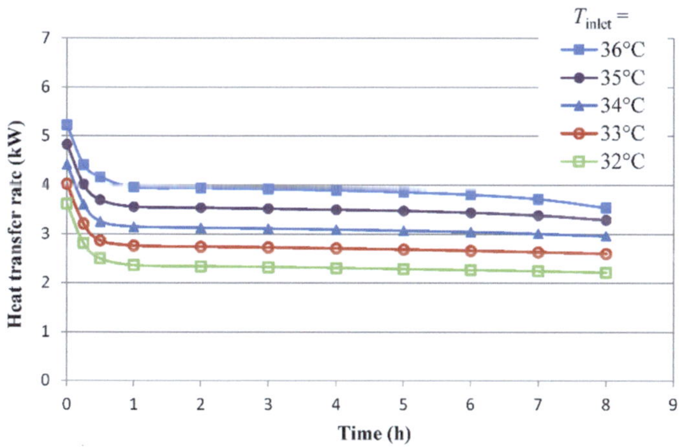

Fig. 9.24 Effect of inlet air temperature on the heat transfer rate during the day ($\dot{V}$ = 1200 m^3/h) [6]

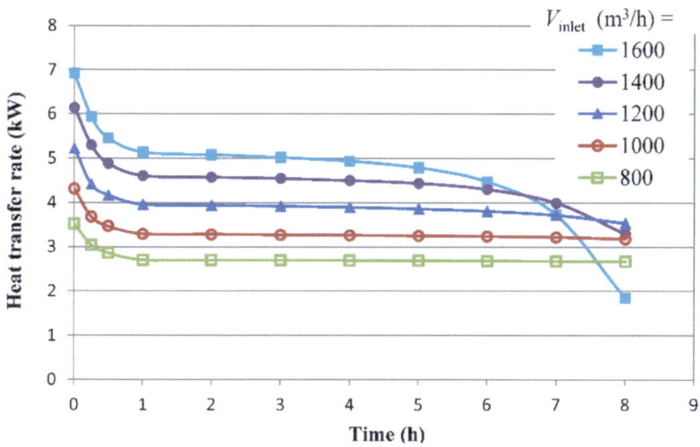

Fig. 9.25 Effect of air flow rate on the heat transfer rate during the day (T_{inlet} = 36 °C) [6]

- *Comparison of single and multiple PCMs*: Table 9.7 compares the exergy efficiencies of the system when using single or multiple PCMs. Multiple-PCMs achieve significantly higher exergy efficiencies due to reduced irreversibility during the charging and discharging processes. The table demonstrates the advantage of using multiple PCMs for improving the thermodynamic performance of the LHTS unit.

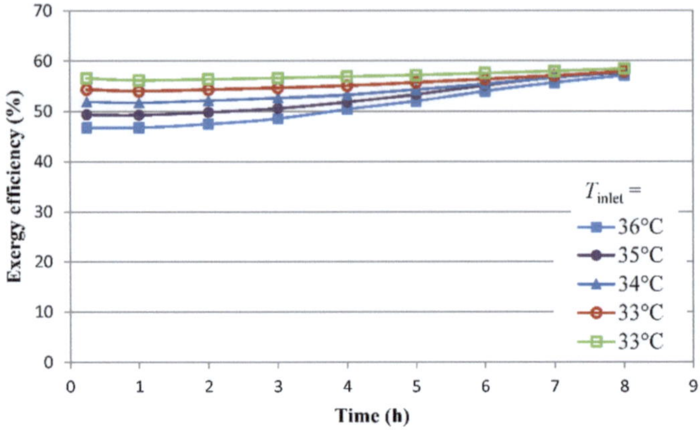

Fig. 9.26 Effect of inlet air temperature on the exergy efficiency during daytime ($\dot{V}$ = 1200 m^3/h) [6]

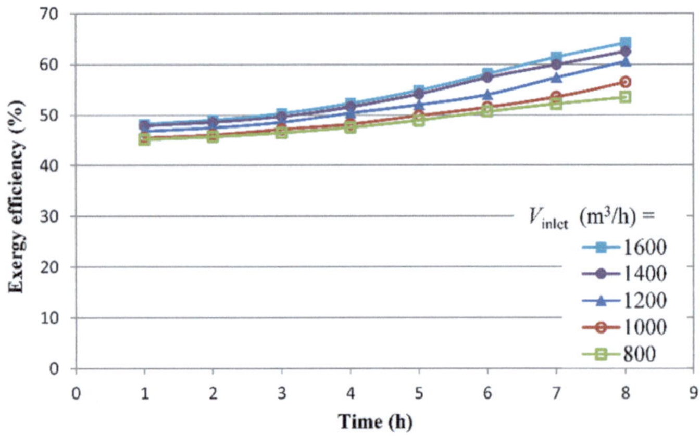

Fig. 9.27 Effect of air flow rate on the exergy efficiency during daytime (T_{inlet} = 36 °C) [6]

9.4.3.3 Optimized Configuration

The optimal system achieves a COP of 7.00 and an exergy efficiency of 15.15% at an inlet air temperature of 36 °C and an air flow rate of 1200 m^3/h. Cooling load increases with higher air flow rates but at the cost of reduced energy efficiency (Fig. 9.28).

Table 9.7 Results for the thermodynamics analyses considering the effects of inlet air temperature and air flow rate [6]

T_{inlet} (°C)	$\dot{V}$ (m^3/h)	$\dot{Q}_c$ (kW)	$t_{extract}$ (h)	W_{fan} (kW h)	$\overline{\eta}_{exe,charging}$ (%)	$\overline{\eta}_{exe,discharging}$ (%)
32	1200	2.534	8.83	3.809	57.0	59.1
33	1200	2.966	9.88	4.118	55.4	59.6
34	1200	3.395	11.43	4.573	53.9	59.9
35	1200	3.821	12.2	4.780	52.5	60.4
36	1200	4.234	12.51	4.890	50.8	60.9
36	800	2.960	9.76	3.576	48.2	60.7
36	1000	3.578	11.14	4.220	49.0	60.8
36	1400	4.803	13.13	5.371	53.0	61.0
36	1600	5.049	14.79	6.188	53.8	61.2

9.4.3.4 Outlet Air Temperature and Exergy Efficiency During a Peak Summer Day

Figure 9.29 presents the variation of outlet air temperature and exergy efficiency over the operating period during a peak summer day. The outlet air temperature remains below 27 °C throughout the day, ensuring thermal comfort. The exergy efficiency varies with the temperature difference between the inlet and outlet air, with higher efficiency observed at lower temperature differences. This highlights the system's ability to provide consistent cooling while minimizing irreversibility.

9.4.4 Conclusions

Energy and exergy analyses of a multiple-PCM LHTS unit for free cooling highlight the system's ability to enhance cooling performance in hot climates. Key findings include:

1. Multiple PCMs outperform single PCMs, achieving higher exergy efficiency by minimizing irreversibility.
2. Optimized configurations maintain outlet air temperatures below 27 °C and achieve a COP of 7.00.
3. Lower inlet air temperatures improve exergy efficiency more effectively than higher air flow rates.

This section provides a robust framework for designing energy-efficient TES systems for diverse climatic conditions.

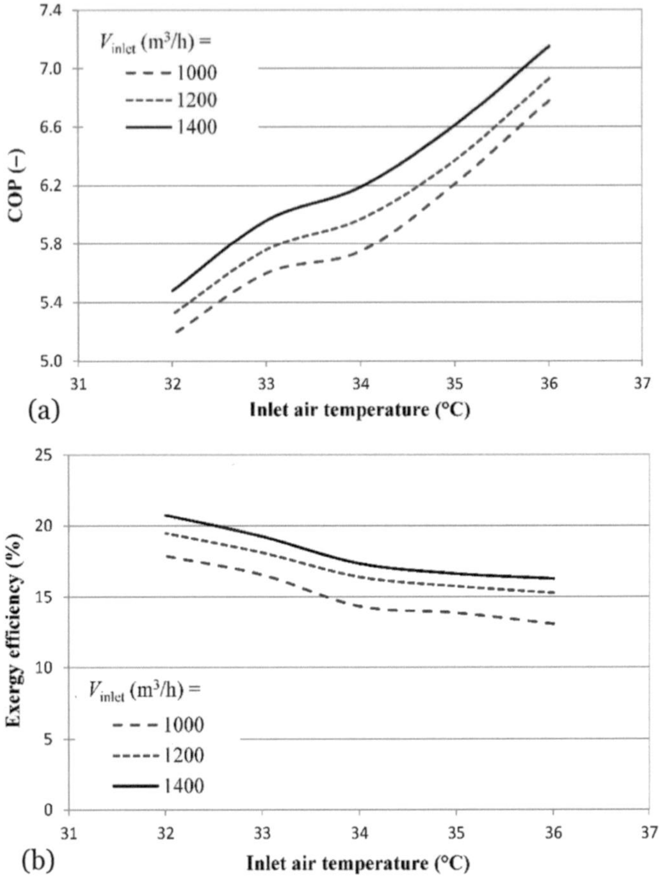

Fig. 9.28 Variation of **a** the COP and **b** overall exergy efficiency with inlet air temperature for various air flow rates [6]

9.5 Advanced Exergy Analysis of Air-Conditioning Systems with Latent Heat Thermal Storage

The integration of thermal storage into air conditioning (AC) systems has gained popularity due to its ability to offset peak cooling loads and lower electricity costs during high-demand periods. Latent heat thermal storage systems, employing PCMs, efficiently store and release cooling energy. While conventional energy analyses focus on energy quantities, they fail to capture the quality of energy conversion. Exergy analysis, grounded in the second law of thermodynamics, provides a comprehensive framework for assessing system inefficiencies and potential improvements.

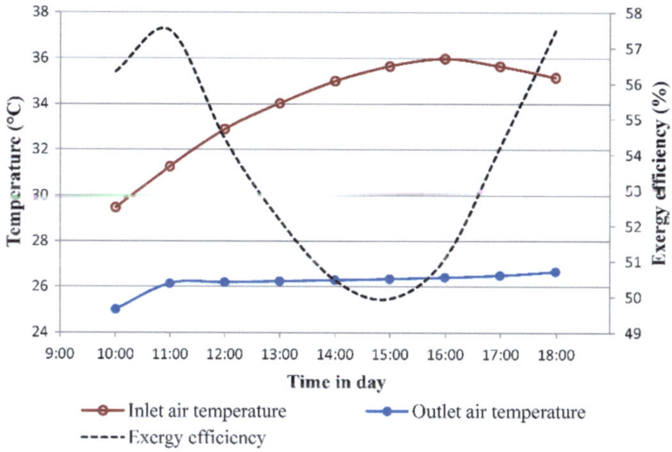

Fig. 9.29 Outlet air temperature and exergy efficiency subject to the variable inlet air temperature during the peak summer day of 2012 ($\dot{V} = 1200$ m^3/h, COP = 7.00 and $\eta_{\text{exe,overall}} = 15.15\%$) [6]

This section applies advanced exergy analysis to an AC system combining an LHTS unit and a VCR cycle [6]. By splitting exergy destruction into various components, this approach identifies key areas for improving system performance, reducing irreversibility, and optimizing energy utilization.

9.5.1 System Description

The system includes a series of PCM slabs arranged in parallel with airflow channels, as illustrated in Fig. 9.30. During nighttime, the VCR cycle solidifies the PCMs by extracting heat, while during daytime, the stored cooling energy is released to cool the air. Key system specifications are provided in Table 9.8.

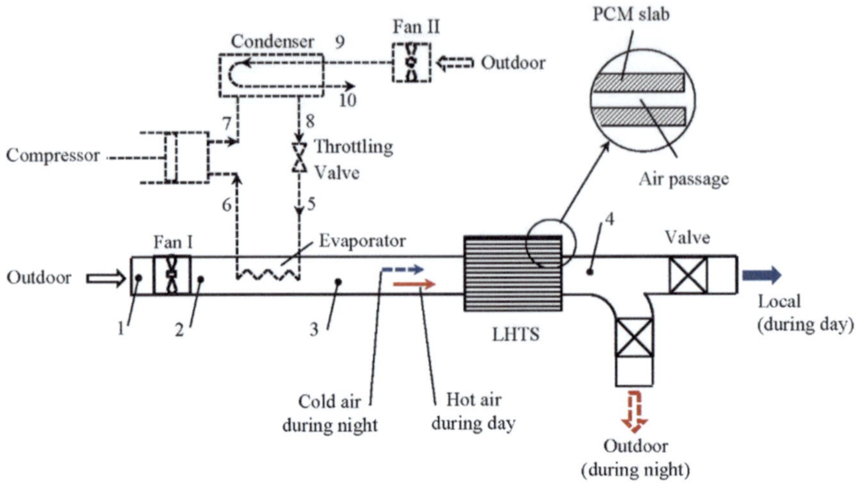

Fig. 9.30 Schematic of the cooling system using a LHTS unit [6]

Table 9.8 LHTS unit specifications used in the calculations [6]

Parameters	Value
Thickness of PCM slabs	10 mm
Thickness of air channels	5 mm
Length of PCM slabs	1.3 m
Width of storage	1.1 m
Number of PCM slabs	34
Operating time during daytime	8 h
Air mass flow rate	1 kg/s

PCM Characteristics

- *Material*: Calcium chloride hexahydrate ($CaCl_2 \cdot 6H_2O$)
- *Melting temperature*: 28–29 °C
- *Latent heat of fusion*: 190,800 J/kg.

The LHTS unit is designed to provide an average cooling load of 8.78 kW during an 8-h daytime operation, maintaining indoor comfort conditions.

9.5.2 Mathematical Formulation

9.5.2.1 Energy Analysis

A comprehensive energy analysis of this LHTS-based air-cooling system has been previously demonstrated. Referring to Fig. 9.30, the time-averaged heat transfer rate within the LHTS unit can be expressed as:

$$\overline{\dot{Q}}_{\mathrm{PCM}} = \frac{1}{\tau}\int_{0}^{\tau} \dot{Q}(t)\mathrm{d}t = \left[\dot{m}c_{\mathrm{p}}\left(\overline{T}_{3} - \overline{T}_{2'}\right)\right]_{\mathrm{air}} \tag{9.35}$$

The time-averaged temperature is defined as:

$$\overline{T} = \frac{1}{\tau}\int_{0}^{\tau} T\mathrm{d}t \tag{9.36}$$

where t represents the operating time.

The system's coefficient of performance (COP) is defined as:

$$\mathrm{COP} = \frac{\overline{\dot{Q}}_{\mathrm{PCM}}}{\dot{W}_{\mathrm{total}}} \tag{9.37}$$

where $\dot{W}_{\mathrm{total}}$ is the total power consumption during the cycle operation (both day and night), which includes the power consumption of the fan and the compressor (if applicable).

9.5.2.2 Exergy Analysis

Exergy analysis provides insights into the thermodynamic inefficiencies of the system by quantifying exergy destruction. The total exergy destruction rate can be determined using an exergy balance:

$$\dot{E}x_{\mathrm{F}} = \dot{E}x_{\mathrm{P}} + \dot{E}x_{\mathrm{D}} \tag{9.38}$$

The exergy efficiency is defined as the ratio of the total product exergy output rate to the total exergy input (fuel) rate:

$$\psi = \frac{\dot{E}x_{\mathrm{P}}}{\dot{E}x_{\mathrm{F}}} \tag{9.39}$$

The relative irreversibility (RI) is a metric used to compare the exergy destruction within a component to the total exergy input to the system:

$$\mathrm{RI} = \frac{\dot{E}x_{\mathrm{D}}}{\dot{E}x_{\mathrm{D,overall}}} \tag{9.40}$$

During the air-cooling (discharging) process, exergy is transferred from the air to the PCM, where a portion is stored, causing a change in the PCM's exergy. The exergy efficiency during discharge can be expressed as:

$$\psi_{\mathrm{discharging}} = \frac{\dot{E}x_{\mathrm{air}}}{\dot{E}x_{\mathrm{PCM}} + \dot{W}_{\mathrm{Fan}}} \tag{9.41}$$

During the PCM solidification (charging) process, the exergy released by the PCM is transferred to the air. The exergy efficiency of the charging process can be expressed as:

$$\psi_{\mathrm{charging}} = \frac{\dot{E}x_{\mathrm{PCM}}}{\dot{E}x_{\mathrm{air}} + \dot{W}_{\mathrm{Fan}} + \dot{W}_{\mathrm{CM}}} \tag{9.42}$$

The exergy destruction rate for the fan can be obtained from an exergy rate balance:

$$\dot{E}x_{\mathrm{D,Fan}} = \dot{E}x_{\mathrm{F,Fan}} - \dot{E}x_{\mathrm{P,Fan}} = \left(\dot{W}_{\mathrm{Fan}}\right) - \left(\dot{E}x_2 - \dot{E}x_1\right) \tag{9.43}$$

where $\dot{W}_{\mathrm{Fan}}$ is the fan power.

9.5.3 *Advanced Exergy Analysis*

Advanced exergy analysis extends conventional exergy analysis by providing a more detailed understanding of exergy destruction within a system. It pinpoints the sources of thermodynamic inefficiencies at the component level. This is achieved by decomposing the total exergy destruction rate of each component into two primary parts:

- *Endogenous exergy destruction*: This represents the exergy destruction that would occur within a specific component if all other components in the system operated ideally (reversibly), while the component under consideration operates at its actual, real-world efficiency. This isolates the inherent irreversibility of the component itself.
- *Exogenous exergy destruction*: This portion of exergy destruction is caused by the inefficiencies (irreversibility) of other components in the system. It quantifies how the performance of other components impacts the exergy destruction within the component being analyzed.

This decomposition allows for a more nuanced understanding of how each component contributes to the overall exergy losses of the system. Further refinement of the analysis involves splitting the exergy destruction into avoidable and unavoidable parts:

- *Avoidable exergy destruction*: This portion represents the exergy destruction that can be potentially reduced through technological improvements, design modifications, or operational changes. It indicates the potential for thermodynamic performance enhancement.
- *Unavoidable exergy destruction*: This represents the exergy destruction that is inherently limited by current technological constraints, material properties, or physical laws. It sets a practical upper limit on the achievable performance improvements.

This further breakdown can be expressed as: $\dot{E}x_{\mathrm{D}} = \dot{E}x_{\mathrm{D}}^{\mathrm{UN}} + \dot{E}x_{\mathrm{D}}^{\mathrm{AV}}$.

9.5.3.1 Application to the LHTS Cooling System

Figure 9.31 illustrates the real, theoretical, and unavoidable thermodynamic processes during the daytime air cooling operation of the LHTS system.

- *Real process (1-2R-a-3)*: This represents the actual operation of the system, including all real-world irreversibility. The fan operates with its actual efficiency, resulting in air leaving the fan at state 2R. The minimum temperature difference in the LHTS unit has its real value, and the air exits the LHTS at state 3.
- *Theoretical process (1-2T-b-3)*: This represents a hypothetical, fully reversible operation of the system, where all components operate with ideal efficiencies ($\eta_{\mathrm{Fan}}^{\mathrm{T}} = 1, \Delta T_{\mathrm{min,\ theoretical}}$). The air exits the fan at state 2T and the LHTS at state 3.
- *Unavoidable process (1-2UN-c-3)*: This process combines the theoretical cycle with the unavoidable irreversibility of each component. It represents the best achievable performance given current technological limitations.

In Fig. 9.31, ΔT_{min} represents the minimum temperature difference within the LHTS unit, calculated as the difference between the time-averaged temperature at the PCM slab's outlet in the airflow direction ($T_{\mathrm{PCM,\ out,\ avg}}$) and the outlet air temperature

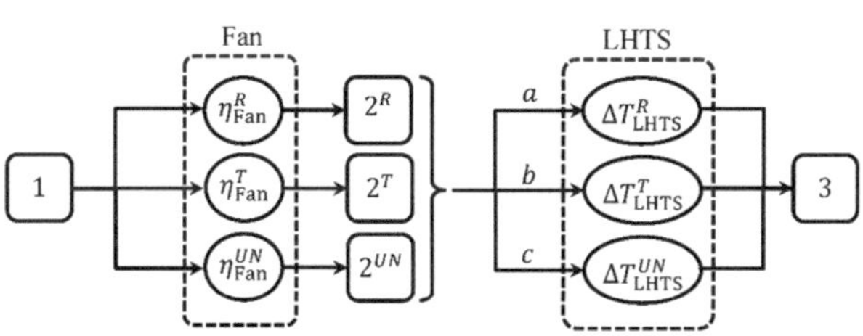

Fig. 9.31 Simplified illustration of various processes for air cooling [6]

(T_{out}). Ambient air (state 1) flows through parallel channels within the LHTS unit, and the cooled air (state 3) provides cooling to the building.

9.5.3.2 Calculating Endogenous Exergy Destruction

To calculate the endogenous exergy destruction of each component, hybrid cycles are analyzed, where only one component operates with its real-world characteristics while all other components operate theoretically. For example:

- *Fan*: The fan operates with its real efficiency ($\eta_{fan, real}$), resulting in air at state 2R. The LHTS unit operates theoretically ($\Delta T_{min, theoretical}$), and the air exits at state 3 (process 1-2R-b-3).
- *LHTS unit*: The fan operates theoretically ($\eta_{fan, theoretical}$), resulting in air at state 2T. The LHTS unit operates with its actual minimum temperature difference ($\Delta T_{min, real}$), and the air exits at state 3 (process 1-2T-a-3).

The unavoidable endogenous exergy destruction is calculated similarly, but each component is assumed to operate at its unavoidable level of irreversibility.

9.5.3.3 Application to the VCR Cycle

Figures 9.32 and 9.33 illustrate the application of this advanced exergy analysis to a Vapor Compression Refrigeration (VCR) cycle, used for both air cooling (Fig. 9.32) and PCM charging (Fig. 9.33). These figures depict the real, theoretical, and hybrid cycles, showing the various states used for calculating the different components of exergy destruction.

For instance, to determine the endogenous exergy destruction of the condenser in the VCR cycle (Fig. 9.32), other components operate under theoretical conditions, while the condenser operates under real conditions. The refrigerant exits the evaporator theoretically at state 4T, is compressed isentropically to the real condenser pressure (state 5^{**}H), and exits the condenser as saturated liquid at state 6R. The throttling valve is replaced by an isentropic expander for the theoretical cycle (state 7^{*}H). A detailed, advanced exergy analysis of a VCR cycle has been conducted by Morosuk and Tsatsaronis [2].

Figure 9.33 shows the processes during the nighttime heat extraction from the PCM using the VCR cycle. The air is cooled by the evaporator (process 1-2T-2′R-3T), with the evaporator operating at its real conditions ($\Delta T_{min, real}$). The refrigerant then undergoes compression, condensation (with $\Delta T_{min, theoretical}$ for the theoretical case), and expansion.

Table 9.9 summarizes the specific conditions analyzed to calculate the endogenous exergy destruction for each component in the VCR cycle.

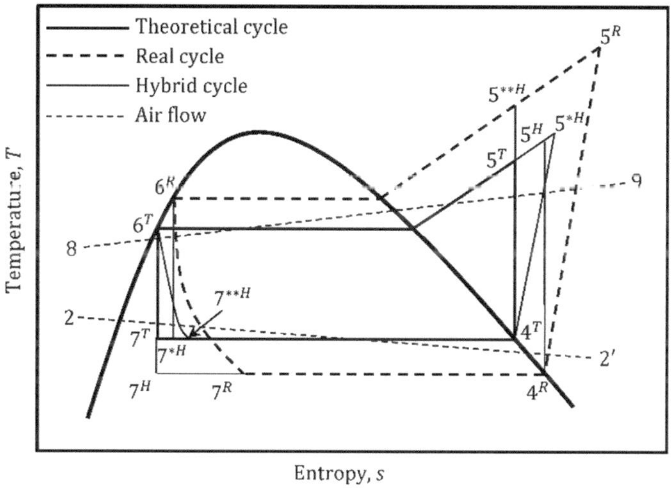

Fig. 9.32 Real, theoretical and hybrid cycles for calculating the endogenous part of the exergy destructions in the components of a VCR cycle [6]

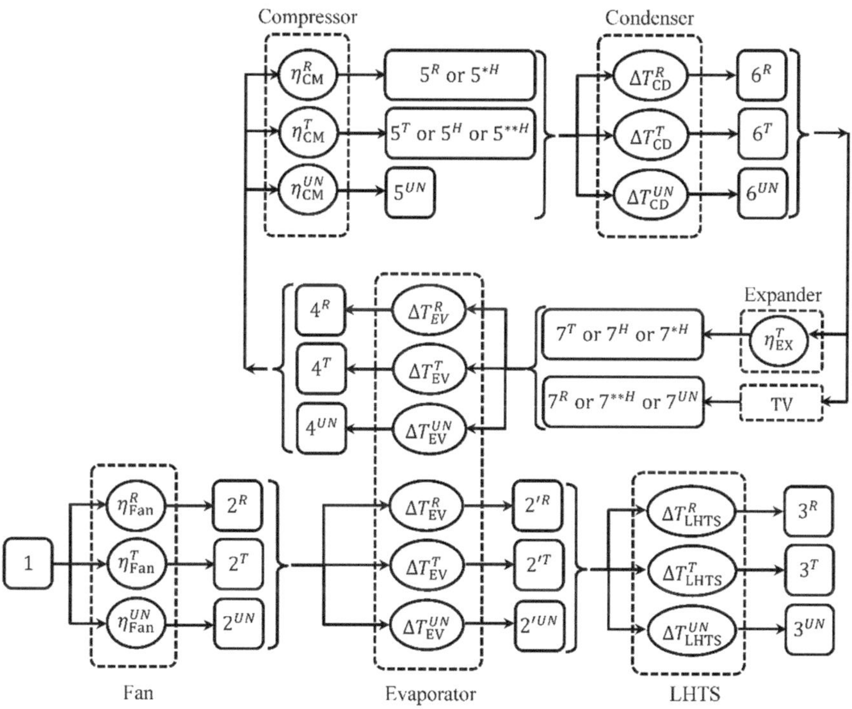

Fig. 9.33 Simplified illustration of various processes for extracting heat from a PCM with a VCR cycle [6]

Table 9.9 Various conditions used to calculate endogenous part of exergy destruction rate for each component [6]

$\dot{E}x_{D}$	Condition
$\dot{E}x^{EN}_{D,Fan}$	$1\text{-}2^{R}\text{-}2'^{T}\text{-}3^{T}$ and $4^{T}\text{-}5^{T}\text{-}6^{T}\text{-}7^{T}$
$\dot{E}x^{EN}_{D,LHTS}$	$1\text{-}2^{T}\text{-}2'^{T}\text{-}3^{R}$ and $4^{T}\text{-}5^{T}\text{-}6^{T}\text{-}7^{T}$
$\dot{E}x^{EN}_{D,EV}$	$1\text{-}2^{T}\text{-}2'^{R}\text{-}3^{T}$ and $4^{R}\text{-}5^{H}\text{-}6^{T}\text{-}7^{H}$
$\dot{E}x^{EN}_{D,CM}$	$1\text{-}2^{T}\text{-}2'^{T}\text{-}3^{T}$ and $4^{T}\text{-}5^{*H}\text{-}6^{T}\text{-}7^{T}$
$\dot{E}x^{EN}_{D,CD}$	$1\text{-}2^{T}\text{-}2'^{T}\text{-}3^{T}$ and $4^{T}\text{-}5^{**H}\text{-}6^{R}\text{-}7^{*H}$
$\dot{E}x^{EN}_{D,TV}$	$1\text{-}2^{T}\text{-}2'^{T}\text{-}3^{T}$ and $4^{T}\text{-}5^{T}\text{-}6^{T}\text{-}7^{**H}$

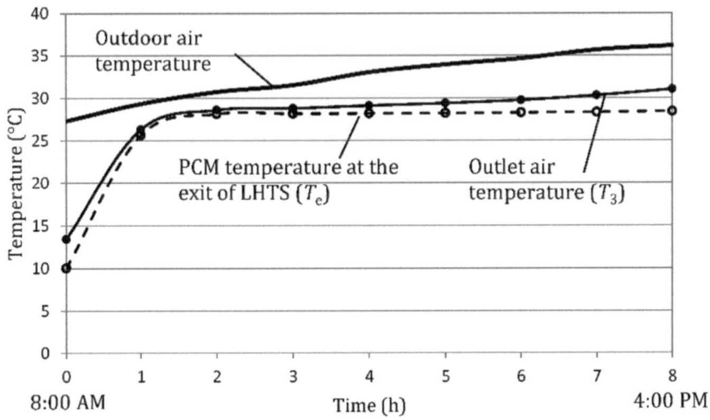

Fig. 9.34 Outdoor temperature, outlet air temperature and PCM temperature at the end of PCM slab profiles for the real process during the day ($\overline{T}_0 = 29.86$ °C, $\overline{T}_1 = 34.11$ °C, $\overline{T}_3 = 25.64$ °C and $\overline{T}_e = 24.69$ °C) [6]

9.5.4 Results and Discussion

9.5.4.1 Air Cooling Process (Day-Time)

Figure 9.34 shows the temperature profiles for outdoor air, outlet air, and the PCM slab. During the day, the PCM absorbs heat and maintains a steady cooling rate. The results indicate a constant heat transfer rate of 8.78 kW, ensuring thermal comfort for 8 h. Exergy analysis reveals that 37.77% of the system's exergy destruction is unavoidable.

9.5.4.2 Charging Process (Nighttime)

Figure 9.35 illustrates the temperature profiles during the charging process. The VCR cycle reduces the outdoor air temperature by 10 °C, enabling effective PCM solidification. The results show that the compressor and condenser exhibit the highest exergy destruction rates, accounting for over 50% of total irreversibility.

Table 9.10 outlines the assumptions used for analyzing the real and unavoidable operation conditions of the air conditioning system components, specifically the fan, compressor (CM), evaporator (EV), condenser (CD), and LHTS unit. These assumptions serve as the basis for splitting exergy destruction into avoidable and unavoidable components during advanced exergy analysis.

- *Fan*:
  - Real condition: Efficiency of 65%, with a minimum temperature difference of 0.95 K during melting and 5.42 K during solidification.
  - Unavoidable condition: Higher efficiency of 75% and reduced irreversibility with a minimum temperature difference of 0.2 K.
- *Compressor (CM)*:

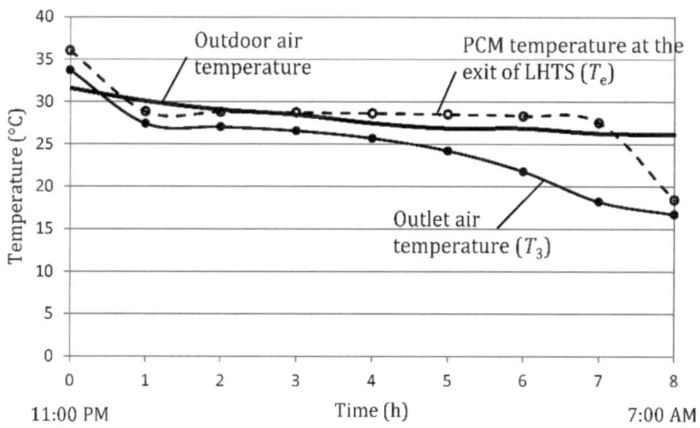

Fig. 9.35 Outdoor temperature, outlet air temperature and PCM temperature at the end of PCM slab profiles for the real process during the night ($\overline{T}_0 = 25.62$ °C, $\overline{T}_1 = 26.06$ °C, $\overline{T}_3 = 25.2$ °C and $\overline{T}_e = 30.62$ °C) [6]

Table 9.10 Assumptions for real and unavoidable conditions [6]

Component	Real condition	Unavoidable condition
Fan	$\eta_{Fan} = 65\%$, $\eta_{el} = 80\%$	$\eta_{Fan} = 5\%$, $\eta_{el} = 98\%$
CM	$\eta_{CM} = 80\%$	$\eta_{CM} = 95\%$
EV	$\Delta T_{min,\,EV} = 10$ K	$\Delta T_{min,\,EV} = 0.5$ K
CD	$\Delta T_{min,\,CD} = 10$ K	$\Delta T_{min,\,CD} = 0.5$ K
LHTS	$\Delta T_{LHTS} = 0.95$ K (melting) $\Delta T_{LHTS} = 5.42$ K (solidification)	$\Delta T_{LHTS} = 0.2$ K

  - Real condition: Efficiency of 80%.
  - Unavoidable condition: Increased efficiency to 95%, representing the theoretical technological limit.

- *Evaporator (EV)*:
  - Real condition: Minimum temperature difference of 10 K.
  - Unavoidable condition: Significantly reduced to 0.5 K, representing ideal performance.

- *Condenser (CD)*:
  - Real condition: Minimum temperature difference of 10 K.
  - Unavoidable condition: Reduced to 0.5 K for theoretical optimal performance.

- *LHTS unit*:
  - Real condition: Minimum temperature differences of 0.95 K for melting and 5.42 K for solidification.
  - Unavoidable condition: Idealized minimum temperature difference of 0.2 K.

Key Insights

1. *Technological Gaps*: The differences between real and unavoidable conditions highlight areas where improvements in material and design can significantly reduce irreversibility.
2. *Fan and LHTS Contributions*: While the fan and LHTS unit show minor gaps between real and unavoidable conditions, the EV and CD exhibit larger inefficiencies, making them prime targets for performance enhancement.
3. *Advanced Exergy Analysis Applications*: These assumptions enable detailed splitting of exergy destruction into avoidable and unavoidable parts, guiding targeted improvements for overall system efficiency.

9.5.4.3 Component Performance

Table 9.11 presents the advanced exergy analysis for daytime operation, while Table 9.12 summarizes nighttime performance. The results reveal that the LHTS unit's exergy destruction is entirely endogenous, indicating that its inefficiencies stem from internal irreversibility. Improvement efforts should focus on reducing these losses through enhanced PCM properties or optimized airflow design.

9.5.5 Conclusions

The advanced exergy analysis of the AC system highlights the potential of integrating LHTS units for efficient cooling. Key findings include:

Table 9.11 Results of advanced exergy analysis for the cooling system using a LHTS unit during the day ($\dot{E}x_{\mathrm{D}}$ in kW) [6]

	$\dot{E}x_{\mathrm{F}}^{\mathrm{R}}$	$\dot{E}x_{\mathrm{P}}^{\mathrm{R}}$	$\dot{E}x_{\mathrm{D}}^{\mathrm{R}}$	ψ (%)	RI (%)	$\dot{E}x_{\mathrm{D}}^{\mathrm{T}}$	$\dot{E}x_{\mathrm{D}}^{\mathrm{UN}}$	$\dot{E}x_{\mathrm{D}}^{\mathrm{AV}}$	$\dot{E}x_{\mathrm{D}}^{\mathrm{EN}}$	$\dot{E}x_{\mathrm{D}}^{\mathrm{EX}}$	$\dot{E}x_{\mathrm{D}}^{\mathrm{EN-UN}}$	$\dot{E}x_{\mathrm{D}}^{\mathrm{EX-UN}}$	$\dot{E}x_{\mathrm{D}}^{\mathrm{EN-AV}}$	$\dot{E}x_{\mathrm{D}}^{\mathrm{EX-AV}}$
Fan	0.352	0.175	0.177	49.66	74.39	0.01	0.075 (42.2%)	0.102 (57.8%)	0.177 (99.83%)	0.0003 (0.17%)	0.074 (99.71%)	0.0002 (0.29%)	0.102 (99.92%)	0.00008 (0.08%)
LHTS	0.0708	0.01	0.061	13.87	25.61	0.007	0.015 (24.91%)	0.046 (75.09%)	0.058 (95.58%)	0. 0026 (4.42%)	0.013 (89.09%)	0.0016 (10.91%)	0.044 (97.74%)	0.00103 (2.26%)
Overall	0.423	0.01	0.238	2.32	100	0.016 (6.88%)	0.090 (37.77%)	0.148 (62.23%)	0.235 (98.74%)	0. 0029 (1.26%)	0.088 (36.98%)	0.0018 (0.79%)	0.147 (61.76%)	0.00111 (0.47%)

Table 9.12 Results of advanced exergy analysis for a cooling system using a LHTS unit during the night ($\dot{E}x_{\mathrm{D}}$ in kW) [6]

	$\dot{E}x_{\mathrm{F}}^{\mathrm{R}}$	$\dot{E}x_{\mathrm{P}}^{\mathrm{R}}$	$\dot{E}x_{\mathrm{D}}^{\mathrm{R}}$	ψ (%)	RI (%)	$\dot{E}x_{\mathrm{D}}^{\mathrm{T}}$	$\dot{E}x_{\mathrm{D}}^{\mathrm{UN}}$	$\dot{E}x_{\mathrm{D}}^{\mathrm{AV}}$	$\dot{E}x_{\mathrm{D}}^{\mathrm{EN}}$	$\dot{E}x_{\mathrm{D}}^{\mathrm{EX}}$	$\dot{E}x_{\mathrm{D}}^{\mathrm{EN-UN}}$	$\dot{E}x_{\mathrm{D}}^{\mathrm{EX-UN}}$	$\dot{E}x_{\mathrm{D}}^{\mathrm{EN-AV}}$	$\dot{E}x_{\mathrm{D}}^{\mathrm{EX-AV}}$
Fan	0.345	0.169	0.175	49.21	8.23	0.01	0.074 (42.73%)	0.100 (57.27%)	0.174 (99.12%)	0.0012 (0.68%)	0.075 (99.76%)	0.00018 (0.24%)	0.099 (98.98%)	0.001 (1.02%)
LHTS	0.144	0.016	0.128	11.32	6.0	0.027	0.029 (22.46%)	0.099 (77.54%)	0.128 (100%)	0	0.029 (100%)	0	0.099 (100%)	0
CD	0.760	0.195	0.565	25.60	26.6	0.081	0.095 (16.79%)	0.470 (83.21%)	0.229 (40.57%)	0.336 (59.43%)	0.089 (93.96%)	0.0057 (6.04%)	0.140 (29.79%)	0.330 (70.21%)
CM	2.161	1.764	0.397	81.63	18.67	0	0.014 (3.48%)	0.383 (96.52%)	0.055 (14.05%)	0.341 (85.95%)	0.011 (86.1%)	0.0019 (13.9%)	0.044 (11.45%)	0.339 (88.55%)
EV	0.678	0.143	0.534	21.15	25.14	0.033	0.046 (8.55%)	0.489 (91.45%)	0.192 (35.9%)	0.342 (64.1%)	0.041 (89.35%)	0.0048 (10.65%)	0.151 (30.9%)	0.338 (69.1%)
TV	0.979	0.653	0.326	66.68	15	0	0.025 (7.71%)	0.301 (92.29%)	0.020 (6.3%)	0.306 (93.7%)	0.021 (81.67%)	0.0046 (18.33%)	0	0.301 (100%)
Overall	2.649	0.016	2.126	0.61	100	0.152 (7.15%)	0.283 (13.32%)	1.843 (86.68%)	0.799 (37.58%)	1.327 (62.42%)	0.266 (12.5%)	0.017 (0.81%)	0.533 (25.08%)	1.309 (61.6%)

1. The fan and LHTS unit exhibit significant exergy destruction during daytime operation, primarily due to internal irreversibility.
2. Nighttime charging performance is limited by high exergy destruction in the compressor and condenser, emphasizing the need for improved component designs.
3. Advanced exergy analysis offers valuable insights into system inefficiencies, guiding targeted improvements for enhancing overall performance.

Future work should explore hybrid cooling systems and alternative PCM materials to further reduce energy consumption and enhance sustainability.

9.6 Exergoeconomic and Environmental Evaluation of Air Conditioning Systems with Latent Heat Thermal Storage

Air conditioning systems experience high energy demands during the daytime, while nighttime usage is minimal. This discrepancy presents an opportunity for incorporating LHTS units to store cooling energy during off-peak hours for use during peak demand. The proposed system reduces energy consumption, operating costs, and CO_2 emissions by utilizing inexpensive off-peak electricity for cooling.

Exergoeconomic analysis combines thermodynamic principles and cost analysis to evaluate system performance and cost-effectiveness. In parallel, environmental analysis estimates CO_2 emissions to quantify the system's ecological impact. This section investigates the impact of PCM choice, evaporator/condenser temperatures, and other operational parameters on the system's energy, exergy, and economic performance [3].

9.6.1 System Description

The system consists of an LHTS unit integrated with a conventional refrigeration cycle, as shown in Fig. 9.30. During the charging process at night, the refrigeration cycle extracts heat from the PCMs, solidifying them. During the discharging process, stored cooling energy is released to cool air during the day. The system specifications include:

- *PCM Types*: RT27, S27, and SP25, with melting points of 27 °C, 27 °C, and 25 °C, respectively.
- *Cooling Capacity*: 10.07–11.8 kW, depending on the PCM configuration.
- *Operating Hours*: 8 h for daytime cooling and nighttime charging.
- *Climate Conditions*: The system is designed for the hot climate of Tehran, with ambient temperatures reaching up to 40 °C during summer.

9.6.2 *Mathematical Formulation*

9.6.2.1 Energy Analysis

The average rate of heat transfer within the LHTS unit during the discharge process (daytime operation) can be determined using the following equation:

$$\overline{\dot{Q}}_{\text{LHTS}} = \frac{Q_{\text{LHTS}}}{t^{\text{dis}}} = \frac{1}{t^{\text{dis}}} \int_{0}^{t^{\text{dis}}} \dot{Q}_{\text{LHTS}}(t)\text{d}t \tag{9.44}$$

The total power consumption of the system encompasses the power consumed during both the charging and discharging phases of the operational cycle:

$$W_{\text{total}} = \left[\dot{W}_{\text{fan I}} \times t\right]^{\text{dis}} + \left[\left(\dot{W}_{\text{fan I}} + \dot{W}_{\text{fan II}} + \dot{W}_{\text{CM}}\right) \times t\right]^{\text{ch}} \tag{9.45}$$

where $\dot{W}^{\text{ch}}$ represents the power consumption during the charging process (e.g., nighttime operation) and $\dot{W}^{\text{dis}}$ represents the power consumption during the discharging process (e.g., daytime operation). This would include fan power and any other auxiliary power consumption.

The system's COP, a key indicator of energy efficiency, is defined as the ratio of the cooling output to the total power input:

$$\text{COP} = \frac{Q_{\text{LHTS}}}{W_{\text{total}}} \tag{9.46}$$

9.6.2.2 Exergy Analysis

Exergy analysis provides a more insightful perspective on thermodynamic performance by quantifying the destruction of exergy, which represents the potential for performing useful work. The rate of exergy destruction within the system can be determined through an exergy balance. The exergy efficiency, a measure of how effectively the exergy input is converted into useful output, is defined as the ratio of the product exergy rate to the fuel exergy rate.

The operation of the proposed LHTS-based air conditioning system constitutes a thermodynamic cycle, comprising a thermal energy storage (charging) process followed by a thermal energy removal (discharging) process. The overall exergy efficiency of the system, considering both charging and discharging, can be expressed as the ratio of the total exergy supplied by the air during discharge to the total exergy input to the system over the entire cycle:

$$\eta_{\text{exe}} = \frac{\left(\dot{E}_{\text{air}} \times t\right)^{\text{dis}}}{W_{\text{total}}} \tag{9.47}$$

where $\dot{E}_{\text{air}}$ is the total exergy transferred to the air stream during the discharging process.

9.6.2.3 Economic Analysis

Economic considerations are crucial in evaluating the feasibility and viability of any engineering system. In this analysis, a cost rate balance is employed to assess the overall economic performance of the LHTS-based air conditioning system:

$$\dot{C}_{\text{P,overall}} = \dot{C}_{\text{F,overall}} + \dot{C}_{\text{env,overall}} + \dot{Z}_{\text{OP}} + \sum_{k} \dot{Z}_k \tag{9.48}$$

The operational cost of the system, encompassing the cost of electricity consumed during both charging and discharging phases, is defined as:

$$\dot{Z}_{\text{OP}} = \left[\dot{W}_{\text{fan I}} \times c_{\text{el}}\right]^{\text{dis}} + \left[\left(\dot{W}_{\text{fan I}} + \dot{W}_{\text{fan II}} + \dot{W}_{\text{CM}}\right) \times c_{\text{el}}\right]^{\text{ch}} \tag{9.49}$$

The investment and maintenance cost rate for each system component k is estimated using the following equation:

$$\dot{Z}_k = \frac{Z_k \times \phi}{N \times 3600}\text{CRF} \tag{9.50}$$

where CRF is the capital recovery factor.

The Capital Recovery Factor (CRF), used to annualize the initial investment cost, is calculated as:

$$\text{CRF} = \frac{i(1+i)^n}{(1+i)^n - 1} \tag{9.51}$$

where i is the annual interest rate and n is the system's economic lifespan (in years).

To determine the investment cost rate of each component, the investment cost (Z_k) can be estimated based on the cost functions listed in Table 9.13.

9.6.2.4 Exergoeconomic Analysis

Exergy destruction and capital investment costs are recognized as the primary cost sources in thermodynamic systems. Exergoeconomic analysis combines exergy analysis with economic principles to evaluate the cost-effectiveness of reducing irreversibilities. A key metric in exergoeconomics is the exergoeconomic factor, which

Table 9.13 The cost functions of various components [3]

Component	Capital cost function (Z)
LHTS	8.67×10^{a} where $a = 2.9211\ \exp(0.1416 \times \log V_{\text{LHTS}})$
EV	$16{,}648.3 \times A_{\text{EV}}^{0.6123}$
CM	$\frac{39.5\,\dot{m}_{\text{r}}}{0.9-\eta_{\text{CM}}}\left(\frac{P_7}{P_6}\right)\ln\left(\frac{P_7}{P_6}\right)$
CD	$268.45 + 516.621 \times A_{\text{CD}}$
TV	$114.5 \times \dot{m}_{\text{r}}$
Fan	$155 \times \left(\dot{V}_{\text{Fan}} + 1.43\right)$
Refrigeration system	$150.2 \times \dot{Q}_{\text{EV}}$

expresses the relative contribution of exergy destruction costs to the total cost of a component:

$$f_k = \frac{\dot{Z}_k}{\dot{Z}_k + c_{\text{F},k}\,\dot{E}_{\text{D},k}} \tag{9.52}$$

The cost rate associated with exergy destruction can be determined using the unit cost of fuel for component k, which is calculated by solving the exergy cost balance equation for each component:

$$\sum_{\text{out}} C_k = \sum_{\text{in}} C_k + \dot{C}_{\text{env},k} + \dot{Z}_k + \dot{Z}_{\text{OP},k} \tag{9.53}$$

and

$$C_k = c_k \dot{E}_k \tag{9.54}$$

In this section, only exergy destruction within each component is considered as a source of thermodynamic inefficiency; exergy losses to the environment are not explicitly accounted for.

A low exergoeconomic factor for a major component suggests that improving its efficiency, even at the cost of increased capital investment, could lead to overall cost savings for the system. Conversely, a high value suggests that reducing the investment cost of that component, even if it results in a slight decrease in exergetic efficiency, might be economically beneficial.

9.6.2.5 Environmental Analysis

The environmental impact of the system is assessed by considering CO_2 emissions. The rate of penalty cost associated with CO_2 emissions is calculated based on the total amount of CO_2 emitted by the system:

$$\dot{C}_{env} = \frac{m_{CO_2}}{N \times 3600} c_{CO_2} \tag{9.55}$$

where c_{CO_2} is the penalty cost per unit mass of CO_2 emitted.

The total CO_2 emission is obtained using:

$$m_{CO_2} = \mu_{CO_2} \times W_{annual} \tag{9.56}$$

In this section, the emission conversion factor for electricity is taken as 0.968 kg/kW h, and the penalty cost of CO_2 emissions is assumed to be 0.09 \$/kg of CO_2.

9.6.3 System Specifications

The investigated system is designed to meet thermal comfort requirements for the climate of Tehran, Iran. The system's operating time is set to 8 h during the daytime (discharge) and is equal to the total solidification time during the nighttime (charge). Figure 9.36 illustrates the summer design conditions on the hottest day and the variation of maximum and minimum ambient temperatures during the hot season in Tehran.

For building applications, PCMs with a phase-change temperature in the range of 18–30 °C are generally preferred to maintain thermal comfort. The PCMs considered in this study are RT27, S27, and SP25. Their thermophysical properties are listed in Table 9.14. The effective heat capacity method was employed to simulate the phase-change process during charging and discharging. The system specifications used in the analyses are provided in Table 9.15.

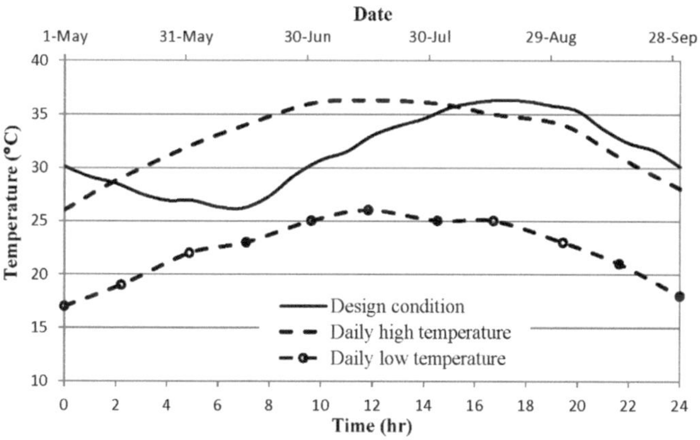

Fig. 9.36 Variation of the ambient temperatures during the summer design condition and the hot season in Tehran [3]

Table 9.14 Thermophysical properties of selected PCMs [3]

Property	RT27	S27	SP25
Density (kg/m^3)	760 (liquid) 880 (solid)	1530	1400 (liquid) 1500 (solid)
Heat capacity at constant pressure, c_p (kJ/kg K)	2.4 (liquid) 1.8 (solid)	2.22 (liquid) 1.5 (solid)	2.5 (liquid) 2.0 (solid)
Thermal conductivity (W/m K)	0.2	0.54	0.6
Latent heat of fusion, L (kJ/kg)	179	183	180
Melting temperature, T_m (°C)	27	27	25
Cost ($/kg)	9	5	8

Table 9.15 System specifications of the storage used in the calculations [3]

Parameters	Value
Number of slabs	34
Width of storage	1.1 m
Thickness of air channels	3.2 mm
Air mass flow rate, $\dot{m}_{air}$	1 kg/s
Reynolds number between slabs	1500
Evaporator heat transfer surface area, A_{EV}	0.2102 m^2
Condenser heat transfer surface area, A_{CD}	0.7124 m^2
Isentropic efficiency of the compressor	80%
Fan efficiency	65%

9.6.4 *Results and Discussion*

9.6.4.1 Effect of PCM Geometry on Outlet Air Temperature

Figure 9.37 illustrates how PCM slab length and thickness affect the outlet air temperature.

- Longer and thicker slabs result in lower outlet temperatures due to increased heat storage capacity.
- RT27 requires larger slabs to achieve cooling due to its low density.
- SP25 performs better in compact configurations, making it suitable for space-constrained systems.

9.6.4.2 Charging Time Versus Inlet Air Temperature

Figure 9.38 shows the effect of the inlet air temperature (T_3) on the charging time (solidification process).

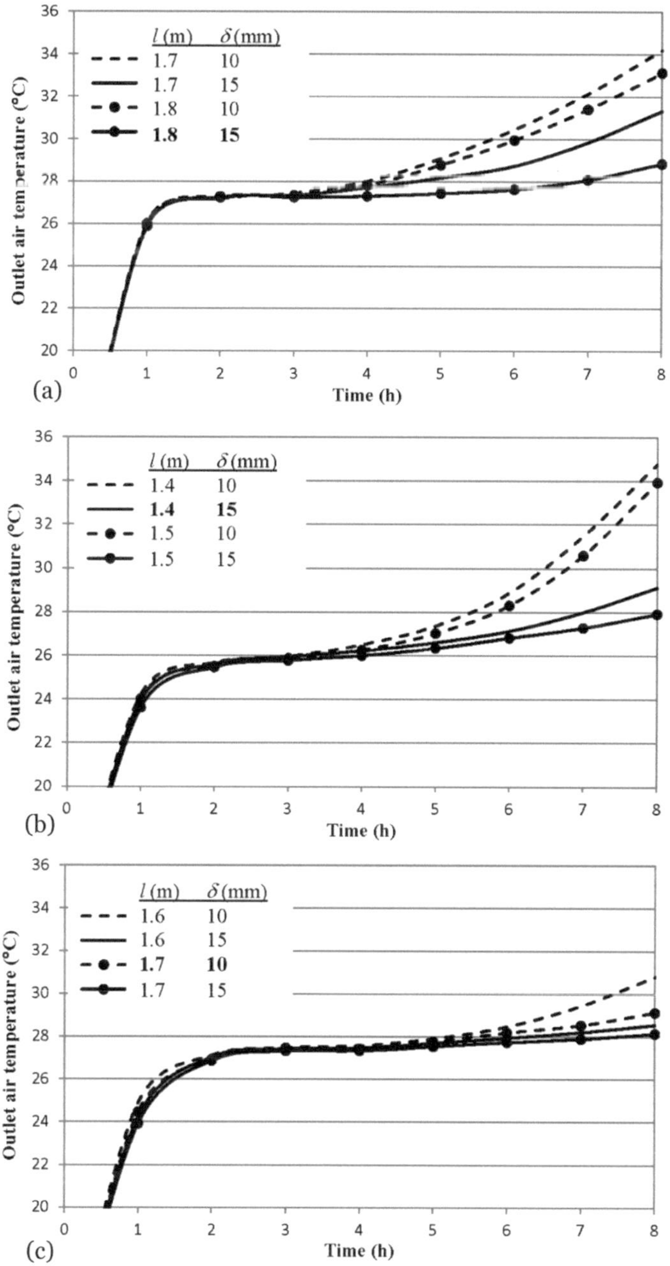

Fig. 9.37 Effect of the length (*l*) and thickness (*δ*) of the PCM slabs on the outlet air temperature from the storage for: **a** RT27, **b** S27, and **c** SP25 [3]

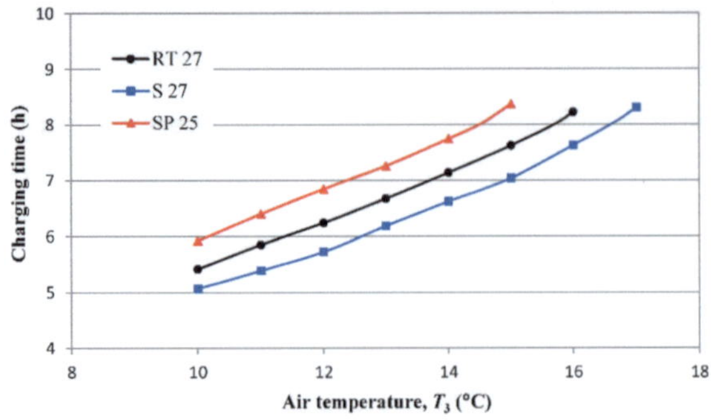

Fig. 9.38 Variation of charging time (solidification of PCM) with inlet air temperature to the LHTS unit for different cases [3]

- Higher T_3 leads to longer charging times due to reduced temperature differences between the air and PCM.
- Case 2 (S27) charges the fastest, highlighting its efficiency for off-peak charging.

9.6.4.3 Effect of T_3 on System COP

Figure 9.39 reveals how T_3 impacts the COP of the system:

- Increasing T_3 reduces the COP due to higher energy consumption during charging.
- Case 3 (SP25) achieves the highest COP (2.66) when operating continuously for 8 h at off-peak conditions.

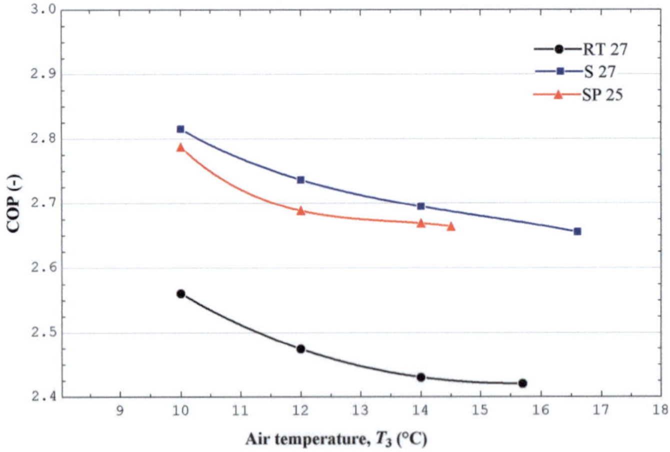

Fig. 9.39 The effect of inlet air temperature on the LHTS unit during night on the COP (T_{EV} = 4 °C and T_{CD} = 48 °C) [3]

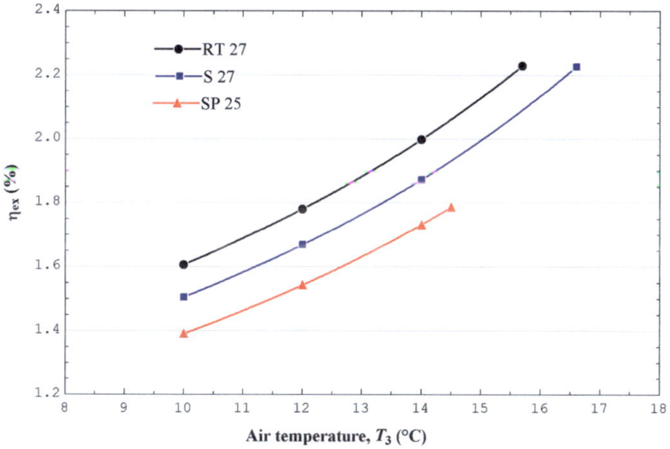

Fig. 9.40 The effect of inlet air temperature on the LHTS unit during night on the exergy efficiency ($T_{EV} = 4$ °C and $T_{CD} = 48$ °C) [3]

9.6.4.4 Effect of T_3 on Exergy Efficiency

Figure 9.40 depicts the variation in exergy efficiency with T_3:

- Higher T_3 improves exergy efficiency by reducing compressor power consumption.
- Case 1 (RT27) achieves the highest exergy efficiency due to its higher product exergy rate during daytime cooling.

9.6.4.5 Economic Analysis

Figure 9.41 shows that CO_2 emissions decrease as the inlet air temperature (T_3) during charging increases. Case 2 produces the lowest emissions due to reduced electricity consumption during charging.

The total cost rate is lowest for Case 2 (S27) at $4094/year due to its reduced investment and operational costs. The exergoeconomic factor (f) for the throttling valve is the lowest, indicating that its costs are dominated by exergy destruction. In contrast, the evaporator has a high f, suggesting that its investment costs dominate.

9.6.4.6 Exergoeconomic Analysis

Table 9.16 compares the performance of the three cases under optimal operating conditions. Key findings include:

- *Case 1 (RT27)*: Highest exergy efficiency (2.93%) due to efficient energy transfer.

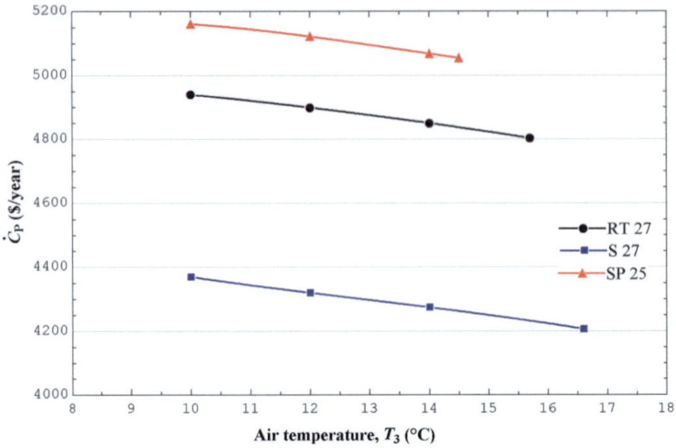

Fig. 9.41 The effect of the inlet air temperature on the LHTS unit during the night on the product cost rate (T_{EV} = 4 °C and T_{CD} = 48 °C) [3]

Table 9.16 The results of energy and exergoeconomic analyses for the best performance conditions of the overall system (T_{EV} = 10 °C and T_{CD} = 44 °C) [3]

PCM	PCM mass in each slab (kg)	COP	η_{exe} (%)	$\dot{C}_p$ ($/year)	T_3 (°C)	f_k (%)						
						Fan I	LHTS	EV	CM	CD	TV	Fan II
RT27	22.57	3.093	2.93	4683	15.7	56.5	19.79	75.58	44.58	20.73	8.64	69.6
S27	28.61	3.378	2.91	4094	16.6	57.63	20.2	76.31	44.53	21.28	8.63	69.6
SP25	32.34	3.434	2.36	4921	14.5	59.14	21.14	75.81	44.56	20.87	9.13	69.6

- *Case 2 (S27)*: Lowest total cost rate ($4094/year) due to reduced investment and operational costs.
- *Case 3 (SP25)*: Highest COP (3.43) and cooling load (11.8 kW), making it ideal for high-demand scenarios.

9.6.4.7 COP Versus Evaporator and Condenser Temperatures

Figure 9.42 shows the impact of evaporator (T_{EV}) and condenser (T_{CD}) temperatures on COP:

- Increasing T_{EV} and decreasing T_{CD} reduce total power consumption, enhancing COP.
- Case 3 consistently outperforms others in terms of COP across temperature ranges.

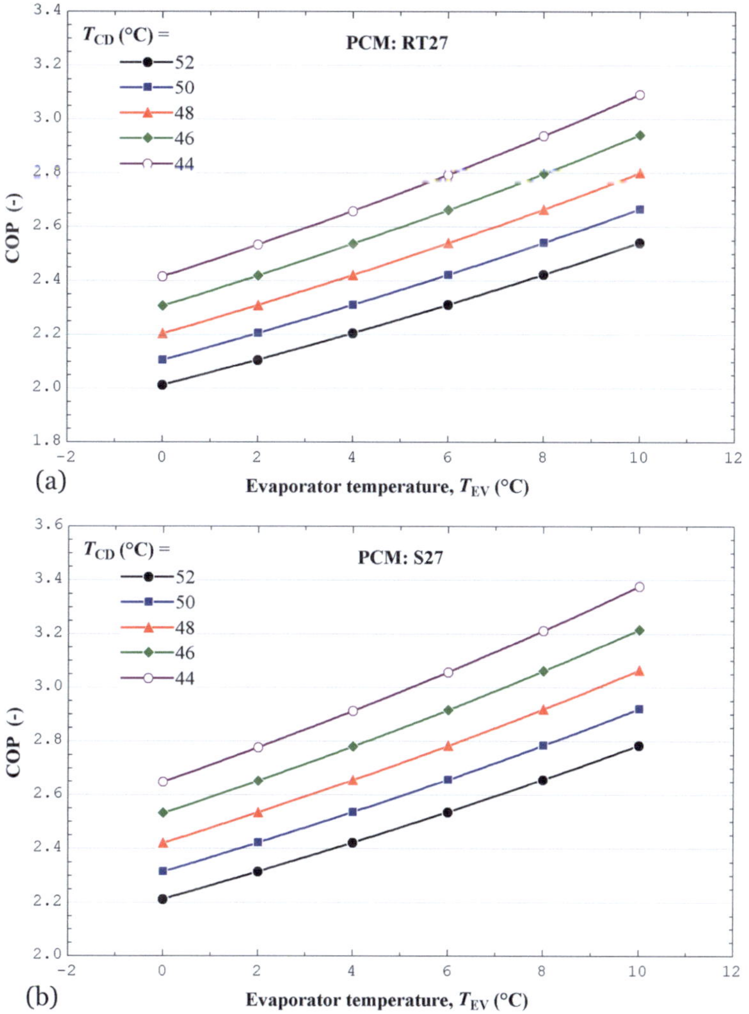

Fig. 9.42 Variation of COP with T_{EV} and T_{CD} for: **a** RT27, **b** S27 and **c** SP25 [3]

9.6.4.8 Exergy Efficiency Versus Evaporator and Condenser Temperatures

Figure 9.43 illustrates the variation of exergy efficiency with T_{EV} and T_{CD}:

- Higher T_{EV} and lower T_{CD} improve exergy efficiency by minimizing irreversibilities.
- Case 1 achieves the highest exergy efficiency due to its optimal thermophysical properties.

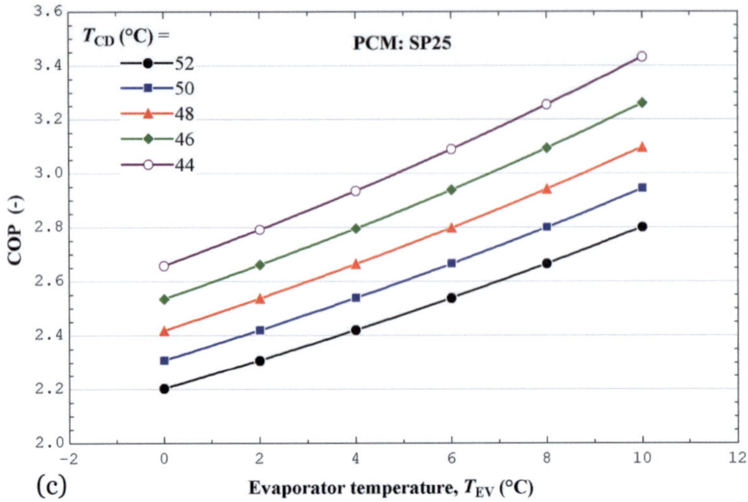

Fig. 9.42 (continued)

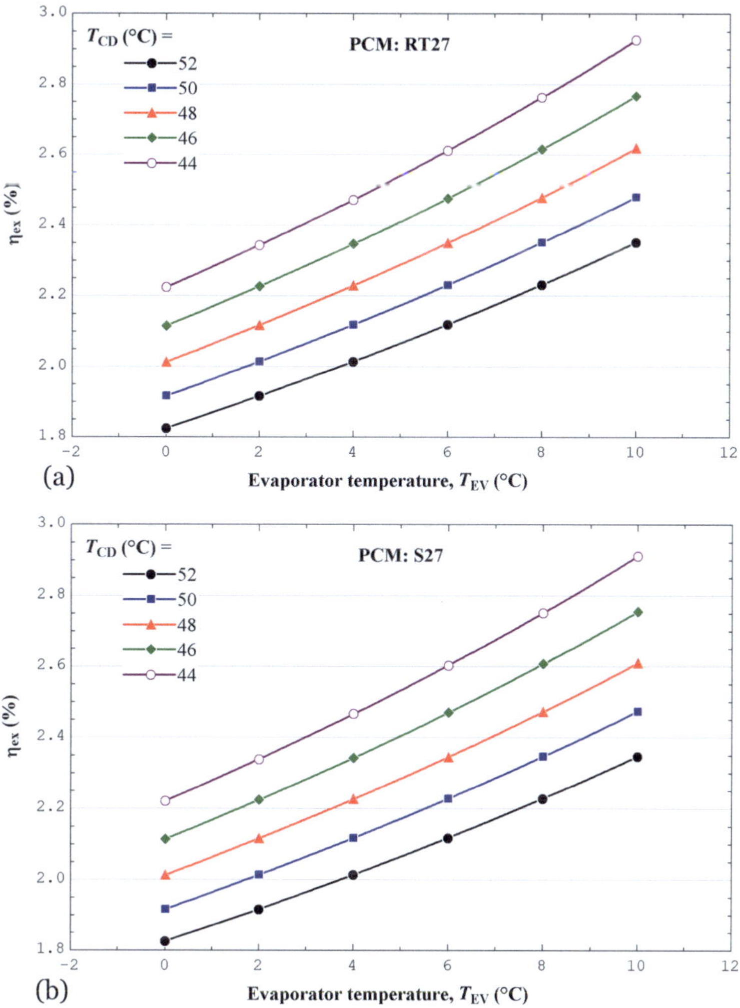

Fig. 9.43 Variation of exergy efficiency with T_{EV} and T_{CD} for: **a** RT27, **b** S27 and **c** SP25 [3]

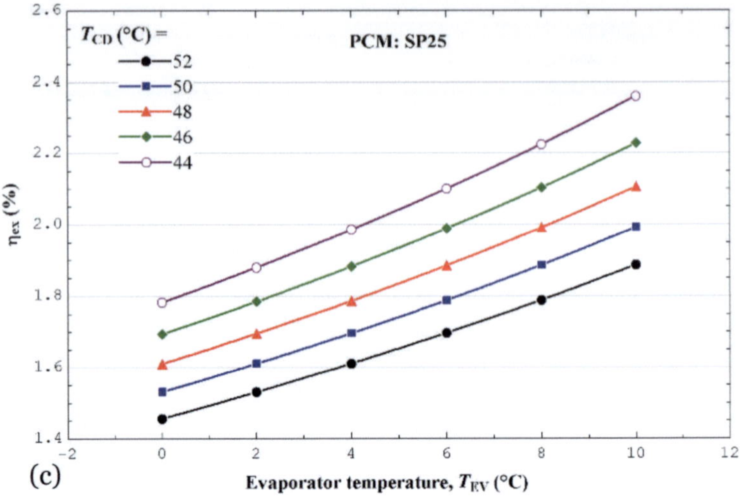

Fig. 9.43 (continued)

9.6.4.9 Product Cost Rate Versus Evaporator and Condenser Temperatures

Figure 9.44 highlights the relationship between evaporator/condenser temperatures and product cost rate:

- Increasing T_{EV} or decreasing T_{CD} reduces the compressor load, leading to lower costs.
- Case 2 offers the lowest product cost rate due to its efficient PCM utilization and lower CO_2 emissions.

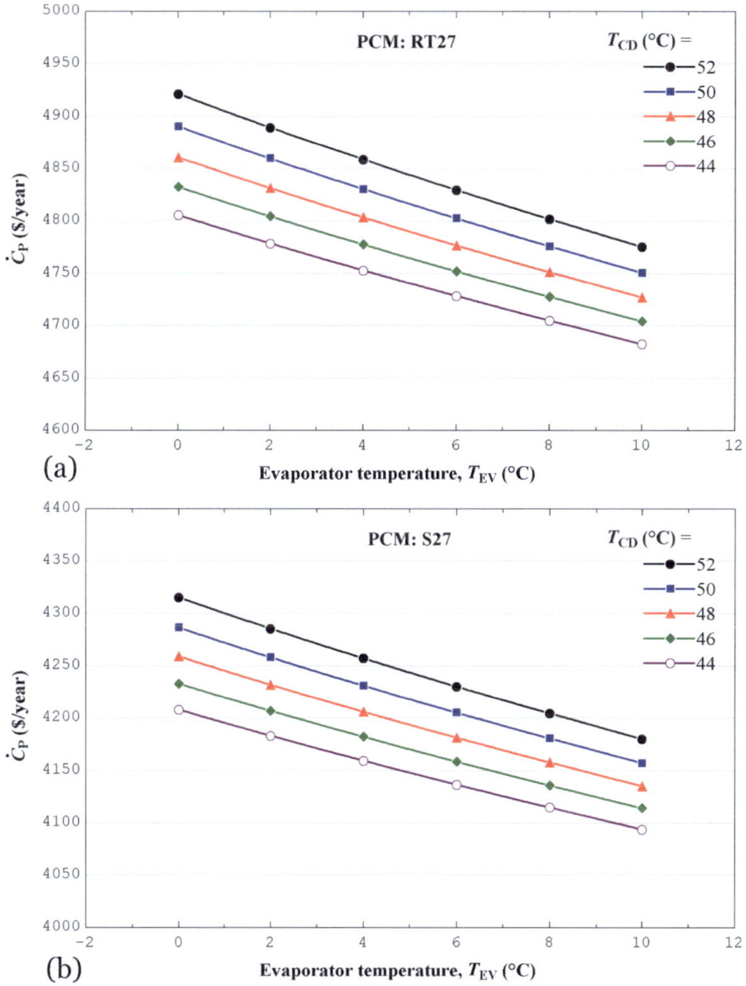

Fig. 9.44 Variation of product cost rate with T_{EV} and T_{CD} for: **a** RT27, **b** S27 and **c** SP25 [3]

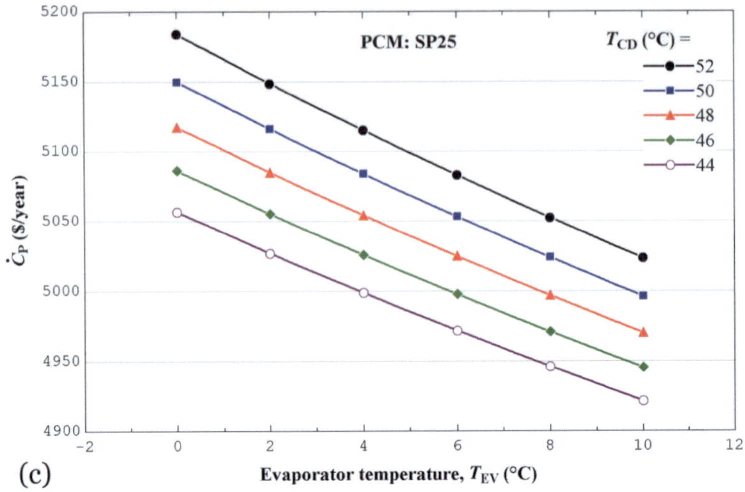

Fig. 9.44 (continued)

9.6.5 *Conclusions*

This chapter presents a comprehensive exergoeconomic and environmental analysis of an LHTS-based air conditioning system. Key findings include:

1. *Case 3 (SP25)*: Achieves the highest COP (3.43) due to its high cooling load.
2. *Case 2 (S27)*: Offers the lowest total cost rate ($4094/year) and the least environmental impact.
3. *Case 1 (RT27)*: Delivers the highest exergy efficiency (2.93%) due to optimal exergy transfer.

The section highlights the importance of PCM selection, operational parameter optimization, and system design for improving energy efficiency and reducing costs and emissions. These insights pave the way for more sustainable cooling solutions in hot climates.

References

1. Laouadi A, Lacroix M (1999) Thermal performance of a latent heat energy storage ventilated panel for electric load management. Int J Heat Mass Transf 42:275–286
2. Morosuk T, Tsatsaronis G (2009) Advanced exergetic evaluation of refrigeration machines using different working fluids. Energy 34:2248–2258
3. Mosaffa AH, Garousi Farshi L (2016) Exergoeconomic and environmental analyses of an air conditioning system using thermal energy storage. Appl Energy 162:515–526
4. Mosaffa AH, Infante Ferreira CA, Rosen MA, Talati F (2013a) Thermal performance optimization of free cooling systems using enhanced latent heat thermal storage unit. Appl Therm Eng 59:473–479

5. Mosaffa AH, Infante Ferreira CA, Talati F, Rosen MA (2013b) Thermal performance of a multiple PCM thermal storage unit for free cooling. Energy Convers Manag 67:1–7
6. Mosaffa AH, Farshi LG, Ferreira CAI, Rosen MA (2014a) Advanced exergy analysis of an air conditioning system incorporating thermal energy storage. Energy 77:945-952
7. Mosaffa AH, Garousi Farshi L, Infante Ferreira CA, Rosen MA (2014b) Energy and exergy evaluation of a multiple-PCM thermal storage unit for free cooling applications. Renew Energy 68:452–458

Chapter 10
Economic and Environmental Dimensions of PCMs

10.1 Introduction

Phase-change materials are new instruments in technology and the economy due to their usefulness in enhancing energy efficiency and sustainability. Their capability to store and release thermal energy during phase changes can lower the energy demand for heating, cooling, and managing thermal control systems. Despite these advantages, there is still a need to strike a balance between cost, economic feasibility and environmental friendliness. Although PCMs may lower energy costs and carbon footprints in the long run, their manufacturing processes, especially paraffins from fossil fuels, have high greenhouse gas emissions. On the other hand, bio-based and inorganic alternatives, like salt hydrates, offer lower environmental impacts but are challenged with costs and scalability. This chapter analyzes global case studies and life-cycle assessments (LCAs) to demonstrate how PCM technologies could be reconceptualized to meet the sustainable developmental goals, focusing on the optimization of PCM systems for life-cycle cost-effectiveness and ecological balance. By resolving issues such as material deterioration, recycling strategies, and market readiness, this chapter illustrates how PCMs can evolve from niche technologies to primary catalysts for a low-carbon future.

10.2 Life-Cycle Assessment of PCM Systems

Life-cycle assessment is a significant approach to determine the impacts of phase-change materials on the environment from the perspective of the entire life cycle, which includes the extraction of raw materials and disposal. It is a methodology that meets the principles of a circular economy by seeking possibilities to reduce resource consumption and waste, while increasing energy efficiency. PCMs are materials that store and release thermal energy when they change from one phase to another,

A. Mosaffa, *Phase Changing Materials*, Materials Horizons: From Nature to Nanomaterials, https://doi.org/10.1007/978-981-95-3742-6_10

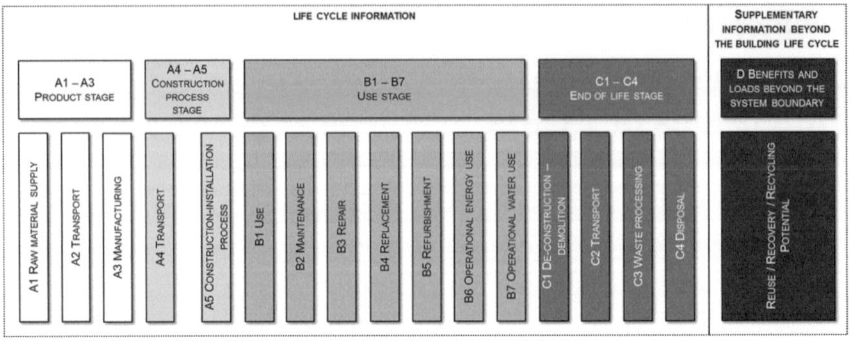

Fig. 10.1 LCA stages for PCM applications [14]

and they are being increasingly integrated into the building envelopes as well as into energy systems to lower operational energy use. However, the environmental advantages must be assessed in the view of impacts from production, encapsulation and end-of-life management.

ISO 14040/44 provides a comprehensive approach to LCA's systematic process, which usually comprises several steps, namely defining the goal and scope, inventory analysis, impact assessment, and interpretation of results. Principal impact categories include GWP (global warming potential), resource depletion, and adverse effects on human health. For PCMs, it is common in LCAs to concentrate on the savings of operational energy versus the cost embodied energy of the paraffin, salt hydrates, and biobased substitutes [22, 32].

LCA evaluates environmental impacts across a product's life cycle, from raw material extraction to disposal (see Fig. 10.1). For PCMs, key stages include:

- *Production (A1–A3)*: Energy-intensive processes for organic PCMs (e.g., paraffin) versus inorganic PCMs (e.g., salt hydrates).
- *Use Phase (B1–B7)*: Operational energy savings versus maintenance/replacement burdens.
- *End-of-Life (C1–C4)*: Recyclability of encapsulation materials (e.g., metals, polymers) [14].

Studies often use generic datasets (e.g., Ecoinvent's "paraffin") rather than manufacturer-specific data, leading to inaccuracies. Methodological inconsistencies in system boundaries and characterization models (e.g., ReCiPe, CML) further hinder comparability [4].

Table 10.1 summarizes critical life-cycle parameters differentiating organic and inorganic PCMs. Organic PCMs, such as paraffins, exhibit higher GWP due to fossil fuel-dependent production processes, while inorganic salt hydrates demonstrate lower embodied carbon. Structural applications of organic PCMs face extended environmental payback periods (> 50 years) due to their high embodied impacts and limited recyclability. In contrast, inorganic PCMs achieve faster payback (25–30 years) owing to higher thermal conductivity and compatibility with recyclable

Table 10.1 Comparative life-cycle parameters of organic and inorganic PCMs in building applications

Aspect	Organic PCMs	Inorganic PCMs
GWP (kg CO_2-eq/kg)	0.649 (paraffin) [38]	0.471 (salt hydrate) [38]
Payback period	> 50 years (structures)	25–30 years (structures)
Thermal conductivity	Low	High
Recyclability	Limited (flammability)	High (metal encapsulation)

metal encapsulation systems. However, inorganic PCMs may require corrosion-resistant containers, adding complexity to their life-cycle management [37].

10.2.1 Operational and Manufacturing Impacts

An ongoing issue with PCM LCAs is the conflict between reduced operational energy use with PCM production's high embodied energy. For example, de Gracia et al. [12] studied cubicles with PCM in Mediterranean climates and observed that while PCM reduced cooling energy consumption by 15%, the global environmental impact over 50 years was equivalent to non-PCM systems because of the high embodied energy of paraffin-based PCMs. Only under unrealistic conditions, such as a hypothetical building lifespan of 100 years or perpetual summer weather, was there an over 10% reduction in overall impact. Kylili and Fokaides [21] have similarly observed that PCMs' operational advantages, such as reduced HVAC load, often offset manufacturing impacts, but only when systems are designed for local climate conditions and frequent phase-change cycles.

Different types of PCM impact the life cycle in very different ways.

- *Paraffins*: These are products of fossil fuels. Their GWP emission is high at 2.1 kg CO_2-eq/kg and so is their primary energy demand at 31.2 MJ/kg. The environmental payback period is between 61 years with heating and 150–260 cycles for cooling [32].
- *Salt Hydrates*: SP-25, as one of the inorganic PCMs, has been shown to have a lower embodied energy of 6.2 MJ/kg and a GWP of 0.5 kg CO_2-eq/kg. Its payback periods are as short as 25 years [22].
- *Bio-Based Esters*: Some bio-PCMs are under development, fatty acid esters for example. While they help reduce the dependence on non-renewable resources, their long-term stability and scalability need to be benzylated [28].

10.2.2 Comparative Analysis of PCM Applications

10.2.2.1 Building Envelopes Versus Active Systems

The LCAs of PCMs indicate that supporting walls and roofs can improve thermal inertia and reduce peak loads. For instance, ventilated double-skin facades (VDSF) with PCMs, although having higher manufacturing impacts, were beneficial as the overall building impact was reduced by 7.7% over 50 years [13]. On the other hand, passive systems, such as solar water heaters with PCM, have greater perceptions of complexity in the design and shorter material life span, leading to steeper challenges [26].

10.2.2.2 Component-Level Considerations

The environmental performance of PCM storage systems will be dependent on how the components are defined. Nienborg et al. [32] analyzed three configurations:

1. **Macroencapsulated PCM**: Very high storage density and encapsulated in aluminum, but has a high impact.
2. **Immersed Heat Exchangers**: Capillary tube systems were lower in material use, over 20–30% lower in GWP than the water storage in the cooling applications (the water was the cooling agent).
3. **Thermochemical Materials (TCMs)**: Closed system sorption materials like silica gel were found to have 260× more GWP than water, which highlights the importance of designs that employ open systems or consider material reuse.

10.2.3 Impact of System Design and Application Scenarios

10.2.3.1 Climate and Usage Patterns

The effectiveness of PCM is greatly determined by the geographic and climatic conditions. For cooler areas, PCMs such as SP-21EK with a lower melting point can reduce cooling energy demand by about 15–30%, which achieves a faster environmental payback [32]. On the other hand, PCMs used for heating applications need higher phase-change temperatures (like SP-58), which have long payback periods due to their low storage density concerning water [12].

10.2.3.2 Lifetime and Recycling

Longer building lifespans are favorable to the PCM value. For instance, hydrated salts paid industrial salts back in 25 years and paraffins achieved payback in 61 years

[22]. Strategies for end-of-life scenarios, like recycling aluminum encapsulation or reusing undegraded PCMs, can further reduce impacts by 20–30% [32].

10.2.4 Challenges and Future Directions

10.2.4.1 Barriers to Circularity

PCMs derived from fossil fuels pose great problems because of the high captured energy. To reduce circularity issues, production must be coupled with renewable energy [22]. In addition, the economic feasibility of PCMs has been stifled due to a lack of industry standards and excessive PCMS's PCMs have suffered from slow adoption rates lack of standardization and high initial investment costs, even though these materials provide savings in the long run [25]. Furthermore, some data gaps hinder meaningful comparisons due to the limited LCA data concerning biobased PCMs and their recyclability [21].

10.2.4.2 Research Priorities

The impact bio-based esters have in the sector greatly improves when the materials' thermal stability increases. Lower impact PCMs must be bio-based to fit the required criteria. Macroeconomically, this enhances other aspects such as energy efficiency in numerous required applications. Integrating the PCMs within the system requires more focus; augmenting PCMs with night ventilation passive design strategies aids in bolstering cyclic efficiency. Policy adoption is instrumental in driving these techniques widespread has to preempt these advancements. Incentives can market the use of PCMs for certification as green buildings while subsidizing circular design advocates for Modified Aided Design (MAD).

LCA reveals that PCM systems can align with circular economy goals when optimized for life cycle performance. While challenges like high embodied energy persist, strategic material selection, climate-responsive design, and recycling protocols offer pathways to net-positive impacts. Future research must prioritize holistic LCAs that account for evolving technologies and circular supply chains.

10.3 Cost Drivers: Material Sourcing, Manufacturing, and Recycling

PCMs have a significant promise to provide reliable support for flamboyant renewable energy and efficient industries, where demand proves to be intermittent. But, all of them are costing drivers, with economic and environmental impacts, related to:

material origins, manufacturing and recycling processes. This article reviews these aspects and to this effect benefits from the recent progress in FSPCM composites.

10.3.1 Material Sourcing

The materials cost and scalability are largely affected by the choice of materials for the PCMs as well as their supporting matrices. Carbon materials, e.g., biochar, expanded graphite (EG), and carbon nanotubes (CNTs), are the most commonly used materials for their good thermal conductivity and structural stability. The biochar prepared from agricultural residues such as wheat straw, soft wood, or water hyacinth is economically viable considering the availability of feedstock and low processing cost [3]. Nonetheless, energy-consuming pyrolysis (550–900 °C) and chemical activation (KOH treatment) are also not cost-effective [10]. Expanded graphite (EG) has better thermal conductivities but needs energy-consuming exfoliation, and therefore it is too costly compared to biochar and not suitable for wide application [45]. Although CNTs can improve the thermal conductivity up to 6600 W/m K, the development and application of CNT tubes are hampered by too expensive manufacturing costs and random interface bonding with the matrix [43]. Mineral-based supports such as diatomite, vermiculite and perlite are abundant and relatively inexpensive; however, are required to undergo pretreatment to develop advanced functionalities. For instance, diatomite and perlite can be subjected to acid treatment or calcination for improved porous structure, which increases the cost by 10–15% for the material [19]. The low-cost clay minerals, such as kaolin or bentonite, might need a surface modification (e.g., acidifying) to enhance the PCM compatibility, which leads to a complicated raw materials sourcing [35]. Polymers, such as high-density polyethylene (HDPE) and polystyrene (PS), provide mechanical stability but rely on fossil fuel-derived feedstocks, leading to price volatility. While recycled polymers can reduce costs, they often compromise thermal performance [31].

10.3.2 Manufacturing Processes

Manufacturing costs are driven by impregnation methods, additive integration, and energy consumption. Vacuum impregnation ensures high PCM loading (75–95%) but requires specialized equipment and energy-intensive vacuum systems, increasing costs by 20–30% compared to simpler immersion techniques [10]. Sol–gel and freeze-drying methods, used for silica and aerogel matrices, produce materials with high surface areas but involve costly reagents, such as tetraethyl orthosilicate, and prolonged processing times [41].

The integration of additives, such as EG, CNTs, or metal powders (e.g., copper), enhances thermal conductivity but raises material expenses. For instance, incorporating 10 wt% EG increases composite costs by 15–20%. Binder systems, including polymers like HDPE or melamine foam, demand precise temperature control during melt-blending or in-situ polymerization, further elevating energy consumption [9]. Scalability remains a challenge, as batch processes for biochar activation or CNT synthesis are less efficient than continuous methods. For example, biochar production at 700 °C for 3 h limits throughput, while EG exfoliation requires specialized reactors [11].

10.3.3 Recycling and Life-Cycle Considerations

End-of-life management significantly impacts both economic and environmental costs. Carbonaceous composites, such as biochar and EG, are chemically stable but difficult to separate from PCMs. While pyrolysis can recover biochar, the energy costs remain prohibitive [3]. Mineral-based composites, like diatomite and vermiculite, are inert and reusable after PCM degradation, but residual organic PCMs may necessitate solvent extraction, increasing recycling expenses. Polymer-based composites, such as HDPE and PS, are technically recyclable, but phase separation from PCMs (e.g., paraffin) requires energy-intensive processes. Cross-linked polymers, such as polymethyl methacrylate (PMMA), are non-recyclable, raising landfill disposal concerns [31].

The durability of FSPCMs under thermal cycling is a critical factor in life-cycle costs. Most composites retain over 90% of their latent heat after 200 cycles, reducing replacement frequency. For instance, softwood biochar/RT28 composites lose only 3% of their capacity after 200 cycles [11]. However, chemical degradation of organic PCMs, such as fatty acids and paraffins, at high temperatures (> 150 °C) necessitates stabilizers, adding 5–10% to material costs [34]. Environmental trade-offs also arise: biochar sequesters carbon, offsetting emissions, while EG and CNTs have high embodied energy due to production processes [3]. Mineral mining, though abundant, causes land disruption, though this risk is mitigated by the widespread availability of materials like diatomite [19].

10.3.4 Economic Feasibility and Future Directions

The economic viability of form-stable phase-change materials (FSPCMs) is directly tied to how performance metrics align with the costs of the material and its manufacturing. There is an existing system cost and performance imbalance. For instance, low-cost composites like biochar-mineral hybrids are moderately priced between \$5–\$10/kg. However, their thermal conductivity range of 0.2–0.5 W/m K makes them unsuitable for high-power applications [5]. In the end, some high-performance composites like EG–CNT–enhanced systems are greater than 10 W/m K of thermal conductivity, but their pricing at \$50–100/kg makes them applicable only in niche fields such as aerospace or advanced electronics [44]. This gap has placed a heavy emphasis on the need for new solutions that genuinely merge inexpensive and effective.

Advancing technology must go after waste valorization to lower feedstock costs. Industrial byproducts such as rice husk ash, sewage sludge, or agricultural residues serve as matrices at a much lower cost than conventionally used ones. For example, biochar from water hyacinth or sugarcane bagasse not only solves waste management problems but has also shown competitive thermal storage capacity [3]. Moreover, the agro-industry byproduct silica-rich rice husk ash allows for a replacement of the synthetic silica matrix, which reduces material costs by 20–30% [42]. Process optimization is equally critical. Transitioning from batch-based pyrolysis to continuous production systems could reduce biochar manufacturing energy consumption by 30–40%, while solvent-free impregnation methods, such as mechanochemical mixing, eliminate costly chemical reagents and streamline production [11].

Incorporating principles of the circular economy will optimize both sustainability and cost-efficiency. Closed-loop recycling systems could recover and reprocess matrices after PCM degradation for polymers like HDPE or PMMA, lessening dependence on virgin resources. For instance, the melt-recycling of composites with paraffin into HDPE has been observed to retain 85–90% of their original thermal capacity over numerous cycles [9]. Biochar is similarly advantageous because its chemical stability enables low-temperature reactivation or regeneration, prolonging its life in thermal storage applications. Policy measures could include subsidizing renewable-integrated PCM systems or offering carbon credits for biochar production, both of which would hasten market adoption.

Standardization and scalability will be critical for long-term competitive advantage. The creation of universal testing criteria for thermal cycling stability and leakage resistance will enable assessment of brand loyalty benchmarks, thus fostering trust with consumers. In addition, the scaling of pilot processes such as freeze drying for aerogel matrices or roll-to-roll fabrication for polymer composite "printing" (also referred to as "manufacturing") necessitates cross-disciplinary resolution of these engineering hurdles alongside academia to industry partnerships [24]. Solving these issues would allow FSPCMs to transition from niche applications into widespread renewable energy technologies, providing a sustainable solution for global decarbonization initiatives.

10.4 Policy and Regulatory Frameworks for PCM Adoption

The integration of phase-change materials in thermal energy storage systems is paramount in improving the energy efficiency and the use of renewable energy. On the other hand, their widespread adoption is strongly dependent on sound policies and regulations dealing with technical, financial, and environmental issues. This part discusses some of the most important policies as well as regional case studies that aim at fostering PCMs adoption, drawing from the policies on renewable energy microgrids and energy storage policy frameworks.

10.4.1 Financial Incentives and Market Mechanisms

Economic policies alter the behavior of individuals toward the enormous initial investment associated with PCM-based systems. Tax rebates and grants tend to lower the capital barriers, making these systems economically viable to both industries and households. Take, for instance, the U.S. Investment Tax Credit (ITC) could be modified to encourage PCM-integrated thermal storage projects just as it has promoted the adoption of solar energy. In the same way, feed-in tariffs (FiTs) promise stable prices to investors for stored thermal energy sold into district heating systems, providing reliable revenue streams. Germany's Renewable Energy Sources Act (EEG) is a case in point. It has been instrumental in promoting decentralized energy systems using FiTs [30].

PCMs are also aided by low-interest loans and green financing. Innovative storage solution loans from Australia's Clean Energy Finance Corporation (CEFC) can be applied to PCM projects. These financial tools not only reduce initial costs but also align with broader climate goals by encouraging low-carbon technologies.

10.4.2 Technical Standards and Grid Integration

Interconnection and interoperability standards guarantee the proper operation of PCM systems within the existing energy framework. As with the IEEE 1547 standard on the integration of distributed energy resources, it is also possible to design standards for voltage control, frequency control, and safety measures for PCM-based TES systems. Also, the IEC 61850 communication protocol standard enables coordination in real-time time pro-active control of PCM storage units and smart grid technologies, increasing reliability of the system [2].

Renewable Portfolio Standards (RPS) requirements have a more indirect effect on stimulating the use of PCM technologies as they mandate the addition of energy storage to the utility's portfolio. In the case of California, RPS requires 60% of renewables by 2030, which raises the need for TES technologies like PCMs to mitigate fluctuations in solar and wind generation [1]. Such policies underscore the need for PCM systems that are grid-compatible and can operate at minimal performance standards.

10.4.3 Environmental and Safety Regulations

Environmental impact assessments (EIAs) are critical for PCM deployment, particularly for inorganic PCMs that may pose toxicity risks. The European Union's Renewable Energy Directive (RED II) mandates EIAs for large-scale TES projects, ensuring minimal ecological disruption. Safety standards, such as California's CPUC regulations for energy storage, enforce fire suppression and thermal management protocols for PCM systems, mitigating risks of thermal runaway.

Compliance with material safety certifications (e.g., REACH in the EU) is essential for PCM market entry. For example, paraffin-based PCMs require rigorous testing to meet flammability and life-cycle environmental impact criteria [40]. These regulations balance innovation with public safety, fostering trust in PCM technologies.

10.4.4 Regional Case Studies and Policy Models

The European Union's Clean Energy for All Europeans package provides a blueprint for PCM adoption. Its emphasis on decentralized energy systems aligns with PCM applications in microgrids, supported by cross-border funding programs like Horizon Europe [15]. Germany's EEG, which prioritizes latent heat storage, demonstrates how FiTs can accelerate PCM use in industrial waste heat recovery.

In the United States, California's Self-Generation Incentive Program (SGIP) has allocated rebates for TES installations, which indirectly aid in promoting PCM-based systems [1]. Japan's Smart Grid Initiative adopts PCMs in its urban energy systems, capitalizing on rigid interoperability requirements to enhance thermal storage effectiveness.

10.4.5 Challenges and Recommendations

There are several challenges faced in the implementation of PCMs.

From a financial point of view, the project is also a challenge (due in part to the cost of materials and manufacturing process, which prevents scalability). From a practical point of view, there's a big jump because there are no standard testing protocols for the long-term performance of PCMs. Regulation also brings additional challenges, not only because safety and environmental frameworks are diverse and varied between jurisdictions and disparate compliance criteria and oversight relations.

Several key steps need to be taken by lawmakers to address the PCM challenges. First, the lack of unified standards should be addressed by creating international performance and safety standards (e.g., ISO) for PCMs. This proposed change would have the benefit of creating a level playing field in the industry. Further, broadening incentives is also a must as similar to the ITC extension for storage, lawmakers could introduce subsidies and research and development grants focused particularly on advancing PCMs. Also, an interdisciplinary collaboration is another crucial aspect. Industry professionals, academics and community members should work in collaboration to develop and conduct pilot experiments that can improve methods of application of these technologies. By doing so, lawmakers can help drive the increased use of PCMs in a real sense.

Efficient policy enablers for the use of PCMs require balanced economic incentives, technical specifications, and strong environmental standards. As for other areas, distraction from PCS to renewable energy policies and adaptation of them to target the integration of PCMs in decarbonization processes may be possible, together with ongoing monitoring and stakeholder involvement. For collection-based PCMs to realize their envisioned role in sustainable energy systems, continued scrutiny and adaptive governance structures are crucial.

10.5 Environmental Impact of Nano-PCMs

The incorporation of nanoparticles (NPs) into PCMs has transformed the TES system by improving thermal conductivity, decreasing phase-change times, and enhancing energy efficiency. However, the environmental impacts of nano-PCMs, specifically their toxicity and difficulties associated with recycling, need scrutiny to guarantee sustainable use. This part critiques these issues, using insights from recent literature.

10.5.1 Toxicity of Nanoparticles in Nano-PCMs

Certain nano-PCMs particles raise risks, which include metals like copper (Cu), silver (Ag), and aluminum (Al), metal oxides like alumina (Al_2O_3), cupric oxide (CuO), titanium (TiO_2), and carbon-based materials such as graphene and carbon nanotubes. In this sense, Metallic NPs like Cu, for example, have proved useful when used in enhancing thermal conductivity, however, their practice is bound by their high tendency to oxidize, leading to the release of toxic ions. For example, silver nanoparticles (Ag NPs) incorporated in polyethylene glycol (PEG) composites are thermally stable but often leach to Ag^+ ions, which are known to be destructive to some aquatic life and soil microbes. Likewise, copper nanoparticles (Cu NPs) embedded in paraffin-based PCMs usually oxidize with time, releasing Cu^{2+} ions, which, Fromelious et al. [16], are linked to cellular damage of living systems.

Metal oxide NPs, such as Al_2O_3 and CuO, are chemically stable but pose risks due to their persistence in ecosystems. Alumina (Al_2O_3) nanoparticles, widely used for their high thermal conductivity, have been shown to accumulate in water bodies, potentially disrupting aquatic life through bioaccumulation. CuO nanoparticles, while effective in reducing phase-change times, exhibit cytotoxicity in mammalian cells, raising concerns about occupational exposure during manufacturing [20].

Carbon-based NPs, including graphene oxide (GO) and multi-walled carbon nanotubes (MWCNTs), are less reactive but not without risks. Graphene oxide, used to stabilize PEG composites, can induce oxidative stress in microorganisms if released into the environment. MWCNTs, despite their superior thermal properties, have been associated with pulmonary toxicity in inhalation studies, necessitating stringent encapsulation protocols [29]. The environmental impact of NPs also depends on their life cycle. For example, during the synthesis of nano-PCMs, improper disposal of nanoparticle-laden waste can lead to groundwater contamination. Studies emphasize the need for life-cycle assessments (LCAs) to evaluate the long-term ecological footprint of nano-PCMs, particularly in industrial applications [6].

10.5.2 Recycling Challenges of Nano-PCMs

Recycling nano-PCMs presents significant technical and economic hurdles due to the complexity of separating nanoparticles from the PCM matrix and ensuring material stability over multiple cycles. Traditional PCM recycling methods, such as melting and reprocessing, are inadequate for nano-PCMs because NPs tend to aggregate or degrade during thermal cycling, reducing their efficacy and increasing contamination risks.

The recovery of NPs from spent nano-PCMs is challenging due to their nanoscale size and strong interfacial interactions with the host material. For instance, metallic NPs like Ag and Cu form stable colloidal suspensions in organic PCMs, making mechanical separation impractical. Advanced techniques such as centrifugation, filtration, and chemical leaching have been explored, but these methods are energy-intensive and often result in nanoparticle loss. Carbon-based NPs, such as graphene oxide, require solvent-based extraction, which generates hazardous waste and negates the environmental benefits of recycling [17].

Thermal cycling worsens the thermal performance of nano-PCMs due to enhanced migration and aggregation of nanoparticles. Studies on paraffin/Al_2O_3 composites show NPs tend to migrate to the container walls, because of which thermal resistance hotspots are formed, after 500 cycles. In the same manner, CuO nanoparticles in fatty acid-based PCM also oxidize with time and give rise to copper carbonates, which decrease the latent heat capacity [7]. The recycling processes, because of modified chemical nano-PCMs, undergo these degradation mechanisms. Figure 10.2 illustrates the factors influencing the thermal conductivity of nano-PCMs.

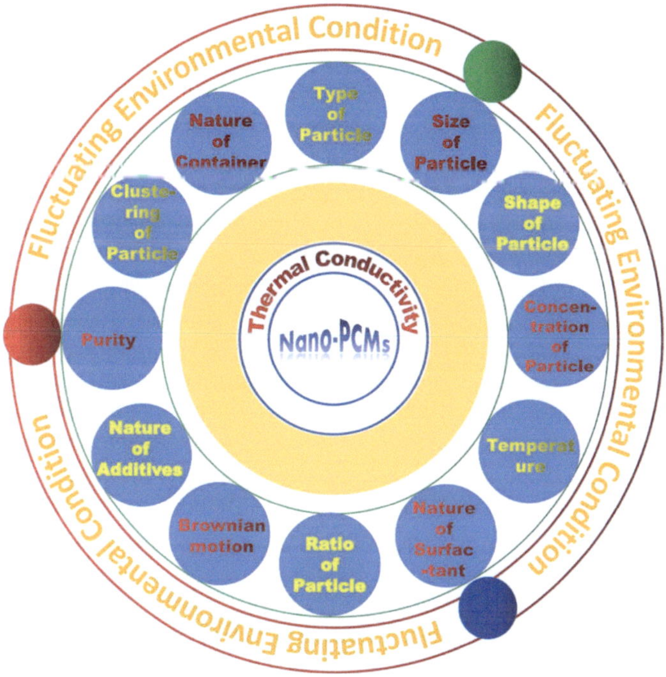

Fig. 10.2 Factors affecting thermal conductivity [39]

10.5.3 Economic Viability

Recycling feasibility is further hindered by the high cost of functionalization and synthesizing the nanoparticles. For instance, Ag NPs incorporated in PEG composites are costly to manufacture and the recovery rates are frequently below 50%, rendering closed-loop recycling economically impractical. The inclusion of several types of nanoparticles, such as metal oxides and carbon nanotubes, into hybrid nano-PCMs results in greater separation difficulty, multi-step recycling processes, which increases the cost [36].

To address toxicity and recycling challenges, researchers propose several mitigation strategies. Surface functionalization of NPs with biocompatible coatings (e.g., silica or polymers) can reduce ion leaching and toxicity. For instance, Al NPs encapsulated in carbon shells exhibit lower oxidation rates and improved environmental compatibility. Similarly, the covalent bonding of graphene oxide with PEG enhances stability and minimizes ecological release [18]. For recycling, developing bio-based nano-PCMs using sustainable nanoparticles (e.g., carbon dots from biomass) offers promise. Li et al. [23] demonstrated that carbonized wood infused with carbon dots retains thermal stability over 1000 cycles, simplifying end-of-life disposal. Additionally, modular TES systems with replaceable nano-PCM cartridges could streamline recycling by isolating spent materials for centralized processing.

10.6 Cost–Benefit Analysis for Building Integration

The integration of PCMs into building systems has emerged as a promising strategy for enhancing energy efficiency and demand management. However, the economic viability of such systems depends on balancing upfront costs with long-term energy savings. This section evaluates the cost–benefit dynamics of PCM-enhanced materials, focusing on active cold storage systems and nano-PCMs, drawing insights from life-cycle assessments and case studies. The initial investment for PCM systems encompasses material costs, installation expenses, and auxiliary equipment. Active cold storage systems, such as those integrating PCM tanks, require a significant capital outlay. For instance, Cui et al. [8] reported a total capital investment of 4350 USD/m^3 for PCM storage tanks, including accessories and space rent, translating to an annualized cost of 435 USD/m^3 over a 10-year lifespan. Nano-PCMs, which incorporate nanoparticles (NPs) like copper, silver, and graphene oxide, further elevate material costs due to the expense of nanoparticle synthesis and functionalization. For example, silver nanoparticles (Ag NPs) used in polyethylene glycol (PEG) composites are costly to produce, with recovery yields often below 50%, complicating cost-effectiveness.

Costs associated with the upfront investment also vary based on the level of PCM deployment. In one of the studies conducted on active cold storage systems, it was found that a small-scale PCM tank (4.2% of daily cooling load) had a storage requirement of 10,680 kWh, which integrated into a commercial building, cost approximately 132,949 USD annually. Moreover, Nano-PCMs have to deal with additional difficulties like the need to guarantee uniform nanoparticle dispersion, preventing aggregation, and advanced manufacturing methods, which further raise the initial costs.

10.6.1 Long-Term Energy Savings and Life-Cycle Cost Analysis

PCM systems offset upfront costs through energy savings from peak shaving, load shifting, and improved thermal efficiency. Active cold storage systems reduced chiller power consumption by 28.9% in a Hong Kong commercial building, yielding annual savings of 168,701 USD through peak load management alone. By shifting cooling demand to off-peak hours under time-of-use pricing, these systems capitalized on lower electricity rates, reducing operational costs by 35% when storage capacity was optimized.

Nano-PCMs enhance energy savings by improving thermal conductivity and reducing phase-change times. For instance, PEG/Ag NP composites achieved a 127% increase in thermal conductivity compared to conventional PCMs, shortening storage/release cycles by 34% and lowering energy consumption. Similarly, carbon-based nanoparticles like graphene oxide (GO) stabilized PCMs, reducing leakage and improving cyclic stability, which prolonged system lifespan and reduced maintenance costs [8].

Life-cycle cost analyses (LCCA) reveal the economic viability of PCM systems. For active cold storage, Cui et al. [8] demonstrated a payback period of 5.4 years for a system with 4.2% daily cooling load capacity, generating net savings of 115,344 USD annually. The study highlighted that 68% of total savings stemmed from peak load management, while 32% originated from demand response incentives.

Nano-PCMs face steeper payback challenges due to higher initial costs. For example, Ag NP-enhanced PCMs required 16.7 years to recoup investments when used solely for demand response, but hybrid applications (e.g., combined with peak shaving) reduced this to 5–7 years [33]. Life-cycle assessments also emphasized the importance of recycling and nanoparticle recovery. Improper disposal of NPs like CuO and Al_2O_3 can lead to groundwater contamination, necessitating costly remediation and offsetting savings.

10.6.2 Challenges in Cost–Benefit Realization

Nano-phase-change materials suffer from critical issues related to recycling and the material's life cycle. One important aspect involves the reduction of efficiency that occurs due to the thermal cycles of nanoparticle migration and the clustering effect known as aggregation. For instance, a study conducted by Mebarek-Oudina and Chabani [27] showed that paraffin/alumina composites had a decline in latent heat capacity by about 30% after 500 thermal cycles due to particle sedimentation.

Along with the mentioned challenges, the grid's varying rates for incentives also create an obstacle for appropriate energy control. Demand response (DR) programs, like the real-time demand response program (RTDRP) offered by ISO New England, reported earnings of $79,654 for a 1443 kW demand reduction, showcasing significant annual savings [8]. Regardless, the infrequency of these events, along with the possible changes in tariff lines, makes such savings unreliable. In addition, the overlapping scope and the degree of integration with the other parts of the system add a layer to the complexity of the overall system. Implementation of PCM systems on a small scale, which constitutes 0.5–4.2% of the total daily cooling load, results in quicker returns on investment. On the other hand, as the systems get larger, the expected return on investment to losses ratio turns unfavorable. This is caused by the growing expenses of materials and space needed for larger systems, as noted by Cui et al. [8].

PCM-enhanced materials offer substantial long-term energy savings, but their economic viability hinges on optimizing upfront costs and addressing life-cycle challenges. Active cold storage systems demonstrate shorter payback periods (2–5.4 years) due to dual benefits in demand response and peak shaving, while nano-PCMs require innovations in recycling and scalable manufacturing to enhance cost-effectiveness. Future research should prioritize hybrid systems combining passive thermal mass with active nano-PCM storage to maximize returns on investment.

References

1. Ajaz W, Bernell D (2021) Microgrids and the transition toward decentralized energy systems in the United States: a multi-level perspective. Energy Policy 149:112094
2. Apostolov A (2022) IEC 61850: digitizing the electric power grid. Artech House Power Engineering Library, Artech House
3. Atinafu DG, Yeol Yun B, Uk Kim Y, Wi S, Kim S (2021) Introduction of eicosane into biochar derived from softwood and wheat straw: influence of porous structure and surface chemistry. Chem Eng J 415:128887
4. Baldassarri C, Sala S, Caverzan A, Lamperti Tornaghi M (2017) Environmental and spatial assessment for the ecodesign of a cladding system with embedded phase change materials. Energy Build 156:374–389
5. Bordoloi U, Das D, Kashyap D, Patwa D, Bora P, Muigai HH, Kalita P (2022) Synthesis and comparative analysis of biochar based form-stable phase change materials for thermal management of buildings. J Energy Storage 55:105801
6. Cabeza LF, Rincón L, Vilariño V, Pérez G, Castell A (2014) Life cycle assessment (LCA) and life cycle energy analysis (LCEA) of buildings and the building sector: a review. Renew Sustain Energy Rev 29:394–416
7. Chen X, Tang Z, Liu P, Gao H, Chang Y, Wang G (2020) Smart utilization of multifunctional metal oxides in phase change materials. Matter 3:708–741
8. Cui B, Gao D, Wang S, Xue X (2015) Effectiveness and life-cycle cost-benefit analysis of active cold storages for building demand management for smart grid applications. Appl Energy 147:523–535
9. Cui W, Li X, Li X, Si T, Lu L, Ma T, Wang Q (2022) Thermal performance of modified melamine foam/graphene/paraffin wax composite phase change materials for solar-thermal energy conversion and storage. J Clean Prod 367:133031
10. Das D, Bordoloi U, Muigai HH, Kalita P (2020) A novel form stable PCM based bio composite material for solar thermal energy storage applications. J Energy Storage 30:101403
11. Das D, Crosby R, Paul MC (2025) A comprehensive review on the form stable phase change materials for storing renewable heat preparation, characterization and application. J Energy Storage 110:115284
12. de Gracia A, Rincón L, Castell A, Jiménez M, Boer D, Medrano M, Cabeza LF (2010) Life cycle assessment of the inclusion of phase change materials (PCM) in experimental buildings. Energy Build 42:1517–1523
13. de Gracia A, Navarro L, Castell A, Boer D, Cabeza LF (2014) Life cycle assessment of a ventilated facade with PCM in its air chamber. Sol Energy 104:115–123
14. EN BS (2012) 15804; sustainability of construction works—environmental product declarations—core rules for the product category of construction products. Br. Stand. Inst.
15. Filho RD, Monteiro ACM, Costa T, Vasconcelos A, Rode AC, Marinho M (2023) Strategic guidelines for battery energy storage system deployment: regulatory framework, incentives, and market planning. Energies 16:7272
16. Frommelius A, Ohlerth T, Noyong M, Simon U (2024) On the application potential of chemically tailored metal oxide and higher chalcogenide nanoparticles for nanoscale resistive switching devices. Phys status solidi 221:2300456
17. Gupta N, Kumar A, Dhasmana H, Kumar V, Kumar A, Shukla P, Verma A, Nutan GV, Dhawan SK, Jain VK (2020) Enhanced thermophysical properties of metal oxide nanoparticles embedded magnesium nitrate hexahydrate based nanocomposite for thermal energy storage applications. J Energy Storage 32:101773
18. Harmen Y, Younes C, Fatima Ezzahrae MA, Fouad B, Charafeddine J, Sophie D, Bensitel M (2021) Thermal performance of PEG-MWCNTs composites as shape-stabilised phase change materials for thermal energy storage. Fullerenes Nanotub Carbon Nanostruct 29:732–738
19. Jeong S-G, Jeon J, Lee J-H, Kim S (2013) Optimal preparation of PCM/diatomite composites for enhancing thermal properties. Int J Heat Mass Transf 62:711–717

20. Kim YH, Ryu H, Lee H-J, Cho Y-K, Nahm S (2012) Transmission electron microscopy microstructure of $(K_{0.5}Na_{0.5})NbO_3$ ceramics with CuO addition. Jpn J Appl Phys 51:035602
21. Kylili A, Fokaides PA (2016) Life cycle assessment (LCA) of phase change materials (PCMs) for building applications: a review. J Build Eng 6:133–143
22. Kyriaki E, Konstantinidou C, Giama E, Papadopoulos AM (2018) Life cycle analysis (LCA) and life cycle cost analysis (LCCA) of phase change materials (PCM) for thermal applications: a review. Int J Energy Res 42:3068–3077
23. Li X, Zhu Z, Yang P, You Z, Dong Y, Tang M, Chen M, Zhou X (2021) Carbonized wood loaded with carbon dots for preparation long-term shape-stabilized composite phase change materials with superior thermal energy conversion capacity. Renew Energy 174:19–30
24. Lin X, Zhang X, Ji J, Liu L, Wu Y, Yang M, Lu D, Zheng H (2021) Development of flexible form-stable phase change material with enhanced electrical resistance for thermal management. J Clean Prod 311:127517
25. Llorach-Massana P, Peña J, Rieradevall J, Montero JI (2017) Analysis of the technical, environmental and economic potential of phase change materials (PCM) for root zone heating in Mediterranean greenhouses. Renew Energy 103:570–581
26. Mahfuz MH, Anisur MR, Kibria MA, Saidur R, Metselaar IHSC (2014) Performance investigation of thermal energy storage system with phase change material (PCM) for solar water heating application. Int Commun Heat Mass Transf 57:132–139
27. Mebarek-Oudina F, Chabani I (2023) Review on nano enhanced PCMs: insight on nePCM application in thermal management/storage systems. Energies 16:1066
28. Menoufi K, Castell A, Farid MM, Boer D, Cabeza LF (2013) Life cycle assessment of experimental cubicles including PCM manufactured from natural resources (esters): a theoretical study. Renew Energy 51:398–403
29. Mishra RK, Mishra P, Verma K, Joseph K (2018) Manipulation of thermo-mechanical, morphological and electrical properties of PP/PET polymer blend using MWCNT as nano compatibilizer: a comprehensive study of hybrid nanocomposites. Vacuum 157:433–441
30. Morris C, Jungjohann A (2016) Energy democracy: Germany's Energiewende to Renewables. Springer International Publishing
31. Mu M, Basheer PAM, Sha W, Bai Y, McNally T (2016) Shape stabilised phase change materials based on a high melt viscosity HDPE and paraffin waxes. Appl Energy 162:68–82
32. Nienborg B, Gschwander S, Munz G, Fröhlich D, Helling T, Horn R, Weinläder H, Klinker F, Schossig P (2018) Life cycle assessment of thermal energy storage materials and components. Energy Procedia 155:111–120
33. Said Z, Pandey AK (2023) Nano enhanced phase change materials: preparation, properties and applications. In: Materials horizons: from nature to nanomaterials. Springer Nature Singapore
34. Sarı A (2004) Form-stable paraffin/high density polyethylene composites as solid–liquid phase change material for thermal energy storage: preparation and thermal properties. Energy Convers Manag 45:2033–2042
35. Sarı A (2015) Fabrication and thermal characterization of kaolin-based composite phase change materials for latent heat storage in buildings. Energy Build 96:193–200
36. Sharma P, Pandey AK, Said Z (2023) Soft computing tools (intelligent techniques) for nano-enhanced PCM. In: Said Z, Pandey, Kumar A (eds) Nano enhanced phase change materials: preparation, properties and applications. Springer Nature Singapore, Singapore, pp 223–246
37. Struhala K, Ostrý M (2022) Life-cycle assessment of phase-change materials in buildings: a review. J Clean Prod 336:130359
38. Struhala K, Ostrý M, Bantová S (2019) Environmental assessment of latent heat storage technology. IOP Conf Ser Earth Environ Sci 290:12087
39. Thirumalaivasan N, Gopi S, Karthik K, Nangan S, Kanagaraj K, Rajendran S (2024) Nano-PCM materials: bridging the gap in energy storage under fluctuating environmental conditions. Process Saf Environ Prot 189:1003–1021
40. Wang Q, Ping P, Zhao X, Chu G, Sun J, Chen C (2012) Thermal runaway caused fire and explosion of lithium ion battery. J Power Sources 208:210–224

41. Wang C, Liang W, Yang Y, Liu F, Sun H, Zhu Z, Li A (2020) Biomass carbon aerogels based shape-stable phase change composites with high light-to-thermal efficiency for energy storage. Renew Energy 153:182–192
42. Yu K, Liu Y, Yang Y (2021) Enhanced thermal properties of polyethylene glycol/modified rice husk ash eco-friendly form-stable phase change material via optimizing support pore structure. J Energy Storage 43:103172
43. Zhang N, Yuan Y, Yuan Y, Li T, Cao X (2014) Lauric–palmitic–stearic acid/expanded perlite composite as form-stable phase change material: preparation and thermal properties. Energy Build 82:505–511
44. Zhang M, Cheng H, Wang C, Zhou Y (2021) Kaolinite nanotube-stearic acid composite as a form-stable phase change material for thermal energy storage. Appl Clay Sci 201:105930
45. Zuo X, Zhao X, Li J, Hu Y, Yang H, Chen D (2020) Enhanced thermal conductivity of form-stable composite phase-change materials with graphite hybridizing expanded perlite/paraffin. Sol Energy 209:85–95

MIX
Papier aus verantwortungsvollen Quellen
Paper from responsible sources
FSC® C105338

If you have any concerns about our products,
you can contact us on
ProductSafety@springernature.com

In case Publisher is established outside the EU,
the EU authorized representative is:
Springer Nature Customer Service Center GmbH
Europaplatz 3, 69115 Heidelberg, Germany

Printed by Libri Plureos GmbH
in Hamburg, Germany